LS006-002-2H

APOLLO LUNAR ROVING VEHICLE OPERATIONS HANDBOOK

APRIL 19, 1971

CONTRACT NAS8-25145

APPROVED: _____
LUNAR ROVING VEHICLE
SYSTEMS ENGINEERING MANAGER

PREPARED BY THE BOEING COMPANY
LRV SYSTEMS ENGINEERING
HUNTSVILLE, ALABAMA

©2012 Periscope Film LLC
ALL RIGHTS RESERVED
ISBN #978-1-937684-89-1

LS006-002-2H
LUNAR ROVING VEHICLE
OPERATIONS HANDBOOK

LIST OF EFFECTIVE PAGES

INSERT LATEST CHANGED PAGES
DESTROY SUPERSEDED PAGES.

TOTAL NUMBER OF PAGES IN THIS PUBLICATION IS 342, CONSISTING OF THE FOLLOWING:

PAGE NO.	ISSUE
Title	19 April 1971
A	19 April 1971
i thru viii	19 April 1971
1-1 thru 1-78	19 April 1971
2-1 thru 2-50	19 April 1971
3-1 thru 3-12	19 April 1971
4-1 thru 4-11	19 April 1971
5-1 thru 5-4	19 April 1971
6-1 thru 6-18	19 April 1971
7-1 thru 7-6	19 April 1971
8-1 thru 8-31	19 April 1971
Title (Appendix A)	19 April 1971
A-i thru A-v	19 April 1971
A-1 thru A-105	19 April 1971
Title (Appendix B)	15 January 1971
B-i	15 January 1971
B-ii	15 January 1971
B-1 thru B-8	15 January 1971

©2012 Periscope Film LLC
ALL RIGHTS RESERVED
ISBN #978-1-937684-89-1

This document has been digitally watermarked
by Periscope Film LLC to prevent illegal duplication.

Mission __J__ Basic Date __12/4/70__ Change Date __4/19/71__ Page __A__

LS006-002-2H
LUNAR ROVING VEHICLE
OPERATIONS HANDBOOK

HANDBOOK CONFIGURATION

This handbook reflects the Lunar Roving Vehicle (LRV) and Space Support Equipment (SSE) delivery review configuration as modified by incorporation of the following:

ECP	DESCRIPTION
LRV 1097	Incorporation of Manual SSE
LRV 1073	Seat Belt Modification
LRV 1075	Rear Steering Recoupling
LRV 1104	Thermal Blanket
LRV 1103	Switch Guards

Mission _____ Basic Date 12/4/70 Change Date 4/19/71 Page i

LS006-002-2H
LUNAR ROVING VEHICLE
OPERATIONS HANDBOOK

TABLE OF CONTENTS

SECTION		TITLE	PAGE
1		General Information	1-1
	1.0	Introduction	1-1
	1.1	Description	1-1
	1.2	Vehicle Systems	1-1
	1.3	Mobility Subsystem	1-5
	1.4	Electrical Power Subsystem	1-28
	1.5	Control and Display Console	1-37
	1.6	Navigation Subsystem	1-42
	1.7	Crew Station	1-49
	1.8	Thermal Control	1-54
	1.9	Space Support Equipment	1-64
2		Normal Procedures	2-1
	2.1	Unloading and Chassis Deployment	2-2
	2.2	Post Deployment Checkout and Drive to MESA	2-22
	2.3	Payload Loading	2-29
	2.4	Pre-Sortie Checkout and Preparation	2-41
	2.5	LRV Configuration for Science Stop	2-45
	2.6	LRV Configuration Prior to Leaving Science Stop	2-46
	2.7	Post Sortie Checkout	2-48
	2.8	Display Reading Sequence and Time Intervals	2-50
3		Malfunction Procedures	3-1
	3.0	Introduction	3-1
4		Auxiliary Equipment	4-1
	4.0	Introduction	4-1
	4.1	Forward Chassis Payload Provisions	4-1
	4.2	Center Chassis Payload Provisions	4-1
	4.3	Rear Chassis Payload Provisions	4-6
5		Operating Limitations	5-1
	5.1	Payload Limitations	5-1
	5.2	Parking Limitations	5-1
	5.3	Sortie Limitations	5-1
	5.4	Navigation System Limitations	5-4

Mission _____ Basic Date 12/4/70 Change Date 4/19/71 Page ii

LS006-002-2H
LUNAR ROVING VEHICLE
OPERATIONS HANDBOOK

TABLE OF CONTENTS
(Continued)

SECTION		TITLE	PAGE
6		Operating Timelines	6-1
	6.0	Introduction	6-1
7		Operating Profiles	7-1
	7.1	LRV Operating Profile	7-1
	7.2	1G Trainer Operating Profile	7-4
8		1G Trainer Non-Crew Procedures	8-1
	8.0	Introduction	8-1
	8.1	General Procedures	8-1
	8.2	Specific Procedures	8-12
	8.3	Preventive Maintenance Assembly Remove and Replace Procedures	8-28

LS006-002-2H
LUNAR ROVING VEHICLE
OPERATIONS HANDBOOK

LIST OF ILLUSTRATIONS

FIGURE NO.	TITLE	PAGE
1-1	Lunar Roving Vehicle	1-2
1-2	1G Trainer	1-4
1-3	Mobility Subsystem	1-6
1-4	LRV Wheel Cross-Section	1-7
1-5	LRV Traction Drive Assembly	1-9
1-6	Traction Drive Installation	1-11
1-7	Suspension Assembly	1-14
1-8	Steering Assembly	1-16
1-9	Steering Control Block Diagram	1-17
1-10	Wheel and Steering Disconnects	1-18
1-11	Hand Controller	1-19
1-12	Torque Required to Rotate Hand Controller for Throttle Control	1-21
1-13	Torque Required to Rotate Hand Controller for Steering Control	1-22
1-14	Brake Control Force Vs. Displacement	1-24
1-15	Drive Control Electronics - Block Diagram	1-26
1-16	LRV Battery Configuration	1-29
1-17	LRV Batteries, Thermal Blanket and Dust Covers	1-30
1-18	Power Distribution System Schematic	1-32
1-19	LRV Monitor Schematic	1-33
1-20	Caution and Warning System	1-34
1-21	Auxiliary Connector Location	1-36
1-22	Control and Display Console	1-38
1-23	Navigation Subsystem Block Diagram	1-43
1-24	Navigation Components on LRV	1-44
1-25	Navigation System Electrical Schematic	1-45
1-26	Vehicle Attitude Indicator	1-46
1-27	Sun Shadow Device	1-47
1-28	Crew Station Components	1-50
1-29	Seat Belts	1-52
1-30	Crew Station Floor Panels	1-53
1-31	Thermal Control Provisions	1-55
1-32	Drive Controller Electronics Thermal Control	1-56
1-33	SPU Electronics and Battery #1 Thermal Control	1-57
1-34	Battery #2 and Directional Gyro Unit Thermal Control	1-58
1-35	Battery Dust Cover Closing Mechanism	1-59
1-36	Forward Chassis Insulation Blanket	1-60
1-37	Space Support Equipment	1-66
1-38	LM/SSE with LRV Installed	1-67
1-39	LRV Deployment Sequence	1-68
1-40	Insulation Blanket	1-69
1-41	LRV Deployment Tapes and Cables	1-70
1-42	D-Handle Release System	1-71

Mission J Basic Date 12/4/70 Change Date 4/19/71 Page iv

LS006-002-2H
LUNAR ROVING VEHICLE
OPERATIONS HANDBOOK

LIST OF ILLUSTRATIONS
(CONTINUED)

FIGURE NO.	TITLE	PAGE
1-43	LRV/SSE Support Structure and Release System	1-72
1-44	Braked Reel	1-73
1-45	LRV Saddle and Forward Chassis Latch System	1-75
1-46	Forward and Rear Chassis Latch	1-76
1-47	Wheel Lock Strut Release	1-77
2-1	Support Arm Latch Mechanisms Latched Configuration	2-3
2-2	LRV Deployment Tapes and Cables	2-4
2-3	LRV Deployment Envelope and Envelope for Deployment Tape Operations	2-6
2-4	Crewman Positioned to Deploy LRV	2-7
2-5	LRV Deployment Sequence	2-9
2-6	LRV Deployment Hardware & Steering Ring Location	2-11
2-7	Foot Rest Deployment	2-15
2-8	Control and Display Console Deployment	2-16
2-9	Seat and PLSS Support Deployment Sequence	2-17
2-10	Crew Position	2-24
2-11	Control and Display Console	2-25
2-12	LCRU/TV/LRV Cable Stowage	2-30
2-13	LCRU, HighGain Antenna, TV Camera Installation	2-31
2-14	16 mm DAC and Low Gain Antenna Installation	2-34
2-15	LCRU Low Gain Antenna Cable Installation on Lunar Surface	2-36
2-16	LRV Rear Payload Pallet Adapters	2-37
2-17	Rear Payload Pallet Installed	2-39
2-18	Buddy SLSS Installation	2-40
4-1	LCRU, High Gain Antenna, TV Camera Installation	4-2
4-2	LCRU/TV/LRV Cable Stowage	4-3
4-3	LCRU Low Gain Antenna Cable Installation on Lunar Surface	4-4
4-4	16 mm DAC and Low Gain Antenna Installation	4-5
4-5	Under-seat Stowage Bag (Left Seat)	4-7
4-6	Passenger Seat Stowage to Create Payload Area on Center Chassis Floor	4-8
4-7	Buddy SLSS Installation	4-9
4-8	LRV Rear Payload Pallet Adapters	4-10
4-9	Rear Payload Pallet Installed	4-11
5-1	Allowable C.G. Envelope for Vehicle Fully Loaded	5-2
5-2	Parking Orientation Constraints	5-3
6-1	LRV Deployment Timeline	6-2
6-2	Post Deployment Checkout Timeline	6-7
6-3	Pre-Sortie Checkout and Preparation Timeline	6-9
6-4	Post Sortie Shutdown Timeline	6-11
6-5	Navigation Update Timeline	6-12
6-6	LRV Traction Drive Decoupling Timeline	6-12

Mission J   Basic Date 12/4/70   Change Date 4/19/71   Page v

LS006-002-2H
LUNAR ROVING VEHICLE
OPERATIONS HANDBOOK

LIST OF ILLUSTRATIONS
(CONTINUED)

FIGURE NO.	TITLE	PAGE
6-7	LRV Steering Decoupling Timeline	6-14
6-8	LRV Rear Steering Recoupling Timeline	6-15
6-9	1G Trainer Battery Changeout Timeline	6-16
6-10	1G Trainer Traction Drive Decoupling Timeline	6-17
6-11	1G Trainer Steering Decoupling Timeline	6-18
7-1	Nominal Operating Profile	7-2
7-2	Nominal Operating Profile for 1G Trainer	7-5
8-1	1G Trainer Basic Vehicle Block Diagram	8-2
8-2	1G Trainer Vehicle Power Distribution Block Diagram	8-3
8-3	1G Trainer Vehicle Front Traction Drive Electrical Signal Routing Block Diagram	8-4
8-4	1G Trainer Vehicle Rear Traction Drive Electrical Signal Routing Block Diagram	8-5
8-5	1G Trainer Vehicle Front Steering Electrical Signal Routing Block Diagram	8-6
8-6	1G Trainer Vehicle Rear Steering Electrical Routing Block Diagram	8-7
8-7	1G Trainer Vehicle Odometer Electrical Signal Routing Block Diagram	8-8
8-8	1G Trainer Temperature Diagnostics Electrical Signal Routing Block Diagram	8-9
8-9	1G Trainer Wheel Decoupling	8-17
8-10	1G Trainer Brake Linkage	8-19
8-11	1G Trainer Steering Arm Clamping	8-22
8-12	1G Trainer Battery Installation	8-24
8-13	1G Trainer Battery Charging Circuit	8-26

Mission _____ Basic Date 12/4/70 Change Date 4/19/71 Page vi

LS006-002-2H
LUNAR ROVING VEHICLE
OPERATIONS HANDBOOK

LIST OF TABLES

TABLE NO.	TITLE	PAGE
1-1	Control and Display Console Controls	1-39
1-2	1G Trainer Thermal Control Device Set Points	1-62
3-1	Malfunction Procedures	3-2
3-2	Malfunction Logic Flow Diagrams	3-3
8-1	1G Trainer Steering Operation Data	8-15

LS006-002-2H
LUNAR ROVING VEHICLE
OPERATIONS HANDBOOK

LRV FLIGHT UNIT PARTIAL DRAWING LIST
(FOR REFERENCE ONLY)

DESCRIPTION	DRAWING NUMBER
Flight Vehicle Assembly	209-35006
Chassis Assembly	209-30006
Tripod	209-30406
Fender	209-31625
Crew Station Installation	209-31006
Suspension and Wheel Installation	209-35206
Steering Installation	209-35306
Electrical and Battery Installation	209-34006
Electrical Cabling Installation	209-35106
Display and Control Console Assembly	209-22200
Gyro and SPU Installation	209-34306
Vehicle Attitude Indicator Installation	209-60050
Thermal Insulation Installation	209-34406
Gyro Thermal Strap	209-70019
SPU Thermal Control Unit	209-70400
DCE Assembly	7553139*
Hand Controller Assembly	7553447*
Traction Drive Assembly	7553102*

*Denotes AC/Delco Electronics Drawings, All Others are Boeing

LS006-002-2H
LUNAR ROVING VEHICLE
OPERATIONS HANDBOOK

SECTION I

GENERAL INFORMATION

1.0   INTRODUCTION

This section contains general information pertaining to the flight operational Lunar Roving Vehicle (LRV). Where applicable, the 1G Trainer differences are noted.

1.1   DESCRIPTION

The LRV system on the lunar surface consists of the LRV, the structure for securing the LRV to the LM stowage bay and the mechanism for deploying the LRV from the LM onto the lunar surface.

1.2   VEHICLE SYSTEMS

The LRV (figure 1-1) is a four-wheeled, self-propelled, manually controlled vehicle to be used for transporting crewmen and equipment on the lunar surface. The vehicle has accommodations for two crewmen and the stowed auxiliary equipment designed for the particular mission.

Control of the LRV during traverses is effected from either of the two crewmen positions by operating the hand controller located between the two crewmen positions. Selection of power supplied to each load, monitoring of key parameters, and operation of the navigation system is effected from the control and display console, which is located for operation by either crewman.

1G Trainer Notes

1. 1G Trainer vehicle systems are shown on figure 1-2.

2. Electrical block diagrams for the 1G Trainer are provided in Section 8.0.

LS006-002-2H
LUNAR ROVING VEHICLE
OPERATIONS HANDBOOK

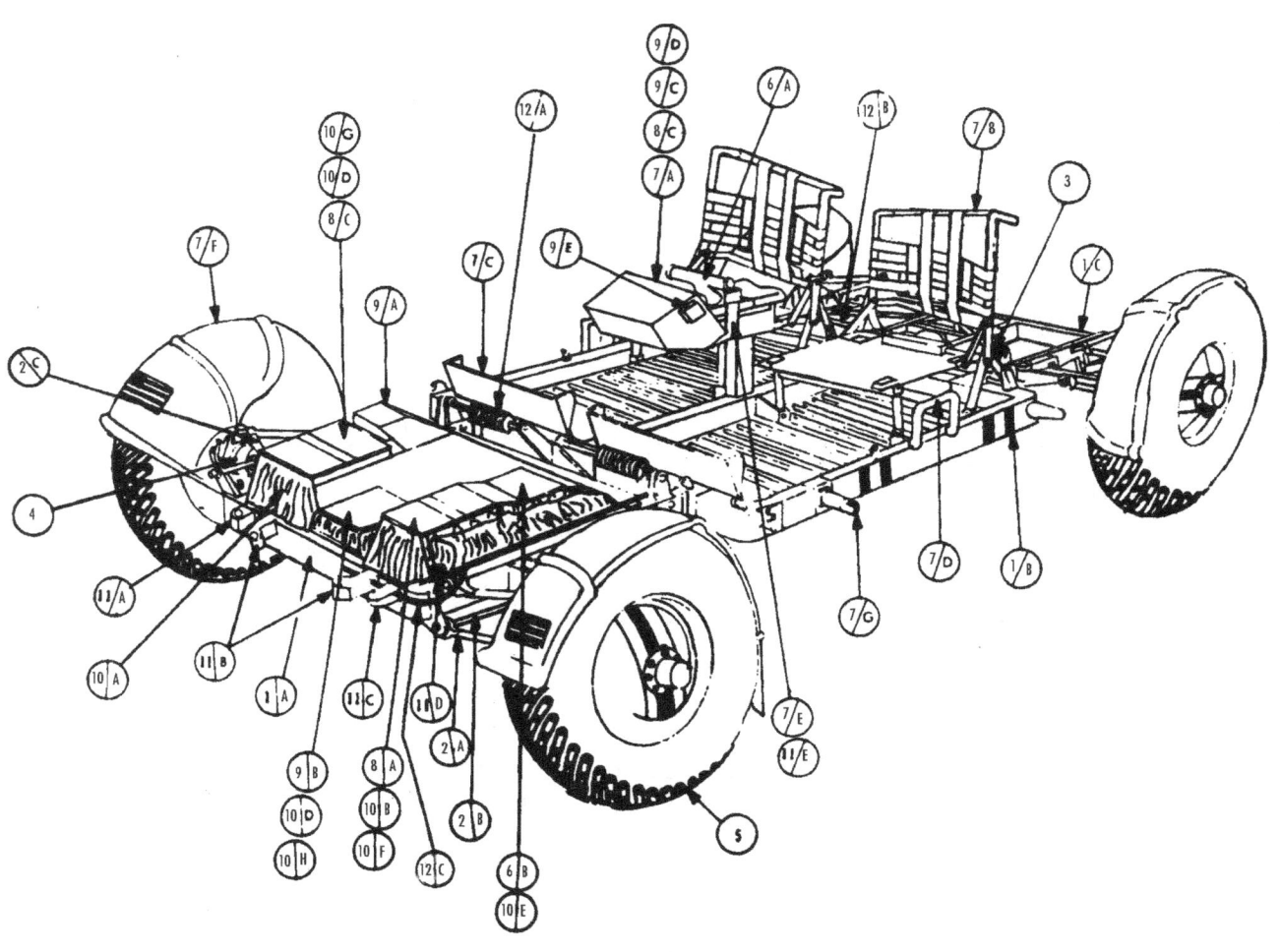

1. CHASSIS
   A. FORWARD CHASSIS
   B. CENTER CHASSIS
   C. AFT CHASSIS
2. SUSPENSION SYSTEM
   A. SUSPENSION ARMS (UPPER AND LOWER)
   B. TORSION BARS (UPPER AND LOWER)
   C. DAMPER
3. STEERING SYSTEM (FORWARD AND REAR)
4. TRACTION DRIVE
5. WHEEL
6. DRIVE CONTROL
   A. HAND CONTROLLER
   B. DRIVE CONTROL ELECTRONICS (DEL)
7. CREW STATION
   A. CONTROL AND DISPLAY CONSOLE
   B. SEAT
   C. FOOTREST
   D. OUTBOARD HANDHOLD
   E. INBOARD HANDHOLD
   F. FENDER
   G. TOEHOLD
   H. SEAT BELT
8. POWER SYSTEM
   A. BATTERY #1
   B. BATTERY #2
   C. INSTRUMENTATION
9. NAVIGATION
   A. DIRECTIONAL GYRO UNIT (DGU)
   B. SIGNAL PROCESSING UNIT (SPU)
   C. INTEGRATED POSITION INDICATOR (IPI)
   D. SUN SHADOW DEVICE
   E. VEHICLE ATTITUDE INDICATOR
10. THERMAL CONTROL
    A. INSULATION BLANKET
    B. BATTERY NO. 1 DUST COVER
    C. BATTERY NO. 2 DUST COVER
    D. SPU DUST COVER
    E. DCE THERMAL CONTROL UNIT
    F. BATTERY NO. 1 RADIATOR
    G. BATTERY NO. 2 RADIATOR
    H. SPU THERMAL CONTROL UNIT
11. PAYLOAD INTERFACE
    A. TV CAMERA RECEPTACLE
    B. LCRU RECEPTACLE
    C. HIGH GAIN ANTENNA RECEPTACLE
    D. AUXILIARY CONNECTOR
    E. LOW GAIN ANTENNA RECEPTACLE
12. DEPLOYMENT COMPONENTS
    A. FWD CHASSIS DEPLOYMENT TORSION SPRINGS
    B. REAR CHASSIS DEPLOYMENT TORSION BARS
    C. SADDLE RELEASE CABLE

FIGURE 1-1 LRV WITHOUT STOWED PAYLOAD (SHEET 1 OF 2)

Mission _____ Basic Date 12/4/70 Change Date 4/19/71 Page 1-2

LS006-002-2H
LUNAR ROVING VEHICLE
OPERATIONS HANDBOOK

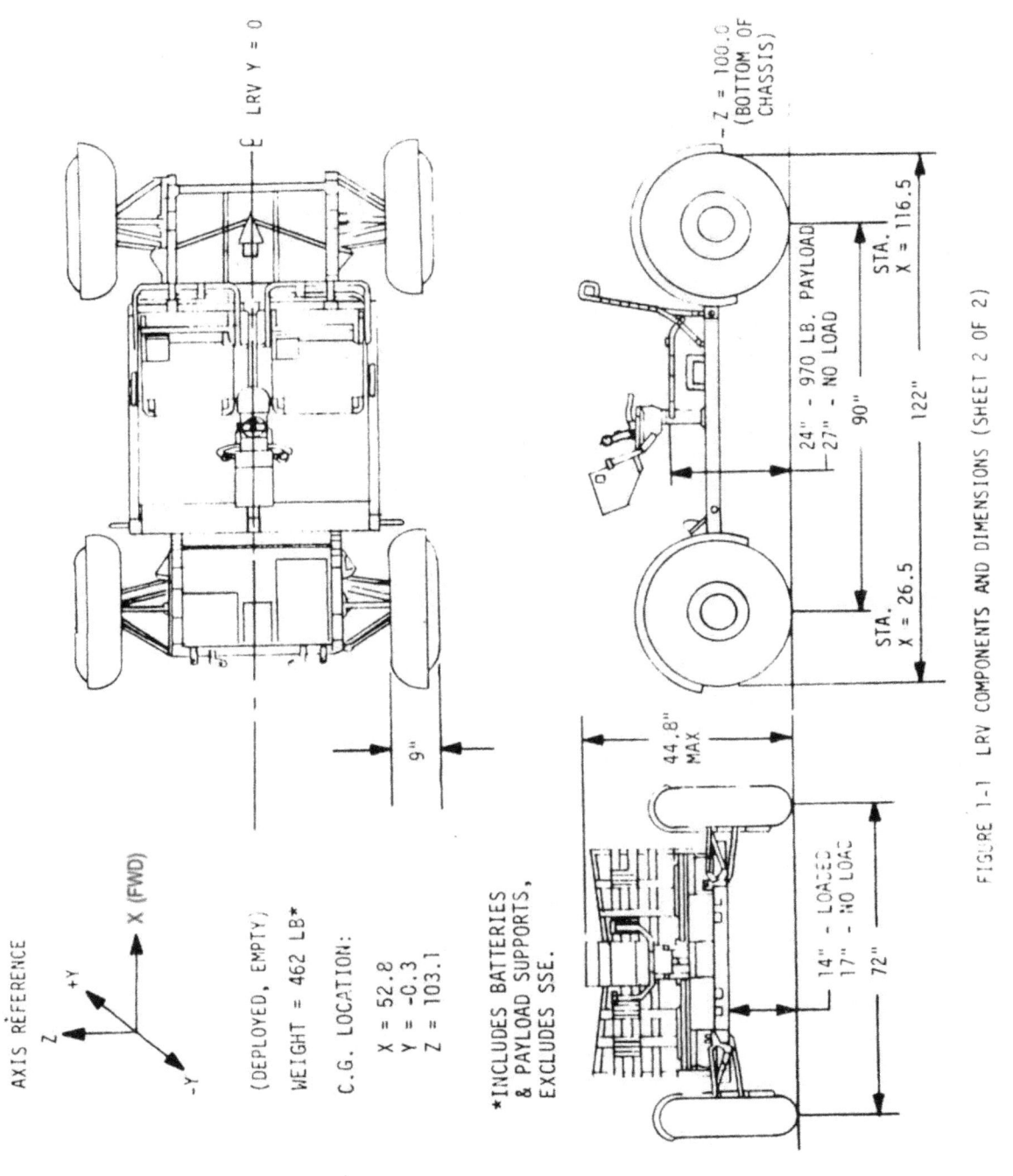

FIGURE 1-1 LRV COMPONENTS AND DIMENSIONS (SHEET 2 OF 2)

LS006-002-2H
LUNAR ROVING VEHICLE
OPERATIONS HANDBOOK

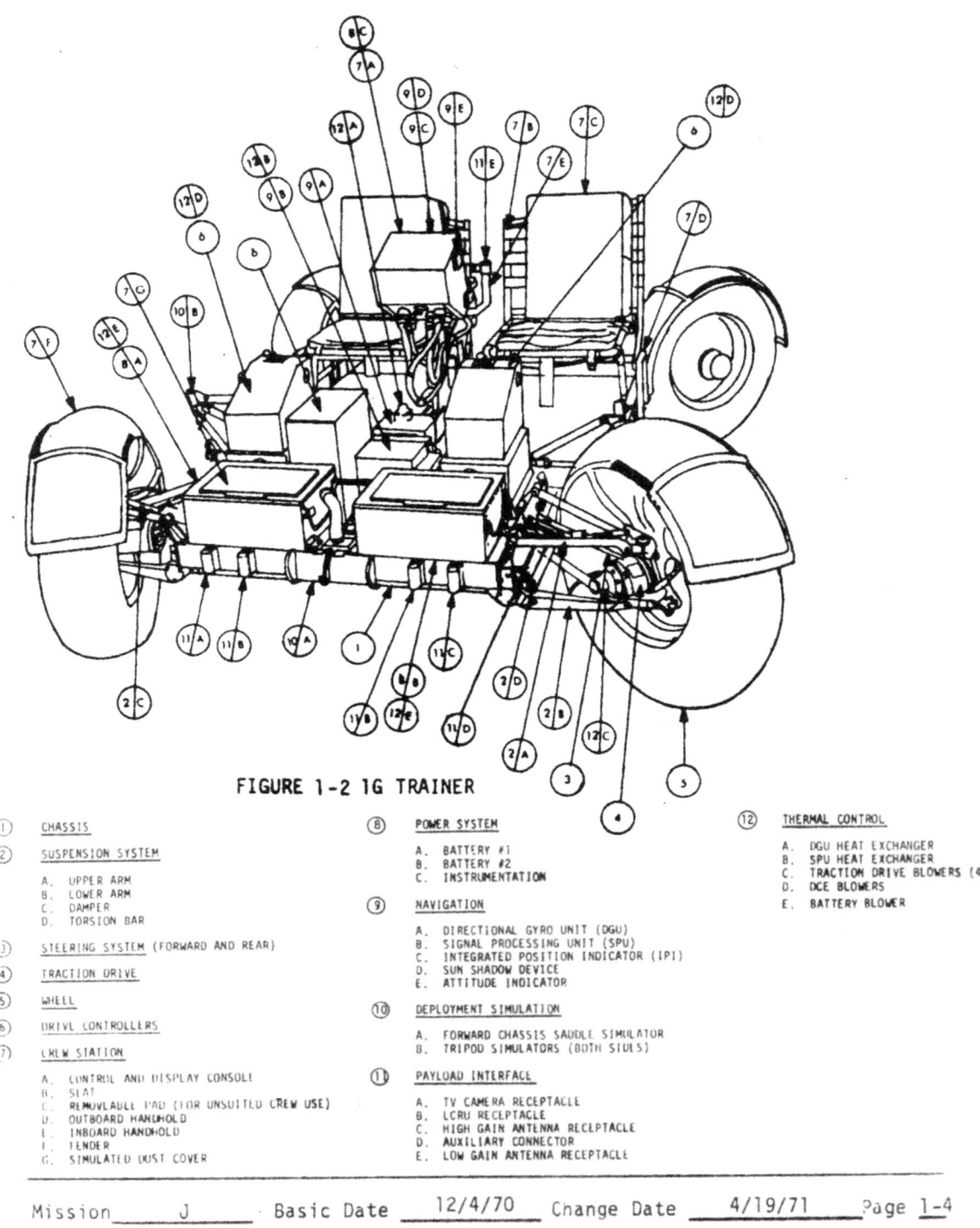

FIGURE 1-2 1G TRAINER

1. CHASSIS
2. SUSPENSION SYSTEM
   A. UPPER ARM
   B. LOWER ARM
   C. DAMPER
   D. TORSION BAR
3. STEERING SYSTEM (FORWARD AND REAR)
4. TRACTION DRIVE
5. WHEEL
6. DRIVE CONTROLLERS
7. CREW STATION
   A. CONTROL AND DISPLAY CONSOLE
   B. SEAT
   C. REMOVABLE PAD (FOR UNSUITED CREW USE)
   D. OUTBOARD HANDHOLD
   E. INBOARD HANDHOLD
   F. FENDER
   G. SIMULATED DUST COVER
8. POWER SYSTEM
   A. BATTERY #1
   B. BATTERY #2
   C. INSTRUMENTATION
9. NAVIGATION
   A. DIRECTIONAL GYRO UNIT (DGU)
   B. SIGNAL PROCESSING UNIT (SPU)
   C. INTEGRATED POSITION INDICATOR (IPI)
   D. SUN SHADOW DEVICE
   E. ATTITUDE INDICATOR
10. DEPLOYMENT SIMULATION
    A. FORWARD CHASSIS SADDLE SIMULATOR
    B. TRIPOD SIMULATORS (BOTH SIDES)
11. PAYLOAD INTERFACE
    A. TV CAMERA RECEPTACLE
    B. LCRU RECEPTACLE
    C. HIGH GAIN ANTENNA RECEPTACLE
    D. AUXILIARY CONNECTOR
    E. LOW GAIN ANTENNA RECEPTACLE
12. THERMAL CONTROL
    A. DGU HEAT EXCHANGER
    B. SPU HEAT EXCHANGER
    C. TRACTION DRIVE BLOWERS (4)
    D. DCE BLOWERS
    E. BATTERY BLOWER

Mission __J__   Basic Date __12/4/70__   Change Date __4/19/71__   Page __1-4__

LS006-002-2H
LUNAR ROVING VEHICLE
OPERATIONS HANDBOOK

1.3   MOBILITY SUBSYSTEM

The mobility subsystem (figure 1-3) consists of the chassis and equipment and controls necessary to propel, suspend, brake and steer the LRV.

1.3.1   Wheel

Each wheel (figure 1-4, Sh 1) includes an open wire mesh tire with chevron tread covering 50 percent of the surface contact area. The tire inner frame prevents excessive deflection of the outer wire mesh frame under high impact load conditions.

Each wheel has a decoupling mechanism (figure 1-5) and can be decoupled from the traction drive by operating the two decoupling mechanisms (figure 1-10) which allows the wheel to "free-wheel" about a bearing independent of the drive train. This decoupling mechanism can also be used to re-engage the wheel with the traction drive. Decoupling disables the brake on the affected wheel.

1G Trainer Notes

1. The 1G Trainer tires for primary use are pneumatic automobile tires (figure 1-4, Sh 2) Special wire mesh wheels are also available for use with the 1G Trainer.

2. The 1G Trainer has simulated wheel decoupling mechanisms to duplicate the LRV-to-Crew interface. Operation of this simulated mechanism, however, will not effect actual decoupling. Procedures for 1G Trainer wheel decoupling are shown in Section 8. Wheel decoupling on the 1G Trainer does not disable the brake on the affected wheels.

1.3.2   Traction Drive

Each wheel is provided with a separate traction drive (figure 1-5, Sh 1) consisting of a harmonic drive reduction unit, drive motor and brake assembly. Each traction drive is hermetically sealed to maintain a 7.5 PSIA internal pressure for improved brush lubrication. Each traction drive also contains an odometer pickup which transmits a pulse to the navigation system signal processing unit at the rate of nine pulses per wheel revolution.

LS006-002-2H
LUNAR ROVING VEHICLE
OPERATIONS HANDBOOK

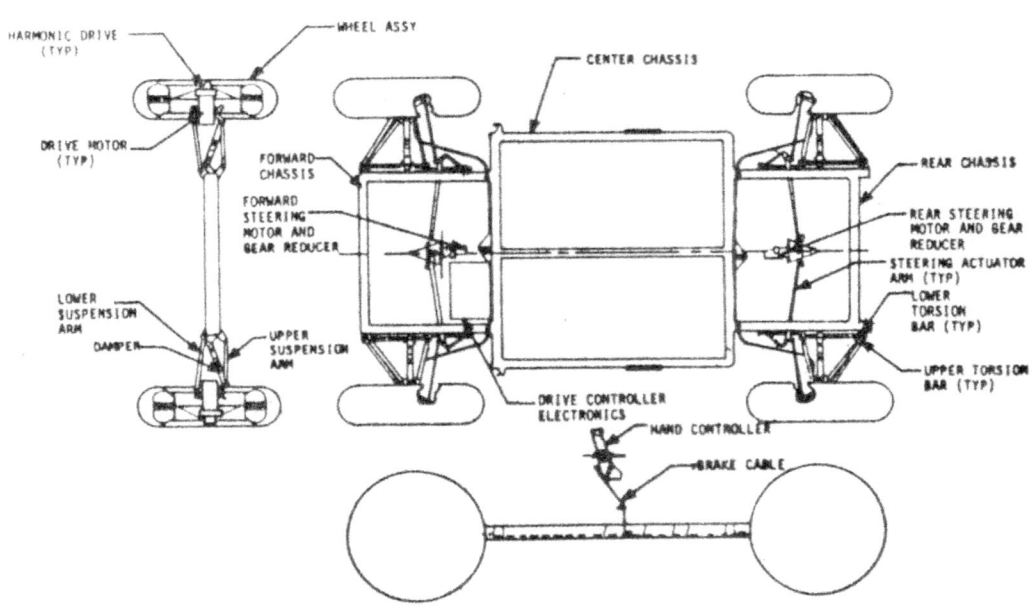

FIGURE 1-3 MOBILITY SUBSYSTEM

LS006-002-2H
LUNAR ROVING VEHICLE
OPERATIONS HANDBOOK

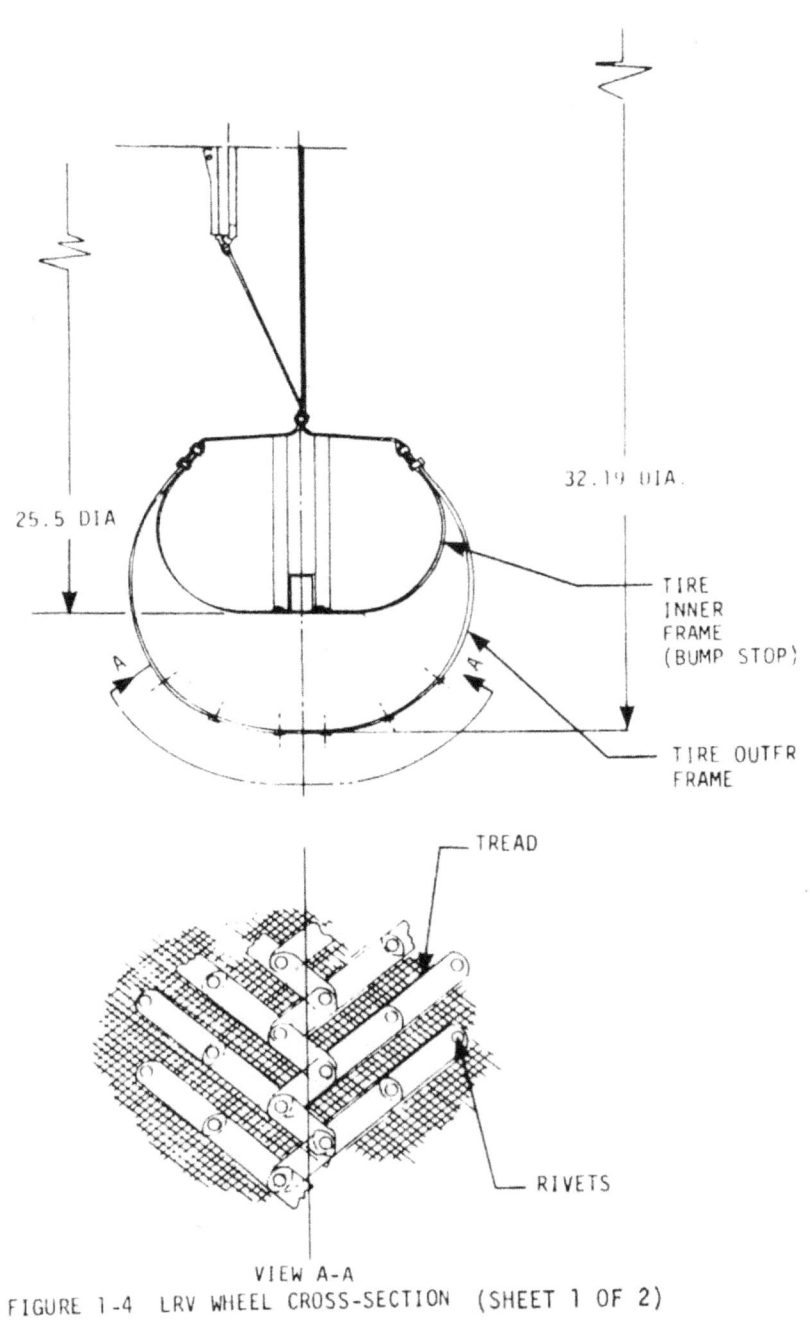

VIEW A-A
FIGURE 1-4 LRV WHEEL CROSS-SECTION (SHEET 1 OF 2)

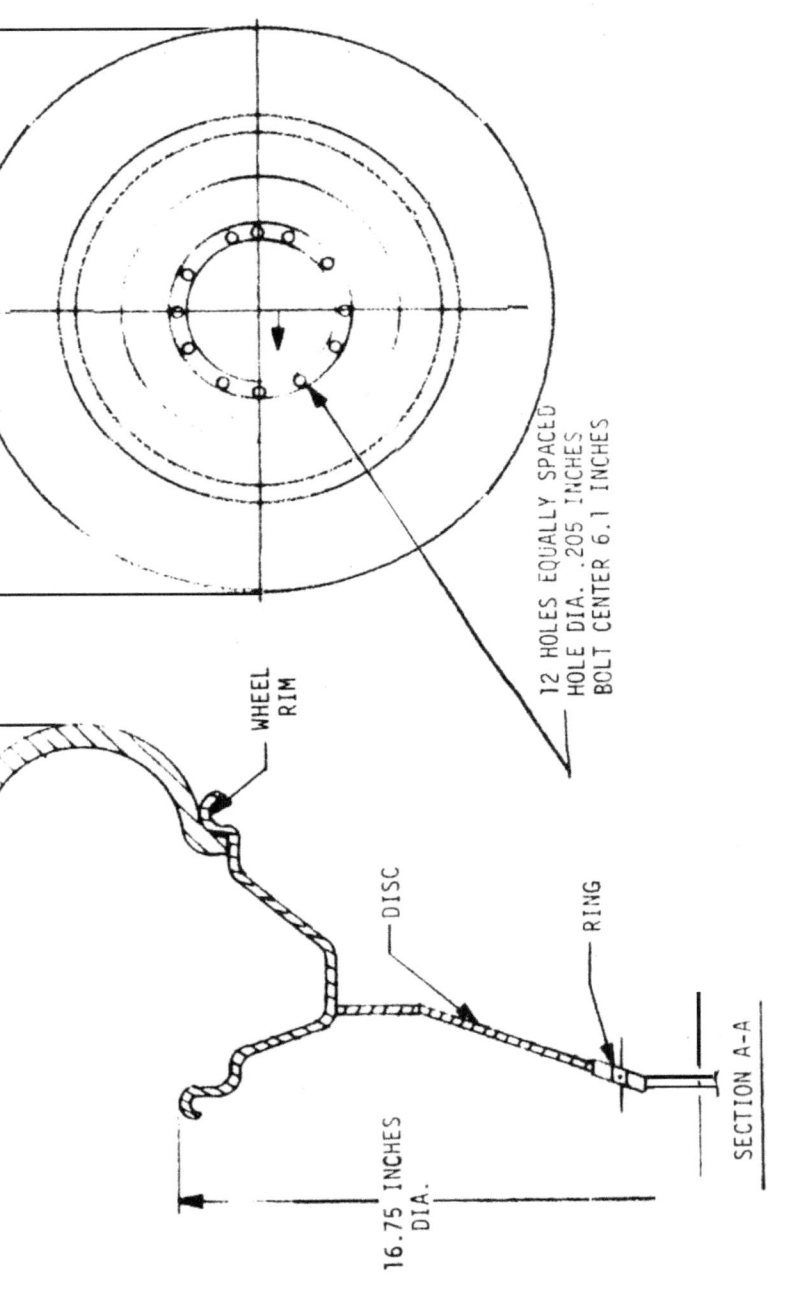

FIGURE 1-4 1G TRAINER WHEEL AND PNEUMATIC TIRE (SHEET 2 OF 2)

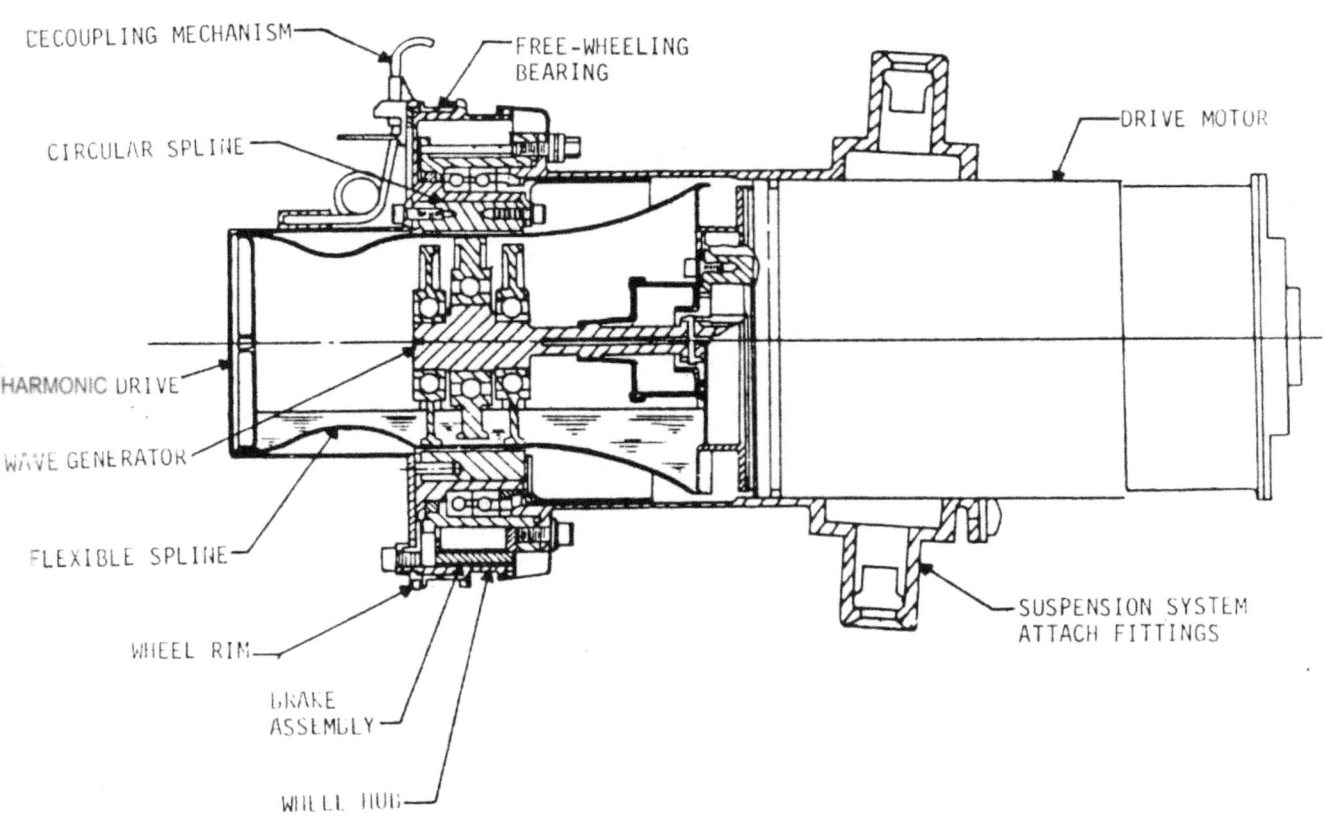

FIGURE 1-5. LRV TRACTION DRIVE ASSEMBLY (SHEET 1 OF 2)

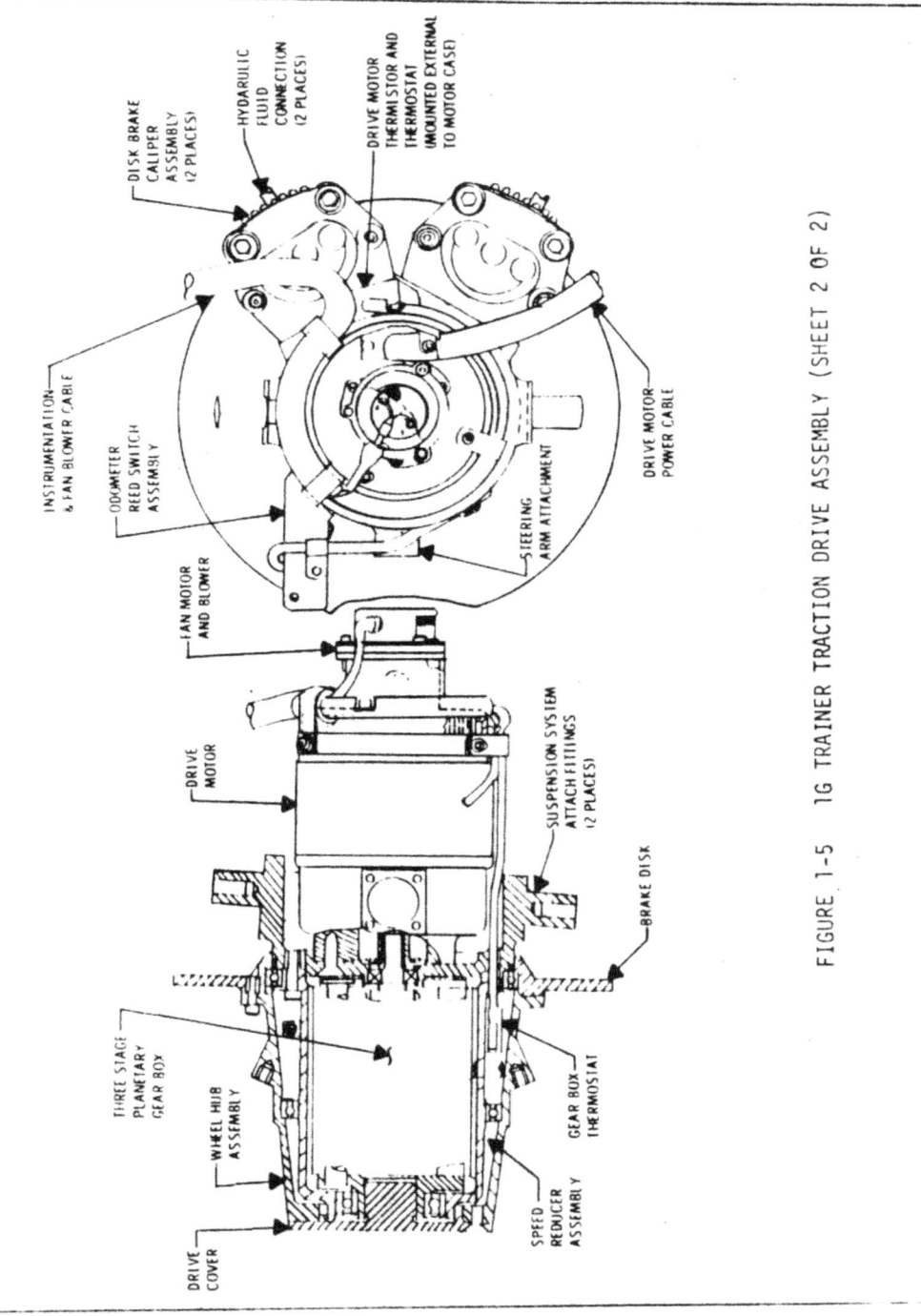

FIGURE 1-5 1G TRAINER TRACTION DRIVE ASSEMBLY (SHEET 2 OF 2)

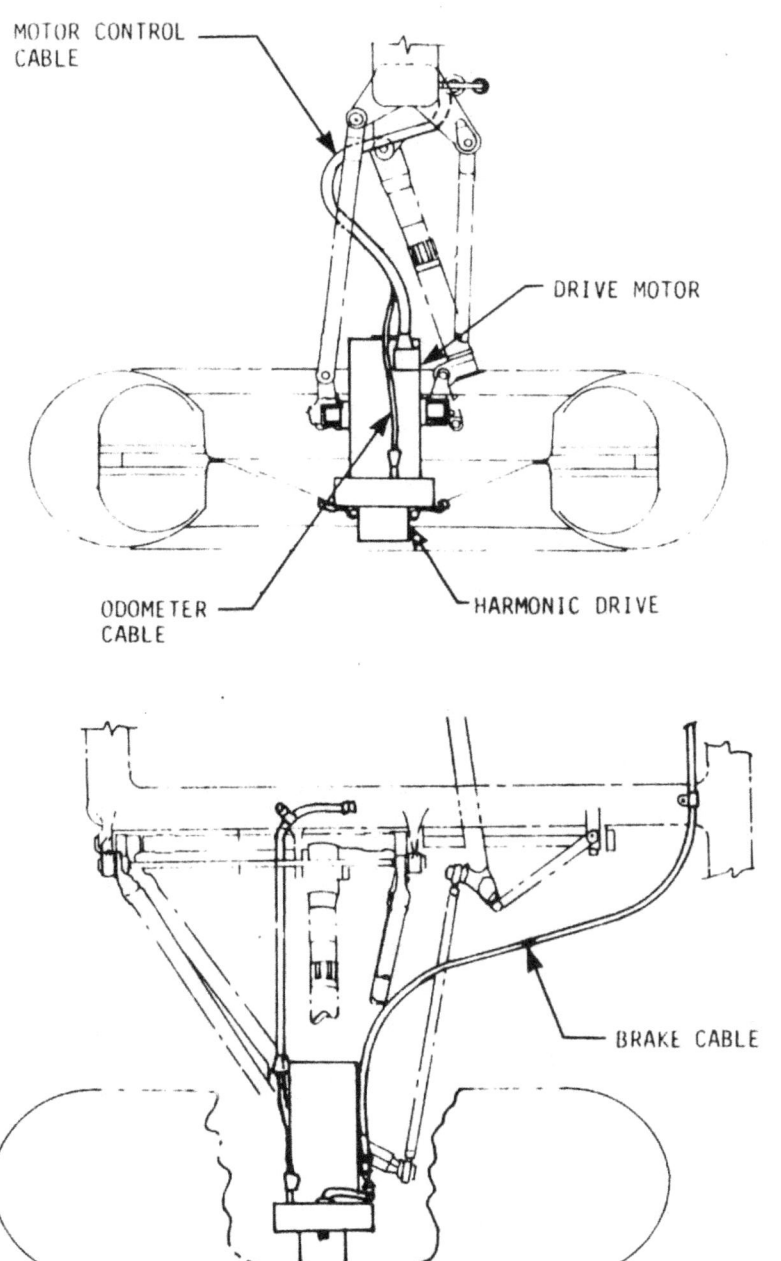

FIGURE 1-6. TRACTION DRIVE INSTALLATION

1G Trainer Notes

1. The traction drive for the 1G Trainer has a 3-stage planetary gear box in lieu of the harmonic drive, (figure 1-5, Sh 2).

2. 1G Trainer traction drives are not hermetically sealed.

1.3.2.1   Harmonic Drive

The four harmonic drive reduction units transmit torque to each wheel. Input torque to the four harmonic drives is supplied by the four electric drive motors. The harmonic drives reduce the motor speed at the rate of 80:1 and allow continuous application of torque to the wheels at all speeds without requiring gear shifting. Speed/torque/efficiency characteristics of the harmonic drive units are shown in Appendix A.

1.3.2.2   Drive Motor

The drive motors are direct current series, brush type motors which operate from a nominal input voltage of 36 VDC. Speed control for the motors is furnished by pulse width modulation from the drive controller electronic package. Performance characteristics for the drive motors are shown in Appendix A. Suspension system attach fittings on each motor also form the king-pin for the LRV steering system. Each motor is instrumented for thermal monitoring. An analog temperature measurement from a thermistor at the stator field is displayed on the control and display panel. In addition, each motor contains a thermal switch which closes on increasing temperature at 400°F and provides an input signal to the caution and warning system to actuate the warning flag.

1G Trainer Notes

1. The 1G Trainer drive motors operate from a nominal input voltage of 34 VDC.

2. The 1G Trainer gear box thermal switch will actuate the warning flag when a gear box temperature reaches 200°F. The indicated temperature, however, will be 450°F to 500°F upon actuation, since the readouts are biased.

3. The 1G Trainer motor temperature switch is set to actuate the flag when the motor external case temperature reaches 225°F. This temperature at the case would correspond to a rotor temperature of about 450°F.

LS006-002-2H
LUNAR ROVING VEHICLE
OPERATIONS HANDBOOK

1.3.2.3   Brakes

Each traction drive is equipped with a mechanical brake actuated by a cable connected to a linkage in the hand controller. Stopping distance capability using these brakes is shown in Appendix A.

Braking is accomplished by moving the hand controller rearward. This operation de-energizes the drive motor and, through a linkage and cable, forces brake shoes against a brake drum which stops the rotation of the wheel hub about the harmonic drive.

1G Trainer Note

> The 1G Trainer brakes are hydraulically actuated disc brakes. Brakes are actuated by the hand controller in the same manner as the LRV mechanical brakes.

1.3.3   Suspension

The chassis (figure 1-7) is suspended from each wheel by a pair of parallel triangular arms connected between the LRV chassis and each traction drive. Loads are transmitted to the chassis through each suspension arm to a separate torsion bar for each arm. Wheel vertical travel and rate of travel is limited by a linear damper connected between the chassis and each traction drive. The deflection of the suspension system and tires combine to allow 14 inches of chassis ground clearance when the LRV is fully loaded and 17 inches when unloaded.

Damping energy heats the fluid in the damper. The heat is conducted from the fluid to the damper walls for dissipation.

The suspension systems can be rotated approximately 135 degrees to allow folding and LRV stowage in the LM.

1G Trainer Notes

1. 1G Trainer suspension is not designed to allow folding for LM stowage.

2. 1G Trainer suspension system contains only a lower torsion bar on each wheel.

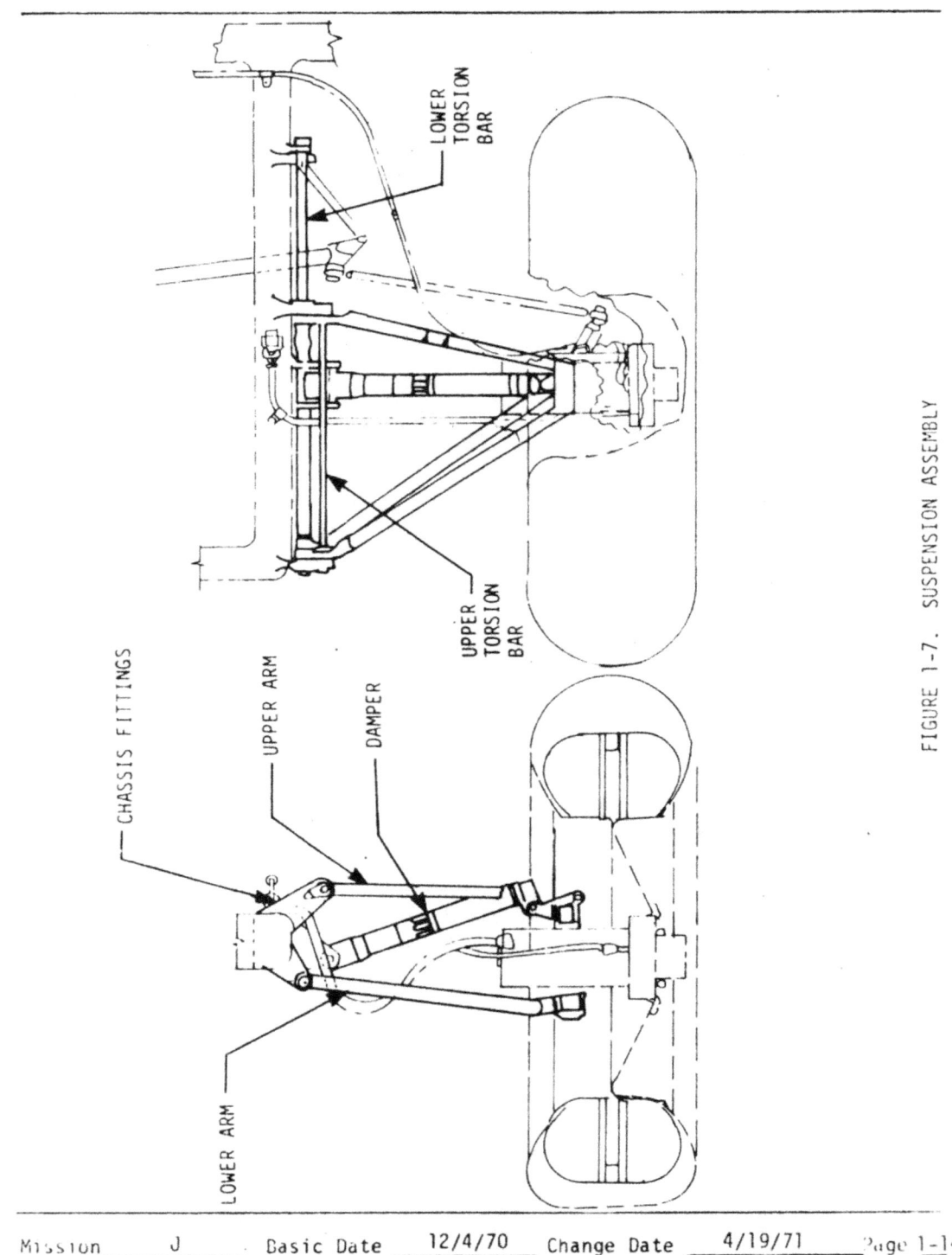

FIGURE 1-7. SUSPENSION ASSEMBLY

LS006-002-2H
LUNAR ROVING VEHICLE
OPERATIONS HANDBOOK

1.3.4   Steering

LRV steering (figure 1-8) is accomplished by Ackermann-geometry steering of both the front and rear wheels allowing a wall-to-wall turning radius of 122 inches. Steering is controlled by moving the hand controller left or right from the nominal position. This operation energizes separate electric motors for the front and rear wheels, and through a servo system, provides a steering angle proportional to the position of the hand controller. (The steering control block diagram is shown in figure 1-9).

Each steering motor is connected to a speed reducer which drives a spur gear sector which, in turn, actuates the steering linkage to accomplish the change in steering angle. Maximum travel position of the sector provides an outer wheel angle of 22 degrees and inner wheel angle of 50 degrees. The steering rate is such that lock-to-lock steering can be accomplished in 5.5 ($\pm$ 0.5) seconds.

The front and rear steering assemblies are mechanically independent of each other. In the event of motor/speed reducer failure, the steering linkage can be disengaged from the sector, the wheels can be centered and locked, and operations can continue using the remaining active steering assembly. Steering disconnect points are shown in figure 1-10. Forward steering reconnection cannot be accomplished by a crewman. The rear steering reconnection can be accomplished by a crewman as described in Section 2.9.

1G Trainer Notes

1. The 1G Trainer steering utilizes continuously operating steering motor. Hand controller movement energizes the appropriate (one of two) counter rotating magnetic particle clutches, thereby engaging the load and effecting steering. A magnetic brake is actuated when the clutches are not engaged.

2. The 1G Trainer has simulated steering decoupling mechanisms to duplicate the LRV-to-Crew interface. Operation of this simulated mechanism, however, will not effect actual decoupling. Procedures for 1G Trainer steering decoupling are shown in Section 8.0.

1.3.5   Hand Controller

The hand controller (figure 1-11) provides the steering, speed, and braking commands to the drive controller electronics. The drive controller electronics then processes these hand controller commands to the appropriate drive motors and steering motors to effect the desired control function. The hand controller is also used as the mechanical brake lever.

Mission   J   Basic Date 12/4/70 Change Date 4/19/71   Page 1-15

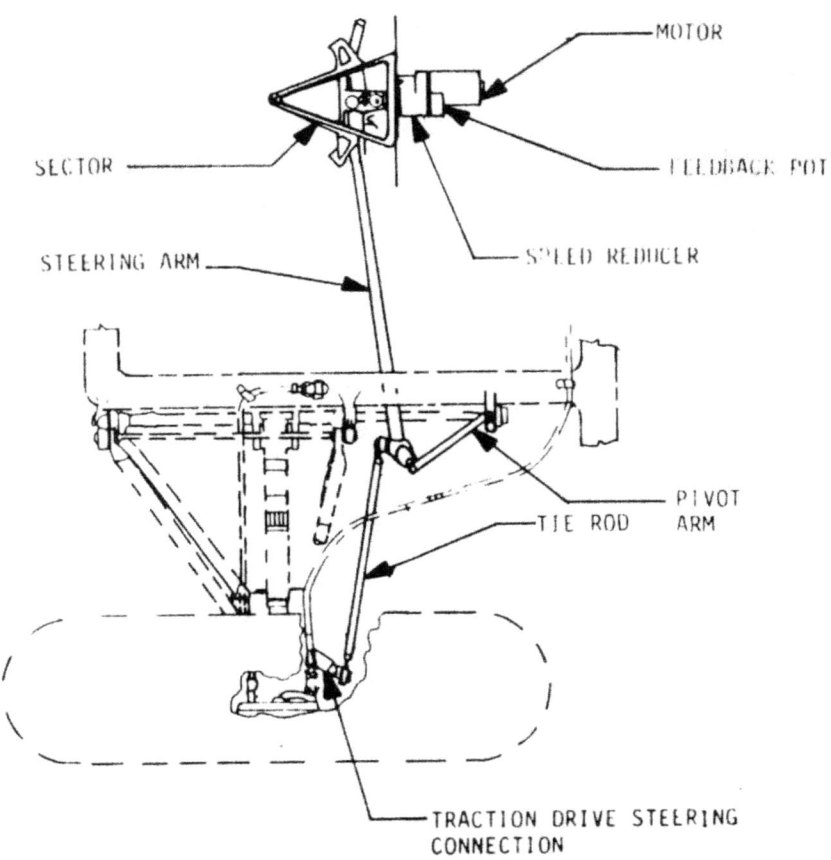

FIGURE 1-8. STEERING ASSEMBLY

LS006-002-2H
LUNAR ROVING VEHICLE
OPERATIONS HANDBOOK

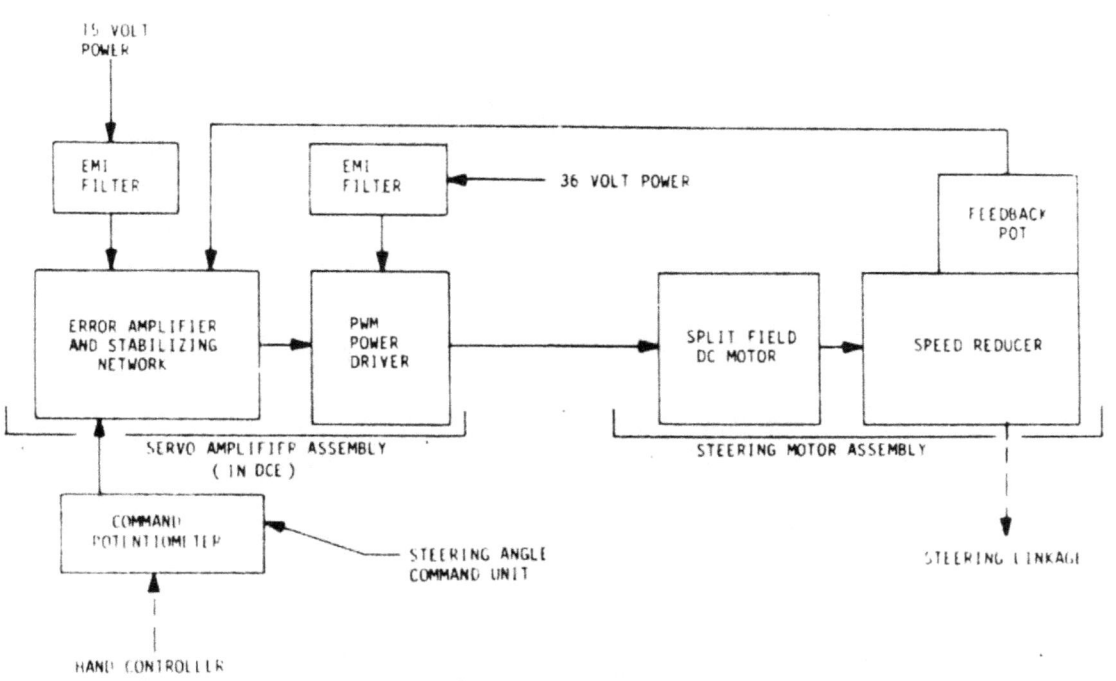

FIGURE 1-9. STEERING CONTROL BLOCK DIAGRAM

LS006-002-2H
LUNAR ROVING VEHICLE
OPERATIONS HANDBOOK

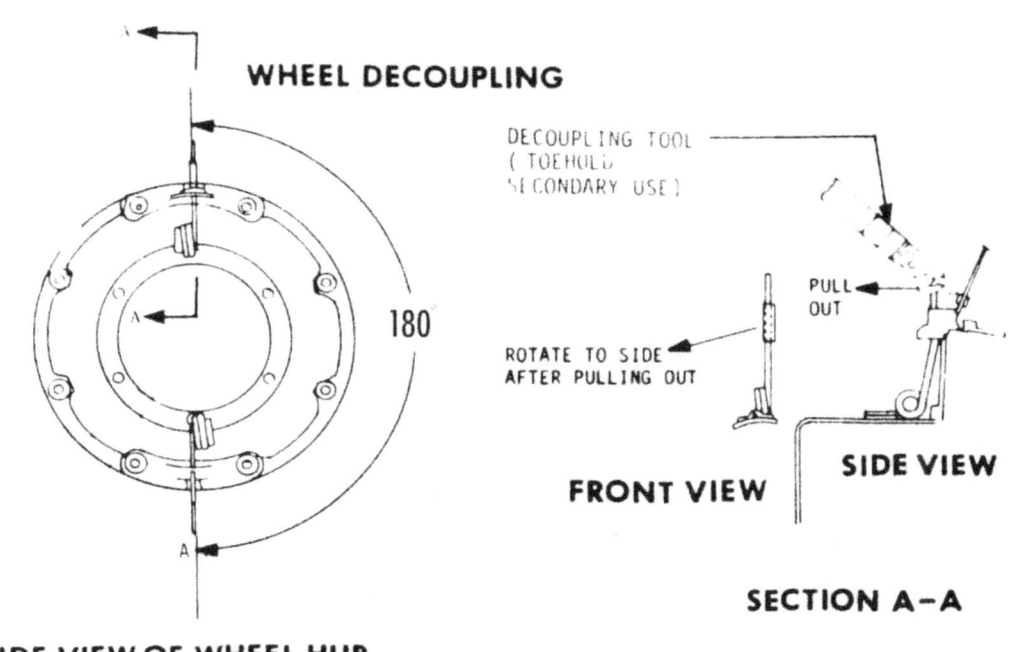

FIGURE 1-10. WHEEL AND STEERING DISCONNECTS

LS006-002-2H
LUNAR ROVING VEHICLE
OPERATIONS HANDBOOK

HAND CONTROLLER OPERATION:

T-HANDLE PIVOT FORWARD - INCREASED DEFLECTION FROM NEUTRAL INCREASES FORWARD SPEED.

T-HANDLE PIVOT REARWARD - INCREASED DEFLECTION FROM NEUTRAL INCREASES REVERSE SPEED.

T-HANDLE PIVOT LEFT - INCREASED DEFLECTION FROM NEUTRAL INCREASES LEFT STEERING ANGLE.

T-HANDLE PIVOT RIGHT - INCREASED DEFLECTION FROM NEUTRAL INCREASES RIGHT STEERING ANGLE.

T-HANDLE DISPLACED REARWARD - REARWARD MOVEMENT INCREASES BRAKING FORCE. FULL 3 INCH REARWARD APPLIES PARKING BRAKE. MOVING INTO BRAKE POSITION DISABLES THROTTLE CONTROL AT 15° MOVEMENT REARWARD.

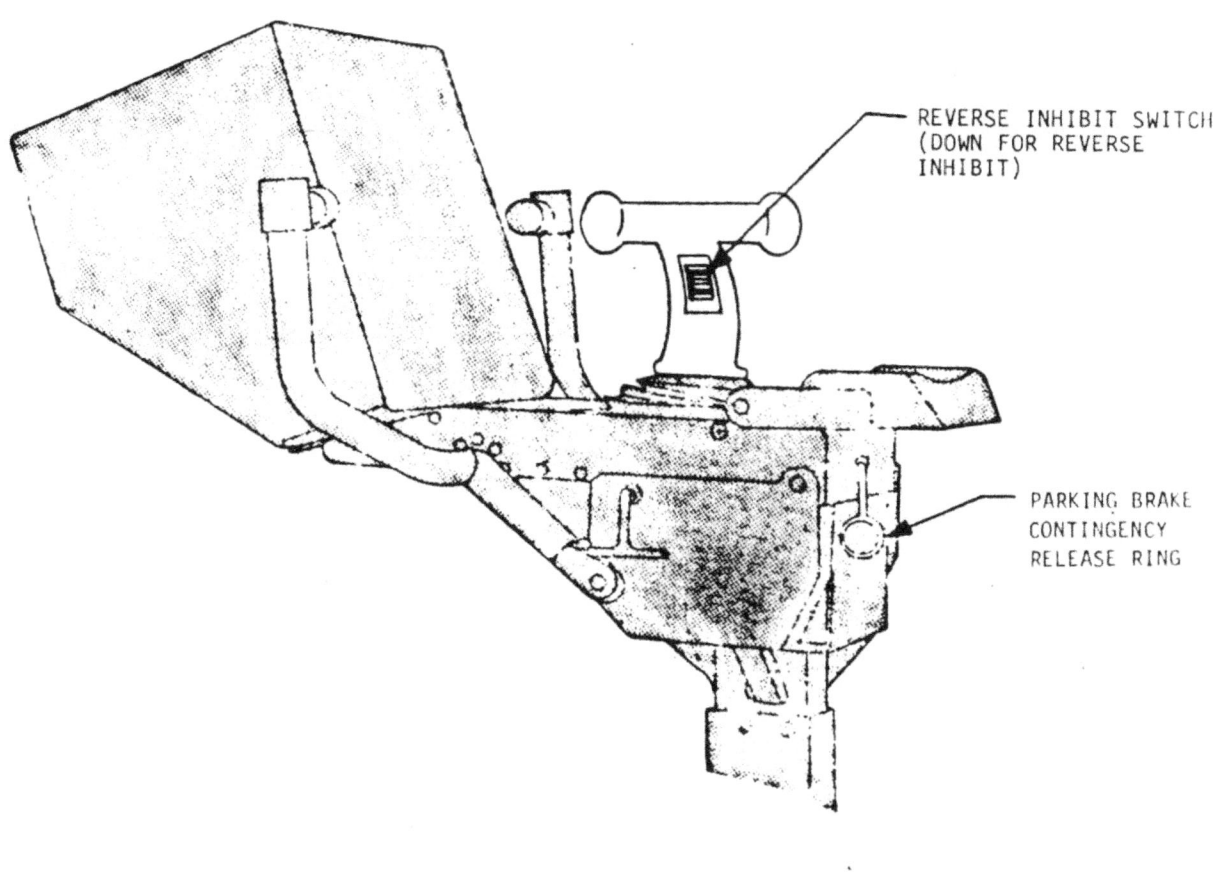

FIGURE 1-11. HAND CONTROLLER

LS006-002-2H
LUNAR ROVING VEHICLE
OPERATIONS HANDBOOK

1.3.5.1   Speed Control

Forward movement of the hand controller about the T-handle throttle pivot axis proportionately increases forward speed. A neutral dead band exists for about the first 1.5 degrees of forward motion. A constant torque of about 6 inch-pounds is required to move the hand controller beyond the limit of the dead band (figure 1-12). The nine degree position corresponds to a pulse duty cycle of approximately 50 percent, at each drive motor, i.e., the motors are at 50 percent of maximum speed condition. The maximum power setting is achieved by pivoting the hand controller to the hard stop (maximum) position at approximately 14 degrees. To decelerate, the hand controller is pivoted rearward. The torques required are shown in figure 1-12. To place the vehicle in neutral, the hand controller is pivoted rearward to the zero ($\pm$ 1/2) degree position.

With the reverse inhibit switch in the down position, the hand controller can be pivoted forward only, thereby preventing inadvertently placing the vehicle in reverse.

To operate the vehicle in reverse, the reverse inhibit switch is placed in the up position and the hand controller pivoted rearward about the throttle pivot point. Torque vs. displacement characteristics for reverse are identical to forward speed operation as shown by figure 1-12.

The vehicle must be brought to a full stop before a direction change is commanded. Direction change is automatically inhibited at vehicle speeds greater than 1 KPH.

The hand controller will remain in the selected forward or reverse speed position in the crewmen "hands off" condition.

1.3.5.2   Steering Control

Pivoting the hand controller left or right about the roll pivot point proportionally changes the wheel steering angle. The steering control, like the throttle control, has a 1/2 degree neutral dead band on either side of zero. (See figure 1-13). A torque of 7 in-lbs. is required to roll the hand controller beyond the neutral position to begin steering angle change. Torque required for increasing the displacement angle about the roll pivot point increases linearly until a displacement of approximately 9 degrees is reached. At the 9 degree position, a soft stop is encountered which requires a step-function torque increase of 5 in-lbs. to pivot the hand controller further outboard for increasing the steering angle. After passing the soft stop

LS006-002-2H
LUNAR ROVING VEHICLE
OPERATIONS HANDBOOK

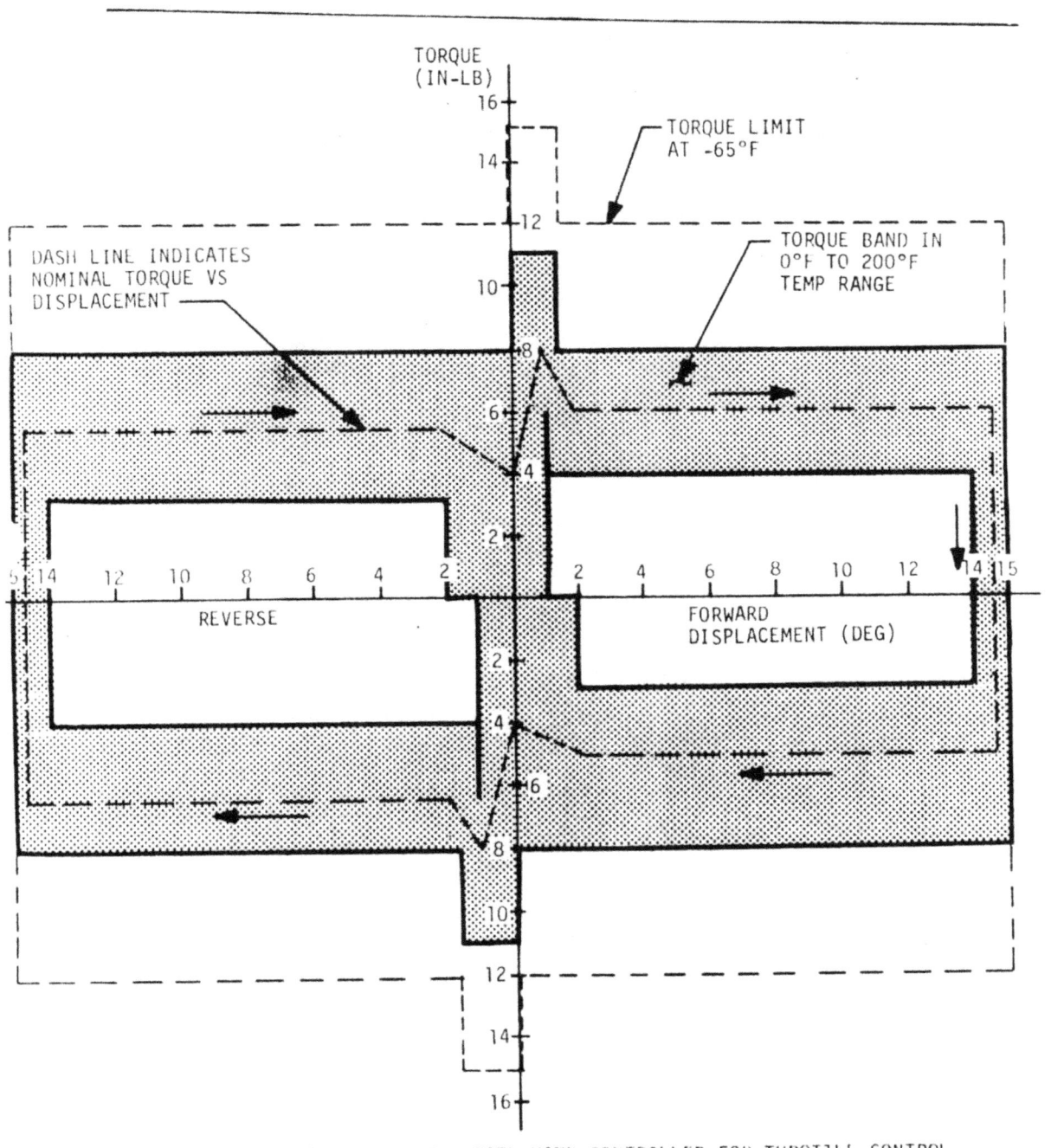

FIGURE 1-12 TORQUE REQUIRED TO ROTATE HAND CONTROLLER FOR THROTTLE CONTROL

LS006-002-2H
LUNAR ROVING VEHICLE
OPERATIONS HANDBOOK

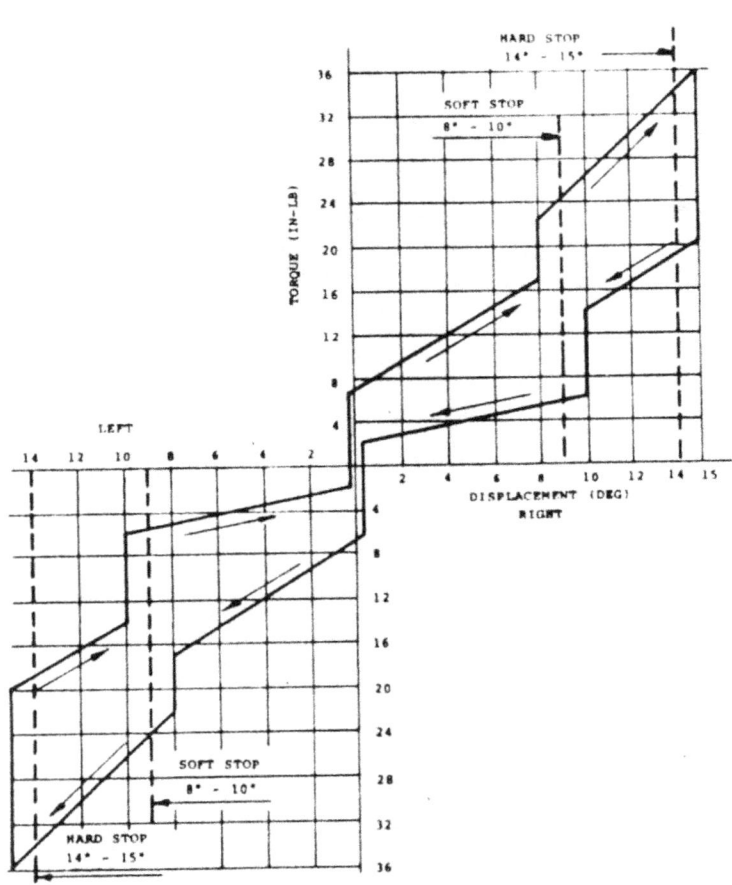

FIGURE 1-13 TORQUE REQUIRED TO ROTATE HAND CONTROLLER FOR STEERING CONTROL

LS006-002-2H
LUNAR ROVING VEHICLE
OPERATIONS HANDBOOK

1.3.5.2 (Continued)

position, the torque required to pivot the hand controller further outboard increases linearly with displacement until the hand controller hard stop limit is reached at the 14 degree outboard hand controller position.

The hand controller is spring loaded to return to the neutral steering position when released.

1.3.5.3 Braking Control

Braking is initiated with the LRV in either forward or reverse by pivoting the hand controller rearward about the brake pivot point. The force required to move the hand controller rearward to increase braking is shown in Figure 1-14. Forward and reverse power is disabled when the brake is displaced 15 degrees.

A three inch rearward displacement of the hand controller engages and locks the parking brake. To disengage the parking brake, the hand controller is placed in the steer left position. A contingency release (figure 1-11) is provided should the brake fail to release when moved to the steer left position. Contingency brake release is effected by moving the brake to full rearward displacement, pulling the release ring, allowing the brake to release and then releasing the ring.

1G Trainer Notes

1. The 1G Trainer hand controller operation (speed, steering and brake) is identical to the LRV hand controller operation, with the exception: If the hand controller is in full throttle position when full brakes are applied, drive power will not be automatically cut-out. This condition (true for both forward and reverse operation) resulted as a consequence of by-passing the DCE logic to eliminate voltage drop and thereby increase trainer top speed.

2. The 1G Trainer brake cables can be adjusted to provide simulated lunar surface stopping characteristics.

Mission   J   Basic Date   12/4/70   Change Date   4/19/71   Page 1-23

LS006-002-2H
LUNAR ROVING VEHICLE
OPERATIONS HANDBOOK

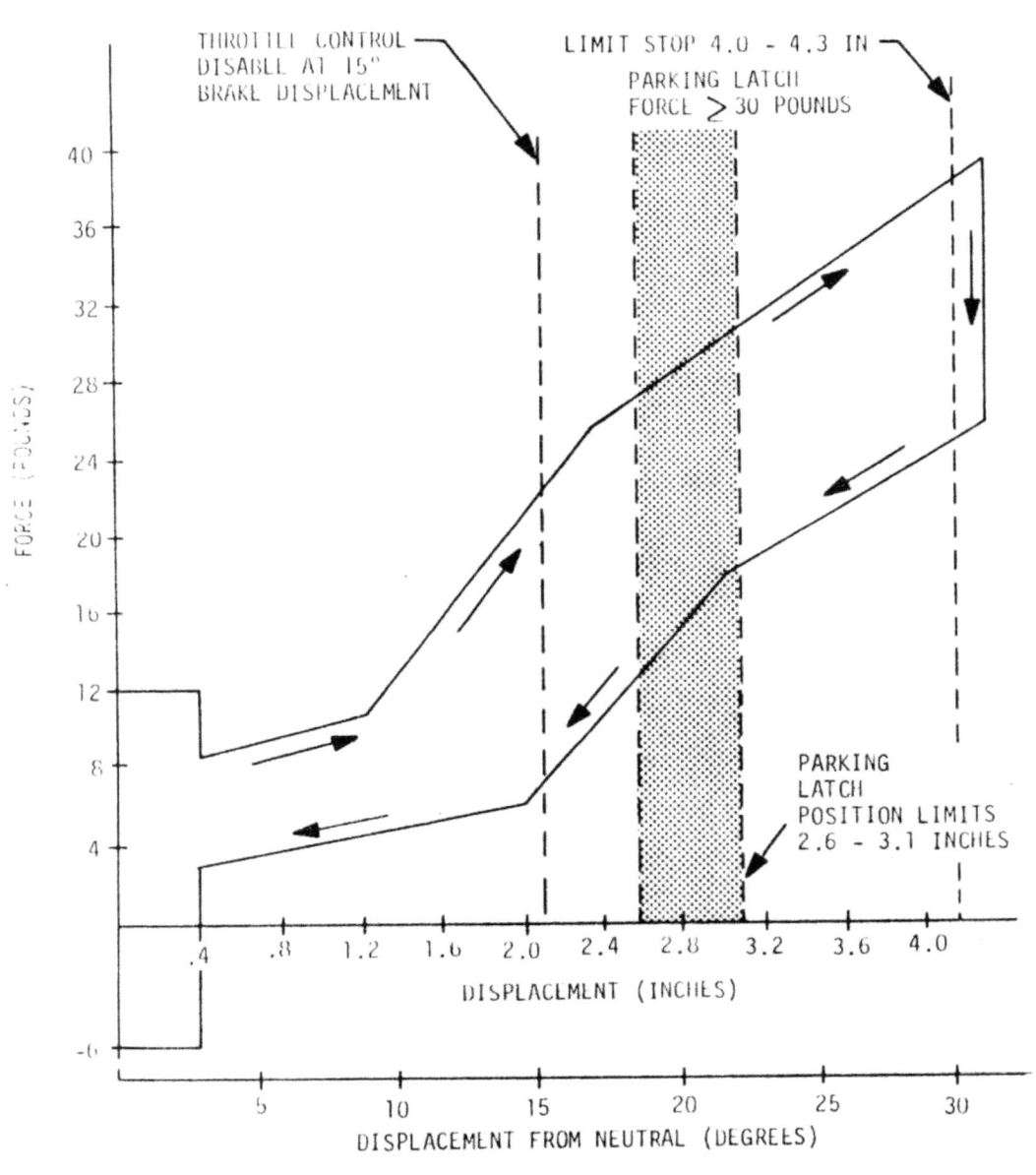

FIGURE 1-14 BRAKE CONTROL FORCE VS DISPLACEMENT

LS006-002-2H
LUNAR ROVING VEHICLE
OPERATIONS HANDBOOK

1.3.7 Drive Control Electronics

The Drive Control Electronics (DCE) accepts forward and reverse speed control signals from the Hand Controller and transmits them to the drive motors in a format which allows drive motor speed control. The steering logic servo amplifier assembly (previously described in paragraph 1.3.4 and figure 1-9) is also contained within the DCE. In addition, the Drive Control Electronics accepts odometer signals from the traction drives and processes the signals for odometer/speedometer readout. The basic manner of operation of the DCE is described below and illustrated in figure 1-15.

When the Hand Controller is actuated in either the forward or reverse positions, two basic signals are generated. One signal is constant voltage signal (A) to the traction drive electronics logic which tells whether the Hand Controller is on the forward or reverse side of neutral, and thus causes relay action to control the direction of drive. The other signal (B) is from the command potentiometers and is a variable voltage which reflects the amount of speed desired. This latter signal is proportional to the position of the Hand Controller and is fed to the Pulse Width Modulators (PWM) where the signal is "chopped" into pulses whose width is proportional to the incoming signal strength. The modulated signal (C) is then fed through the Drive Enable switches (astronaut operated) for each Drive Motor whose function is to determine whether the command signal for each drive motor is to be derived from PWM #1 or PWM #2. The position of the PWM select switch -- astronaut operated -- allows disabling of a defective PWM if desired.

1G Trainer Note

The 1G Trainer Drive Enable PWM 1 and PWM 2 positions are common "ON" positions. PWM 1 is an integral part of Drive Controller No. 1 which powers only the two front drive motors and similarly PWM 2 is an integral part of Drive Controller No. 2 which powers only the two rear motors.

After the modulated signal (C) has passed through the Drive Enable switch for each traction drive, it enters a gating switch which serves several purposes. First, it inhibits drive power if the brake is on (D). Second, if drive current becomes excessive, it inhibits drive power until the current level falls to an acceptable value (E). Third, it inhibits drive power momentarily while the Hand Controller is being switched from forward to reverse or reverse to forward (F). If none of these three inhibits is present, then the gate passes the modulated signal (C) on to the power switching driver and the power switch, which produce the proper power levels for motor control (G).

The last step prior to application of power to the motor is selection of forward or reverse motor drive. This is accomplished by the reversing relay and relay driver. The relay driver determines the position of the reversing relays and is actuated by position of the Hand Controller (forward or reverse)

Mission ___J___   Basic Date __12/4/70__   Change Date __4/19/71__   Page __1-25__

LS006-002-2H
LUNAR ROVING VEHICLE
OPERATIONS HANDBOOK

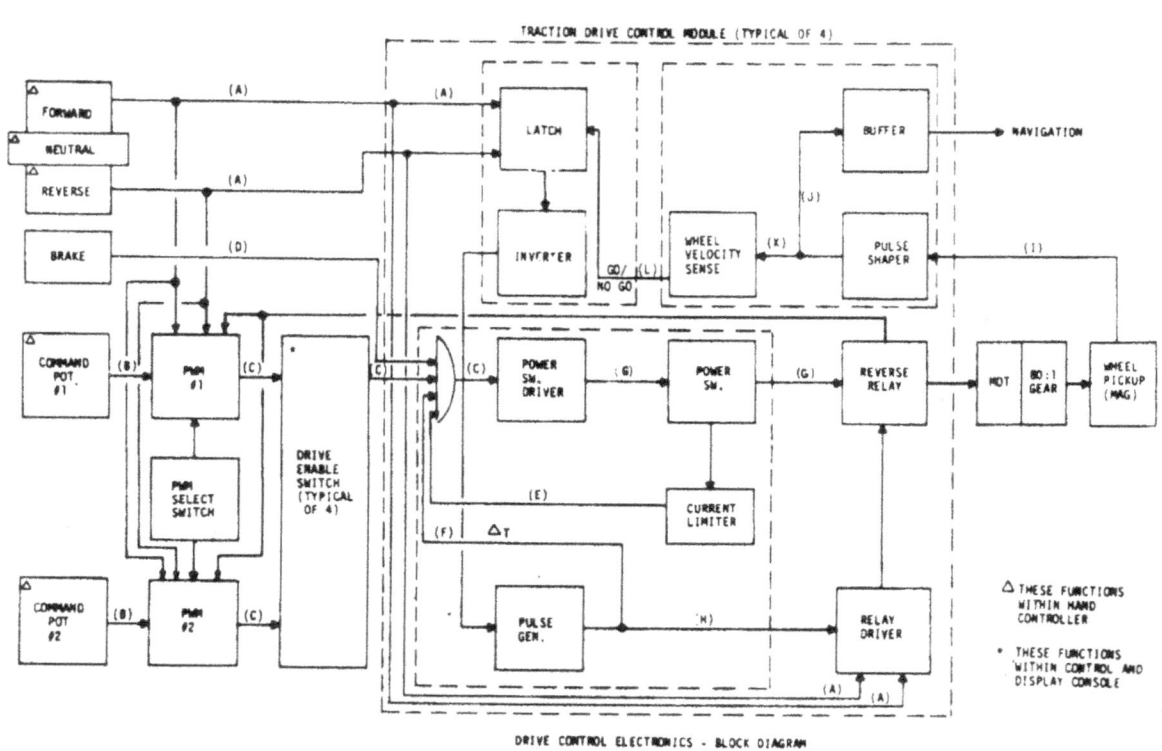

FIGURE 1-15 DRIVE CONTROL ELECTRONICS - BLOCK DIAGRAM

1.3.7 (Continued)

and a signal from the pulse generator (H) which indicates the power to the drive motor has been inhibited and switching can take place. The condition of the reversing relays determine the direction of current through the motor armature and thus the direction of rotation of the motor. The wheels are driven through 80:1 ratio harmonic drive units as explained in paragraph 1.3.3.

Each traction drive has a magnetic pickup for sensing the rotational motion of each wheel. This signal (I) is picked up as a series of pulses and transmitted back into the Traction Drive Control Module where it is properly pulse shaped for use and then used for two purposes. One, it is fed (J) through a buffer to the navigation subsystem for the odometer. Two, it is fed (K) to a wheel velocity sensing circuit which determines if the wheel velocity is greater than 1 KPH. If it is, a signal (L) is sent to the traction drive electronics logic to disallow switching from forward to reverse or reverse to forward until wheel speed drops below 1 KPH. From 1 KPH to full stop the state of the inhibit circuit may be indeterminate, thus it is imperative that the vehicle be brought to a full stop before a direction change is commanded.

LS006-002-2H
LUNAR ROVING VEHICLE
OPERATIONS HANDBOOK

1.4   ELECTRICAL POWER SUBSYSTEM

The electrical power subsystem consists of two batteries, distributing wiring, connectors, switches, circuit breakers and meters for controlling and monitoring electrical power.

1.4.1   Batteries

The LRV contains two primary silver zinc batteries (figure 1-16) each having a nominal voltage of 36 (+5/-3) VDC and each having a capacity of 115 ampere hours. Both batteries are normally used simultaneously on an approximate equal load basis during LRV operation by selection of various load-to-bus combinations through circuit breakers and switch settings on the control and display console.

The batteries are located on the forward chassis enclosed by the thermal blanket and dust covers (figure 1-17). Battery No. 1 (on the left side) is connected thermally to the navigation Signal Processing Unit (SPU), and serves as a partial heat sink for the SPU. Battery No. 2 (on the right side) is thermally tied to the navigation Directional Gyro Unit (DGU) and serves as a heat sink for the DGU.

The batteries are installed in the LRV on the pad at KSC in an activated condition and are monitored for voltage and temperature on the ground until approximately T-18 hours in the countdown. On the lunar surface, the batteries are monitored for temperature, voltage, output current, and remaining ampere-hours. These displays are located on the control and display panel.

Each battery is protected from excessive internal pressure by a pressure relief valve that is set to open at 3.1 to 7 PSI differential pressure. The relief valve closes when the differential pressure is below the valve's relief pressure. Each battery is capable of carrying the entire LRV electrical load, and the circuitry is designed such that in the event one battery fails, the entire electrical load can be switched to the remaining battery. LRV range capability is shown in Appendix A.

1G Trainer Notes

1. 1G Trainer uses two rechargeable nickel cadmium batteries having a voltage output of 34 VDC and a capacity of 24 ampere hours each. Both batteries must be used for 1G Trainer operation.

2. 1G Trainer estimated operation time before recharge for a set of batteries (two batteries per set) is 63 minutes on smooth level ground (800 pound payload configuration and 10 KPH). This estimate includes 19 minutes of stand-by time.

(Continued)

LS006-002-2H
LUNAR ROVING VEHICLE
OPERATIONS HANDBOOK

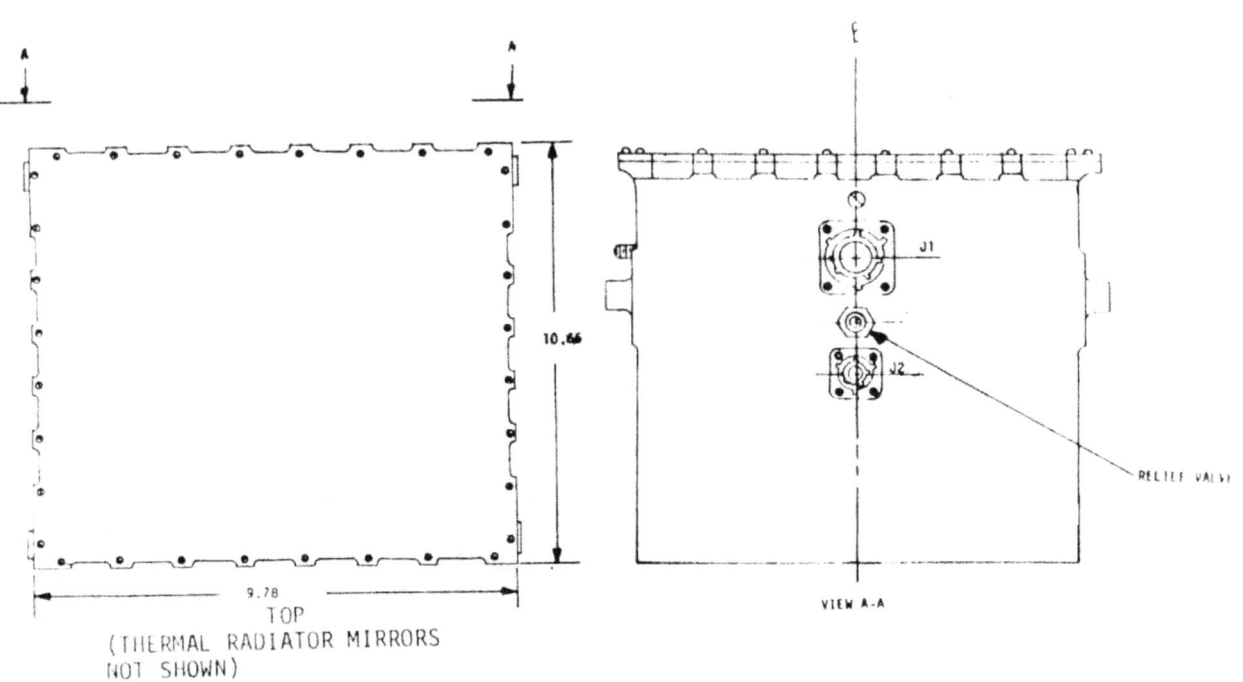

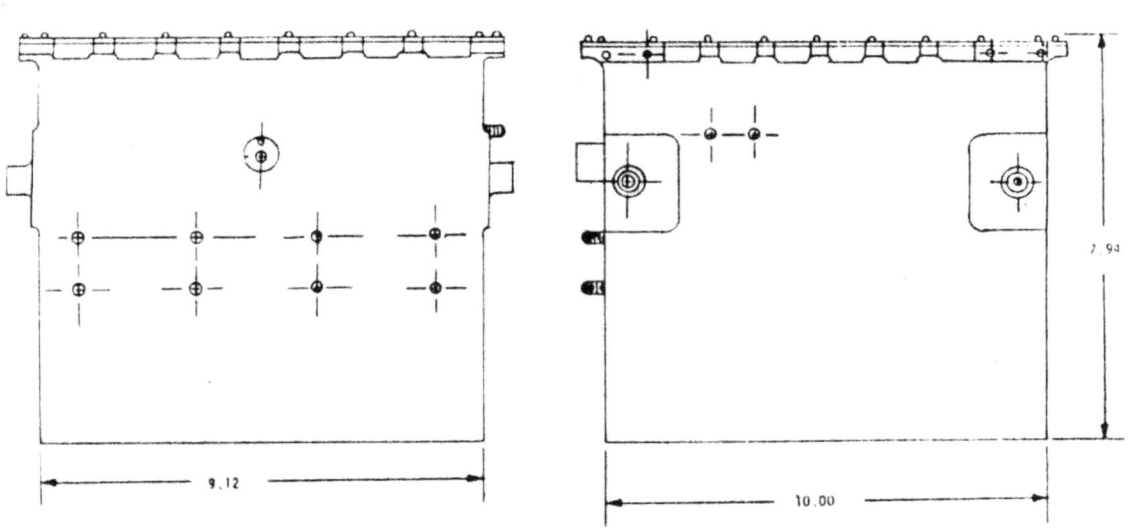

FIGURE 1-16. LRV BATTERY CONFIGURATION

Mission ___J___ Basic Date _12/4/70_ Change Date _4/19/71_ Page 1-29

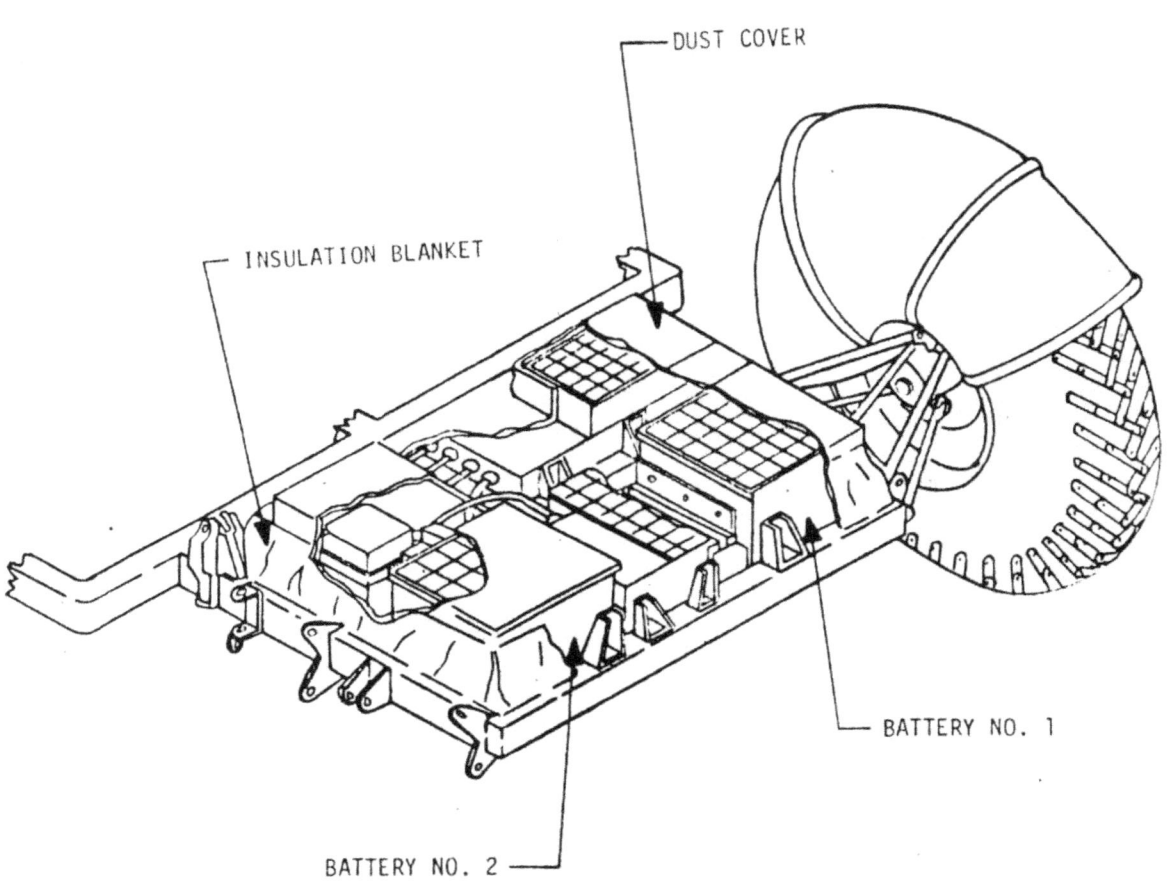

FIGURE 1-17 LRV BATTERIES, THERMAL BLANKET AND DUST COVERS

1G Trainer Notes
(Continued)

3. The 1G Trainer batteries are fan cooled when internal temperatures rise above a preset thermostatic switch value. The batteries are not covered by a thermal blanket.

1.4.2   Distribution and Monitoring System

The electrical distribution and monitoring schematics for the LRV are shown in figures 1-18 and 1-19. The switch and circuit breaker arrangement is designed to allow switching any electrical load to either battery.

During normal LRV operation, the navigation system power remains on during the entire sortie. To conserve power for increased range, all mobility elements (i.e., traction drives, steering motors, electronic controller, and PWM power supplies) are turned off if a stop is to exceed 5 minutes duration.

1G Trainer Note

The 1G Trainer has alternate provision for utilizing an external power source by means of a rotary switch selector (figure 8-3) and umbilical connector.

1.4.3   Caution and Warning System

Refer to figure 1-20 for the caution and warning system schematic. The normally open temperature switches in the batteries and drive motors close on increasing temperatures. When either battery reaches 125°F or any drive motor reaches 400°F, the temperature switch closes, energizing the "OR" logic element and the driver. The driver then sends a 10 millisecond 36V pulse to the coil of the electromagnet which releases the magnetic hold on the indicator at the top of the console and a spring loaded flag flips up. The astronaut can reset the flag by pushing it down even though the cause has not been eliminated. The flag will not flip up again unless an overtemperature occurs on another battery or traction drive or the initial overtemperature subsides and then re-occurs.

1G Trainer Notes

1. The 1G Trainer traction drive gear box thermal switches will actuate the warning flag when gear box temperature reaches 200°F. The gear box temperature readout is biased so a reading of 450°F to 500°F will exist when the thermal switch actuates.

2. The 1G Trainer motor temperature switches are set to actuate the flag when motor external case temperature reaches 275°F. This temperature at the case would correspond to a rotor temperature of about 450°F.

LS006-002-2H
LUNAR ROVING VEHICLE
OPERATIONS HANDBOOK

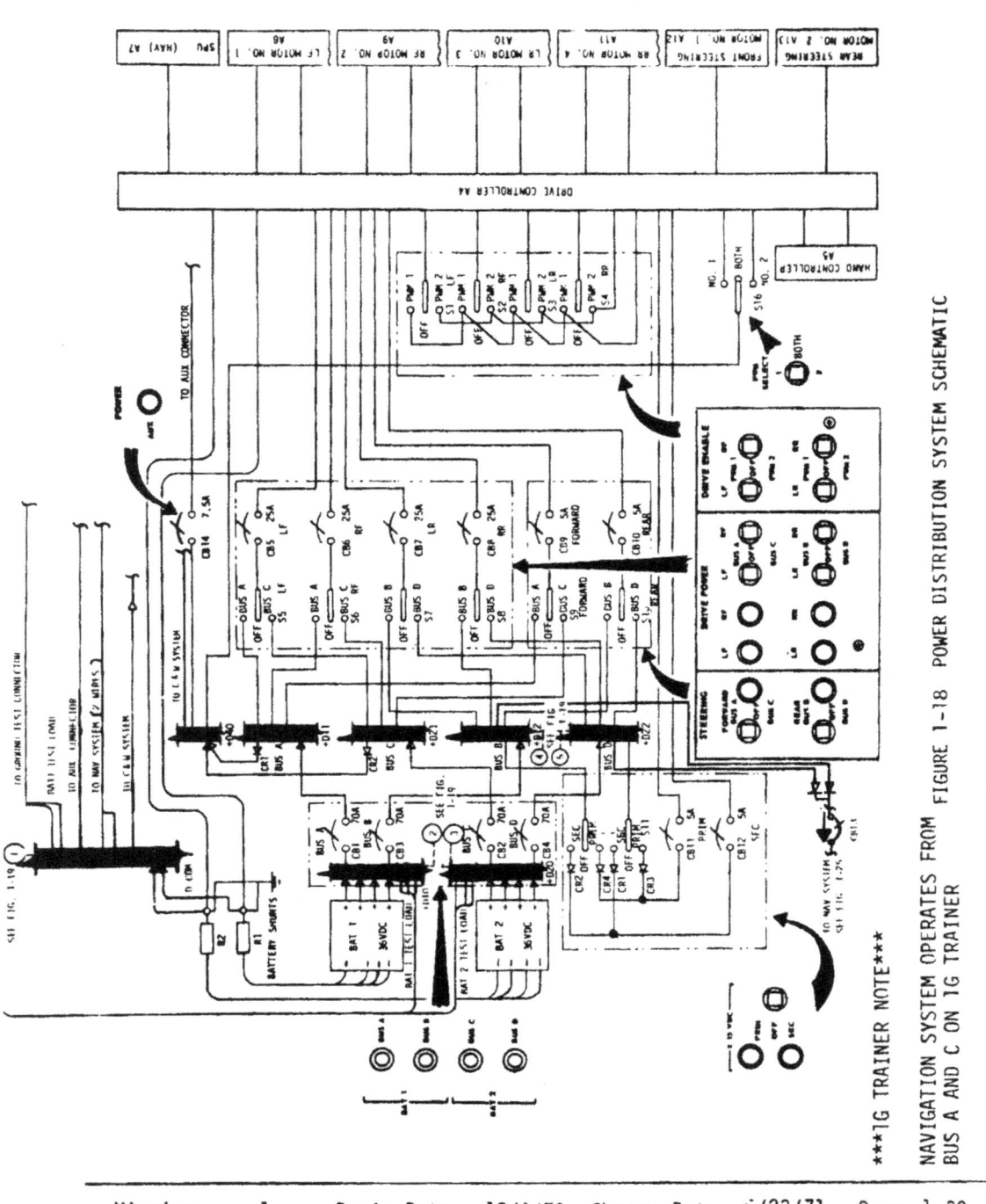

Figure 1-18 POWER DISTRIBUTION SYSTEM SCHEMATIC

1G TRAINER NOTE

NAVIGATION SYSTEM OPERATES FROM FIGURE 1-18
BUS A AND C ON 1G TRAINER

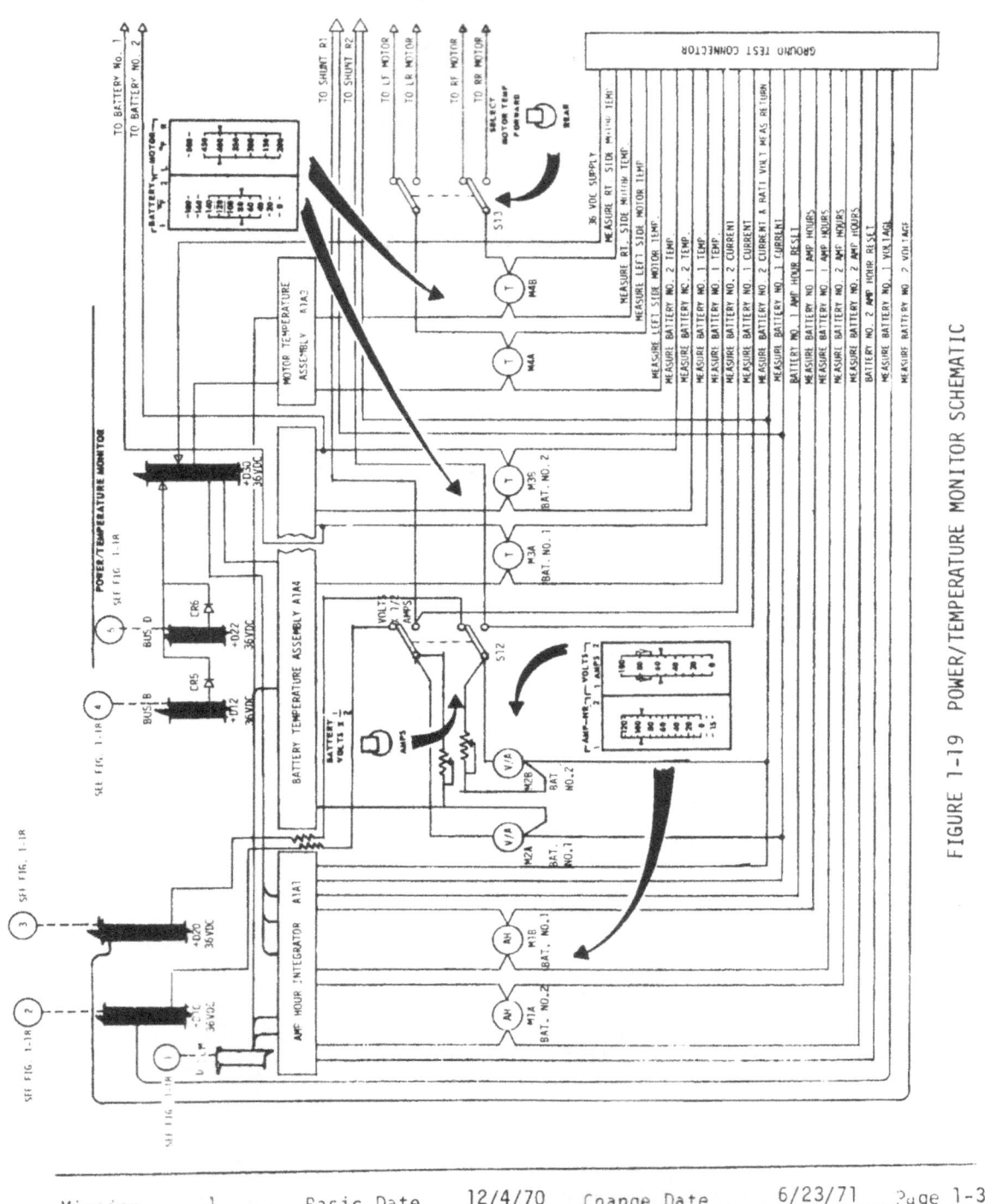

FIGURE 1-19 POWER/TEMPERATURE MONITOR SCHEMATIC

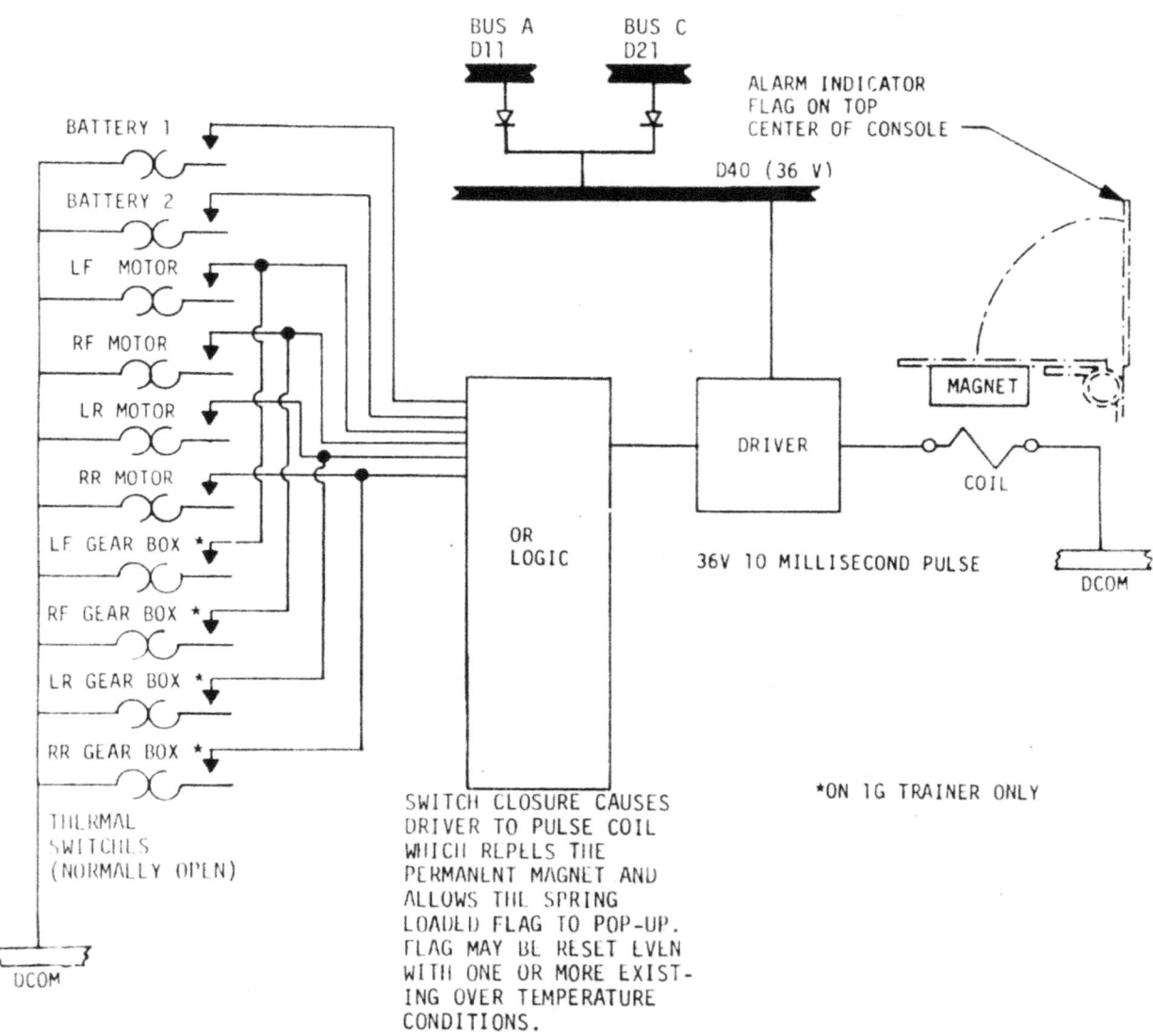

FIGURE 1-20. CAUTION AND WARNING SYSTEM

1.4.4   Auxiliary Connector

The auxiliary connector (figure 1-21) provides power for the Lunar Communications Relay Unit (LCRU). Power at the connector is furnished at 36 (+5/-3) VDC through a 7.5 ampere circuit breaker. Source impedance at the connector is less than 0.4 ohms shunted by a 440 micro-farad capacitor. Prior to launch, the LCRU power cable is attached to the auxiliary connector.

1G Trainer Note

The 1G Trainer auxiliary connector is not electrically functional.

LS006-002-2H
LUNAR ROVING VEHICLE
OPERATIONS HANDBOOK

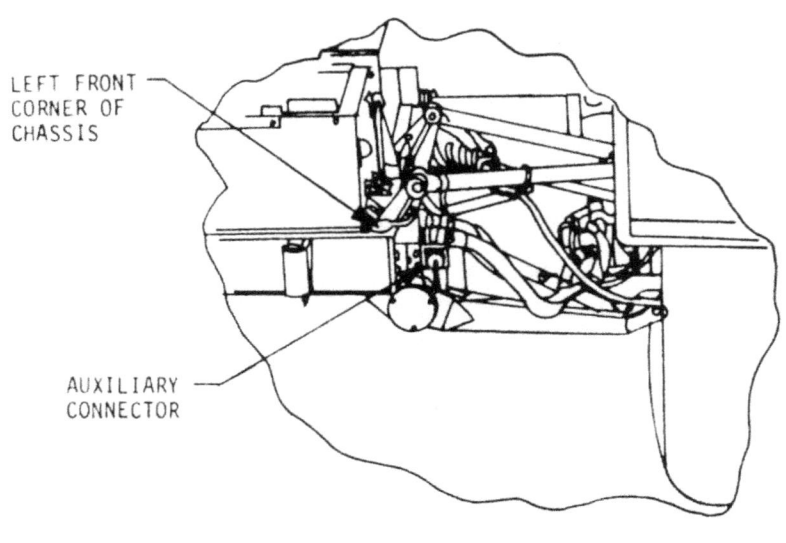

1G TRAINER

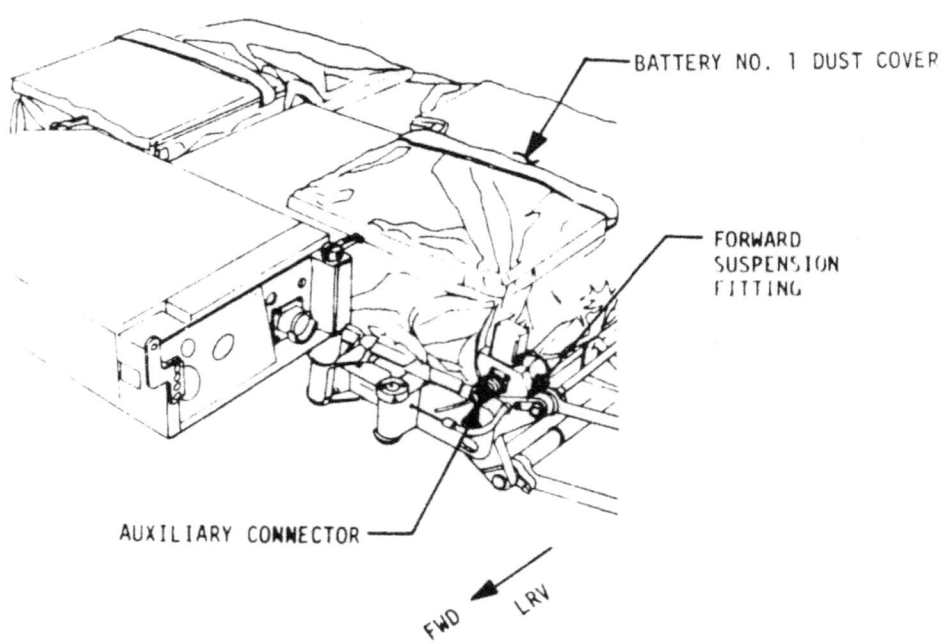

FIGURE 1-21. AUXILIARY CONNECTOR LOCATION

LS006-002-2H
LUNAR ROVING VEHICLE
OPERATIONS HANDBOOK

1.5   CONTROL AND DISPLAY CONSOLE

The Control and Display Console (figure 1-22) is separated into two main functional parts: Navigation on the upper part of the panel and monitoring and controls on the lower part of the panel. Refer to Table 1-1 for controls and use. The Control and Display Panel is activated with a radioluminescent material (Promethium) which provides visibility of displays even under lunar shadow conditions.

1.5.1   Attitude Indicator

This instrument (figure 1-26) provides indications of LRV pitch and roll. It indicates PITCH upslope (U) or downslope (D) within a range of plus 25 to minus 25 degrees in five degree increments and indicates ROLL within a range of 25 degrees left to 25 degrees right in one degree increments. The damper on the side of the indicator can be lifted to damp out oscillations. The pitch indication is readable in the stowed position of the indicator. The indicator is rotated outward which exposes the ROLL scale to the left side crewman. The pitch and roll reading is transmitted to MCC for navigation update computation.

1.5.2   Heading Indicator

This instrument displays the LRV heading with respect to lunar north. The initial setting and updating of this instrument is accomplished by operating the GYRO TORQUING switch LEFT or RIGHT.

1.5.3   Bearing Indicator

This instrument displays bearing to the LM in one degree digits. In the event of power loss to the navigation system, the bearing indication will remain displayed. The indications are lost when power is reapplied to the navigation system, however.

NOTE

Insufficient data is available for bearing computation until the LRV has moved about 50 meters from the point of nav initialization, therefore, the display indication should be disregarded until the vehicle is at least 50 meters from the point of nav initialization.

1.5.4   Distance Indicator

NOTE

Operating the LRV in reverse will add to the distance displayed on this instrument.

Mission   J   Basic Date   12/4/70   Change Date   4/19/71   Page 1-37

LS006-002-2H
LUNAR ROVING VEHICLE
OPERATIONS HANDBOOK

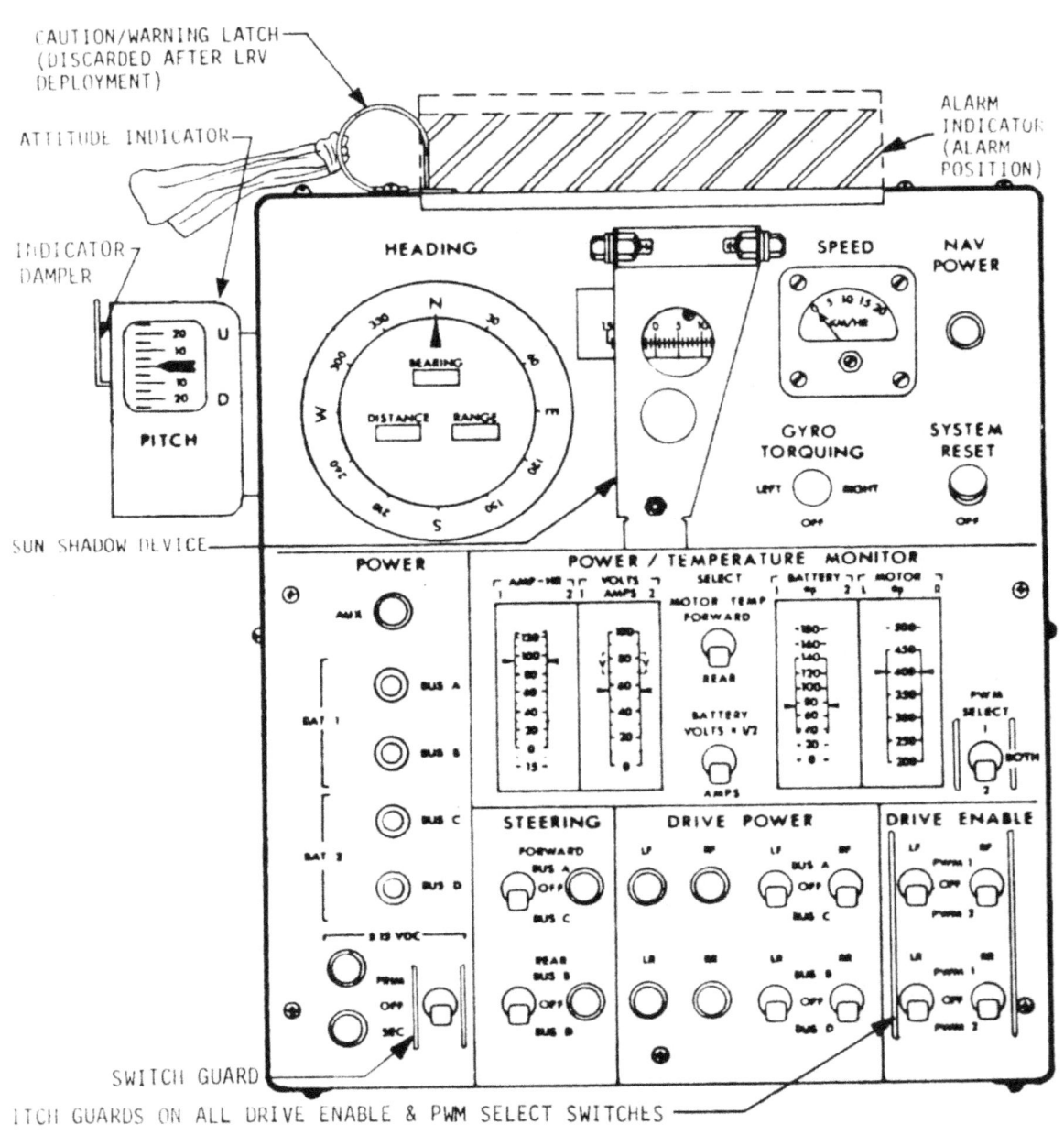

FIGURE 1-22 CONTROL AND DISPLAY CONSOLE

LS006-002-2H
LUNAR ROVING VEHICLE
OPERATIONS HANDBOOK

CONTROL	USE
GYRO TORQUING Switch	Adjusts heading indication during navigation updating. Switch is locked OFF in center position.
NAV POWER CB (5.0 Amps)	Routes power from main busses to navigation system. NOTE When Nav Power CB is open SPEED METER does not function.
AUX CB (7.5 Amps)	Energizes LRV Auxiliary Connector for LCRU power.
BAT 1 BUS A CB (70.0 Amps)	Energizes Bus A from LRV Battery No. 1.
BAT 1 BUS B CB (70.0 Amps)	Energizes Bus B from LRV Battery No. 1.
BAT 2 BUS C CB (70.0 Amps)	Energizes Bus C from LRV Battery No. 2.
BAT 2 BUS D CB (70.0 Amps)	Energizes Bus D from LRV Battery No. 2.
$\pm$ 15 DC PRIM CB (5.0 Amps)	Routes power from $\pm$ 15 DC PRIM/SEC Switch to 15 VDC power supplies and protects primary $\pm$ 15 VDC power supply from overload.
$\pm$ 15 DC SEC CB (5.0 Amps)	Redundant to $\pm$ 15 DC PRIM CB and protects secondary $\pm$ 15 VDC power supply from overload.
$\pm$ 15 DC PRIM/SEC Switch	Routes 36 VDC power from Bus B or D through $\pm$ 15 VDC CB's and on to 15 VDC Power Supplies in Motor Controller.
SYSTEM RESET Switch	Resets BEARING, DISTANCE and RANGE indicators to zero.
AMP-HR Indicator	Monitors battery residual capacity.
VOLTS/AMPS Indicator	Monitors battery volts or LRV current (amps) consumption.

TABLE 1-1 CONTROL AND DISPLAY CONSOLE CONTROLS

Mission __J__ Basic Date __12/4/70__ Change Date __4/19/71__ Page __1-39__

LS006-002-2H
LUNAR ROVING VEHICLE
OPERATIONS HANDBOOK

CONTROL	USE
MOTOR TEMP SELECT Switch	Selects Forward or Rear drive motors to be monitored on MOTOR °F indicator.
Battery Select Switch	Selects battery voltage or current to be displayed on VOLTS AMPS indicator.
Battery °F Meter	Monitors temperature of batteries. Allowable temperature is 40°F to 125°F for each battery.
MOTOR °F Meter	Monitors temperature of each motor. Allowable temperature is 400°F for each drive motor.
PWM SELECT Switch	Energizes Pulse Width Modulators (PWM's) in Motor Controller. With Switch in position 1 or 2, only corresponding PWM is energized. With switch in position BOTH, both PWM's are energized.
FORWARD STEERING Switch	Selects Bus A or C to supply power to Forward Steering Motor.
REAR STEERING Switch	Selects Bus B or D to supply power to Rear Steering Motor.
FORWARD STEERING CB (5.0 Amps)	Protects Forward Steering Motor from overload current.
REAR STEERING CB (5.0 Amps)	Protects Rear Steering Motor from overload current.
DRIVE ENABLE Switches	Switches LF, RF, LR and RR - Select PWM 1 or 2 for control of drive motors.
DRIVE POWER Switches	Switches LF, RF, LR and RR - Select desired bus to supply power to drive motors.
DRIVE POWER CB's (25.0 Amps each)	Protects the four drive motors from overload damage.

TABLE 1-1   CONTROL AND DISPLAY CONSOLE CONTROLS
(Continued)

Mission __J__   Basic Date __12/4/70__   Change Date __4/19/71__   Page __1-40__

1.5.4 (Continued)

This instrument displays distance traveled by the LRV in increments of 0.1 kilometer. This display is driven from the navigation signal processing unit which receives its inputs from the third fastest traction drive odometer. Total digital scale capacity is 99.9 km. In the event of power loss to the navigation system the distance indicator at time of power loss will remain displayed.

1.5.5 Range Indicator

This instrument displays the distance to the LM, and is graduated in 0.1 km increments with a total digital scale capacity of 99.9 km. In the event of power loss to the navigation system the range indicated at time of power loss will remain displayed.

1.5.6 Speed Indicator

NOTE

When the NAV POWER circuit breaker is open, no speed indication will be attained.

The instrument shows LRV velocity from 0 to 20 km/hr. This display is driven from the odometer pulses from the right rear wheel, through the SPU.

1.5.7 Sun Shadow Device

This device is used to determine the LRV heading with respect to the sun azimuth. When deployed, the device casts a shadow on a graduated scale when the vehicle is facing away from the sun. The point at which the shadow intersects the scale is transmitted by the crew to MCC for navigation update. The scale length is 15 degrees either side of zero with one degree divisions. The sun shadow device can be utilized at sun elevation angles up to 75 degrees.

LS006-002-2H
LUNAR ROVING VEHICLE
OPERATIONS HANDBOOK

1.6   NAVIGATION SUBSYSTEM

Refer to figure 1-23 for the Navigation Subsystem Block Diagram; figure 1-24 for hardware locations, and figure 1-25 for electrical schematic.

The power supply converts the vehicle battery voltage to the AC and DC voltages required for operation of the navigation subsystem components. Signal outputs to the subsystem are: direction (obtained from a directional gyro) and distance (obtained from odometer pulses from each traction drive unit). These signals are operated on by the navigation subsystem which displays the results as: heading with respect to lunar north, bearing back to the LM, range back to the LM, total distance traveled and velocity.

NOTE

The Navigation System is initialized by pressing the SYSTEM RESET button, which resets all digital displays and internal registers to zero. Initialization is only performed at the start of each EVA.

Alignment of the directional gyro is accomplished by measuring the pitch and roll of the LRV using the attitude indicator (figure 1-26), and measuring the LRV orientation with respect to the sun using the sun shadow device (figure 1-27). This information is relayed to MCC where a heading angle is calculated. The gyro is then adjusted by slewing with the torquing switch until the heading indicator reads the same as the calculated value. Slew rate is approximately 1.5 degrees per second.

The heading angle of the LRV is implicit in the output from the gyro, which is generated by a three wire synchro transmitter. The heading indicator in the IPI contains a synchro control transformer and an electromechanical servo system which drives the control transformer until a null is achieved with the inputs from the gyro.

NOTE

The odometer logic cannot distinguish between forward and reverse wheel rotation. Therefore, reverse operation of the LRV adds to the odometer reading.

There are four odometers in the system, one for each traction drive unit. Nine odometer pulses are generated for each revolution of each wheel. These signals are amplified and shaped in the motor controller circuitry and enter the line receiver in the SPU. The odometer pulses from the right rear wheel enter the velocity processor for display on the LRV SPEED indicator.

Mission   J   Basic Date 12/4/70 Change Date 4/19/71 Page 1-42

LS006-002-2H
LUNAR ROVING VEHICLE
OPERATIONS HANDBOOK

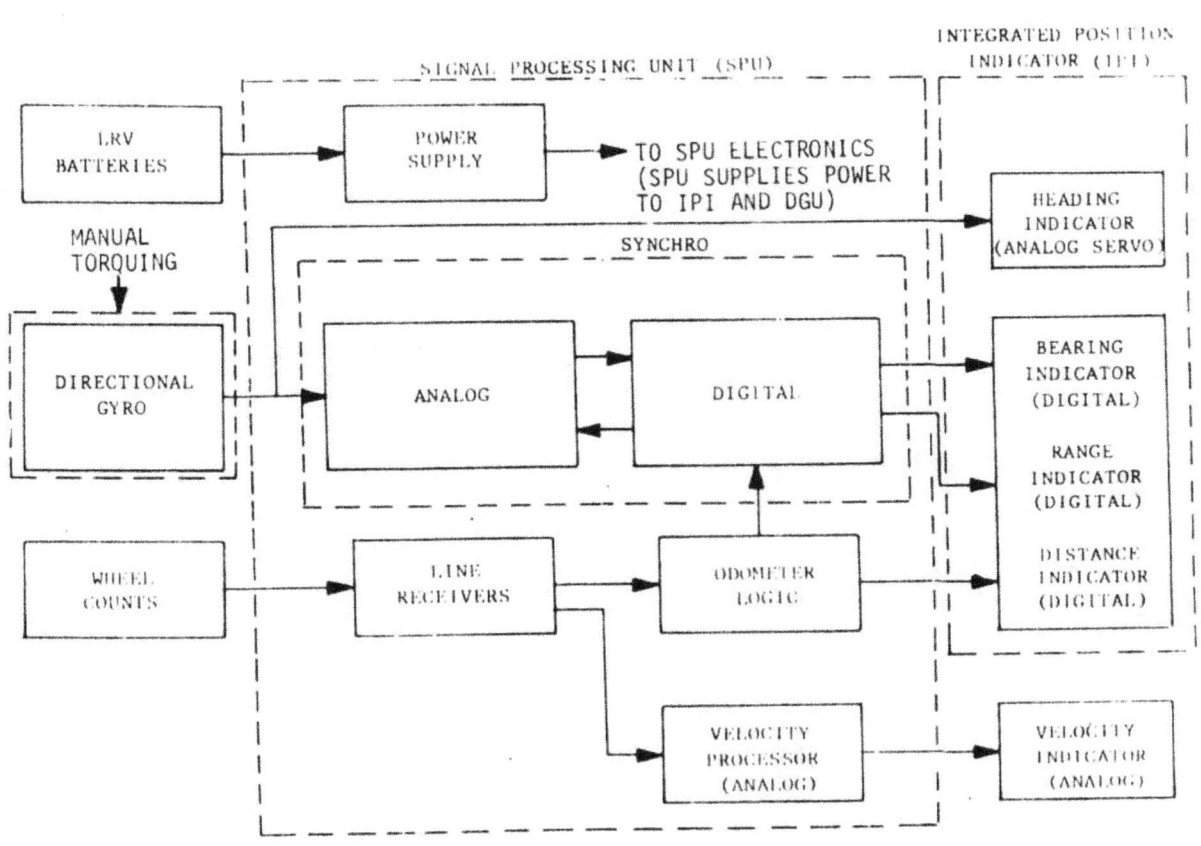

FIGURE 1-23. NAVIGATION SUBSYSTEM BLOCK DIAGRAM

LS006-002-2H
LUNAR ROVING VEHICLE
OPERATIONS HANDBOOK

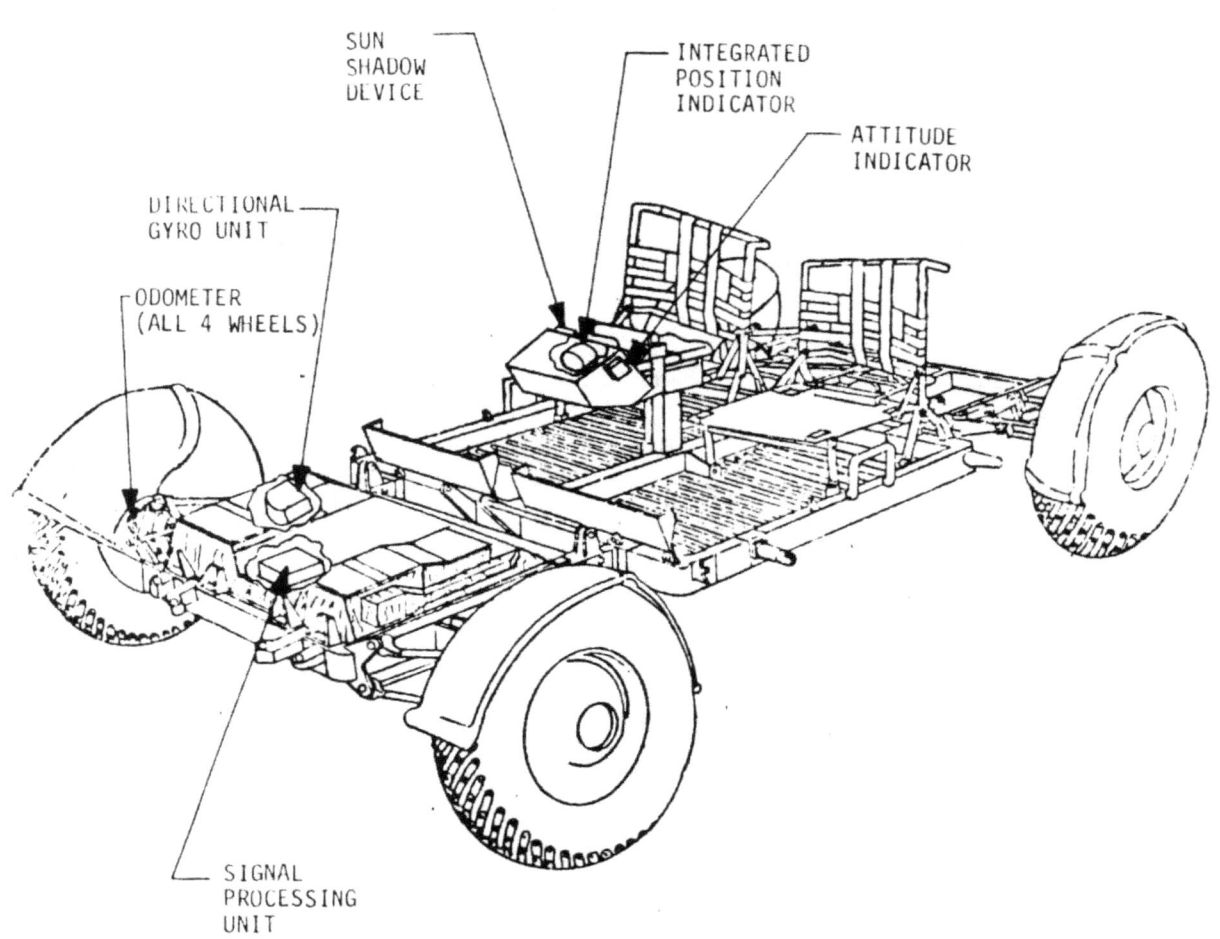

FIGURE 1-24. NAVIGATION COMPONENTS ON LRV

LS006-002-2H
LUNAR ROVING VEHICLE
OPERATIONS HANDBOOK

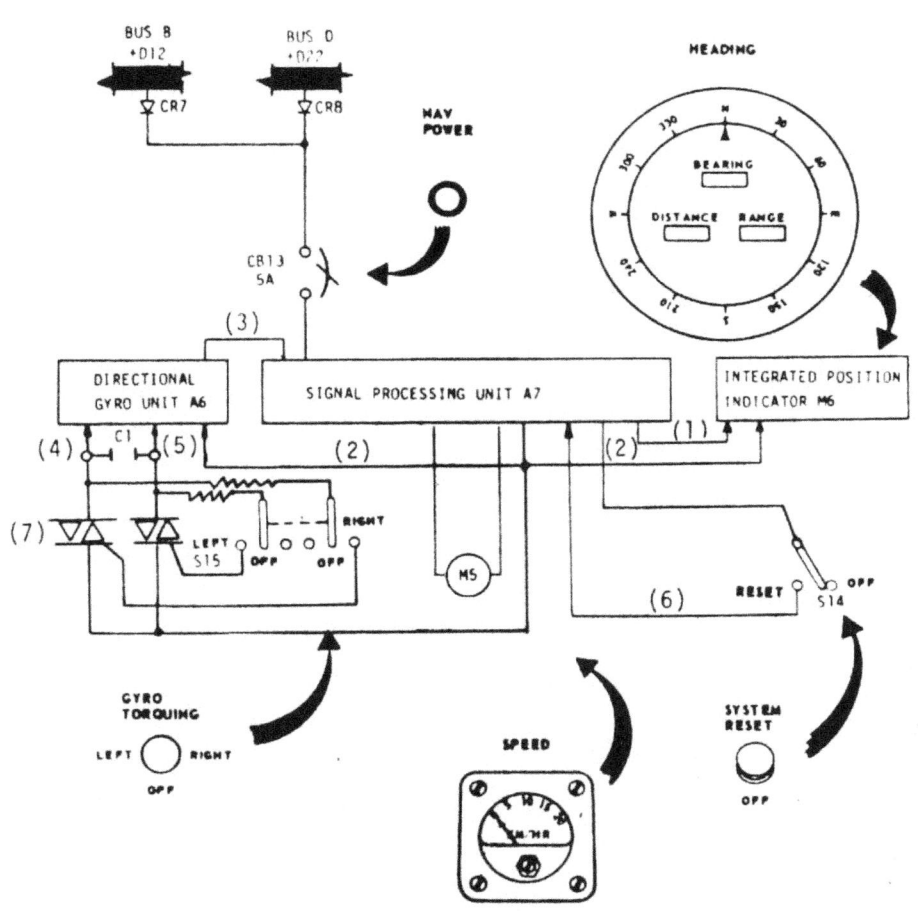

LEGEND

1. 3 WIRE SYNCHRO POSITIONING SIGNAL AND 3 WIRES TO INCREMENT COUNTERS.
2. 400 Hz ELECTRICAL POWER.
3. POSITION SYNCHRO OUTPUT FROM GYRO.
4. TORQUE RIGHT INPUT SIGNAL.
5. TORQUE LEFT INPUT SIGNAL.
6. DIGITAL DISPLAY RESET SIGNAL.
7. ⚡ SYMBOL INDICATES "TRIAC" (BI-DIRECTIONAL TRIODE THYRISTOR).

FIGURE 1-25   NAVIGATION SYSTEM ELECTRICAL SCHEMATIC

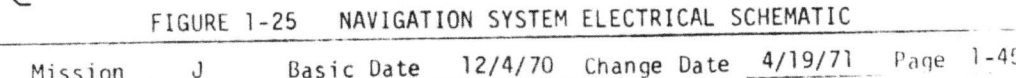

LS006-002-2H
LUNAR ROVING VEHICLE
OPERATIONS HANDBOOK

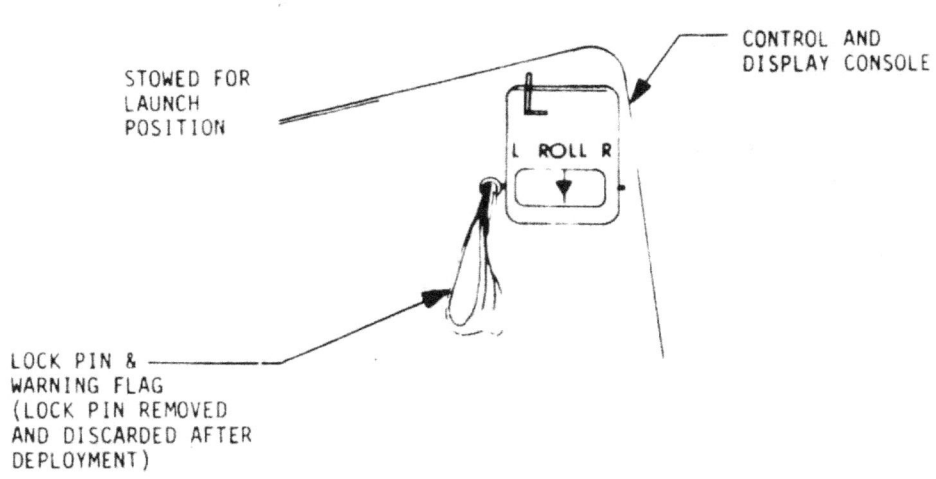

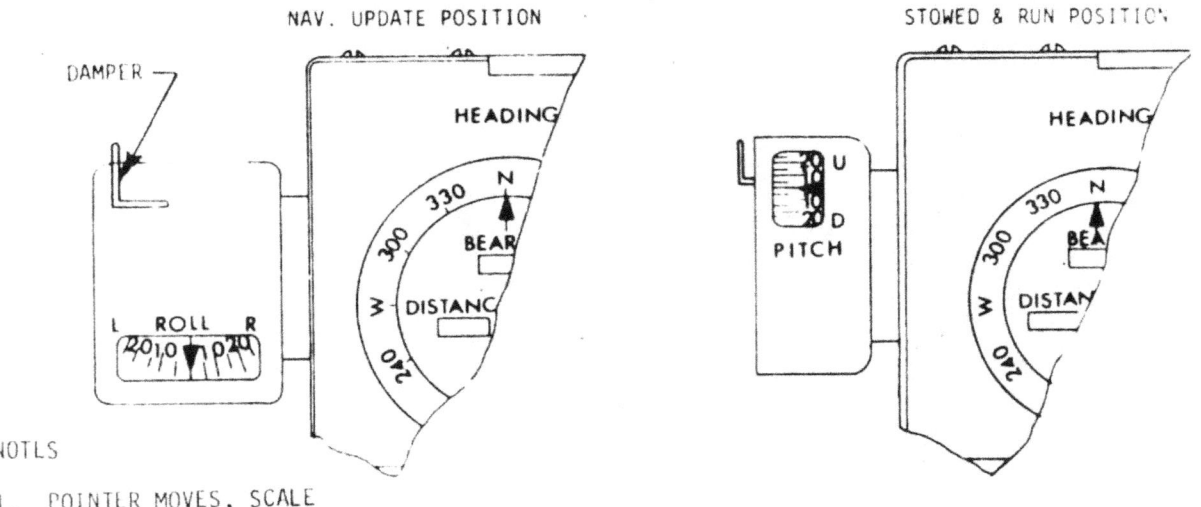

NOTES

1. POINTER MOVES, SCALE IS FIXED
2. NOMENCLATURE ON CASE EXTERIOR IS NOT ACTIVATED WITH RADIOLUMINESCENT MATERIAL

FIGURE 1-26 VEHICLE ATTITUDE INDICATOR

Mission __J__ Basic Date __12/4/70__ Change Date __4/19/71__ Page __1-46__

LS006-002-2H
LUNAR ROVING VEHICLE
OPERATIONS HANDBOOK

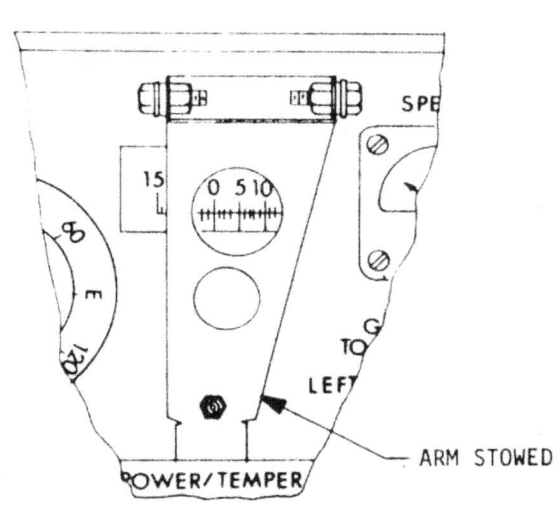
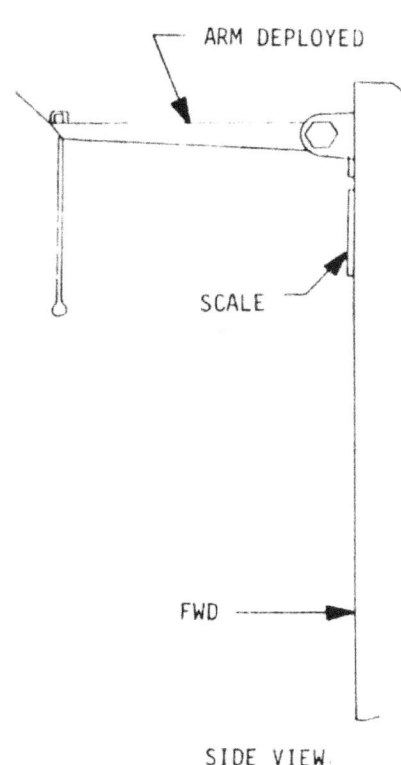

FIGURE 1-27 SUN SHADOW DEVICE

1G Trainer Note

The 1G Trainer navigation system is calibrated for use with the wire wheels, however, the navigation errors incurred when pneumatic tires are used are small. This is because the effective radius of both types of wheels are approximately the same, (pneumatic tires inflated to 30 psig) and the pneumatic wheel has essentially no slip. The only error incurred is in the range and distance calculation. The distance display when pneumatic wheels are used is estimated to be about 3.5% low. The ability to navigate back to the starting point during training should not be affected as all other errors cancel on a closed course.

Odometer pulses from all four wheels enter the odometer logic via the SPU line receivers. This logic selects the third fastest wheel for use in the distance computation. This insures that the odometer output pulses will not be based on a wheel which is locked, nor will they be based on a wheel that has excessive slip.

NOTE

Loss of Nav information occurs if vehicle is operated with more than one traction drive shut off.

The odometer logic sends outputs to the digital distance indicator in the IPI and to the range/bearing processor in the SPU. Upon entering the range/bearing processor, the outputs initiate selection, and conversion of heading, sine and cosine to digital numbers.

The effect of conversion of heading, sine and cosine, at distance increments is equivalent to entering (distance increment x sine heading) and (distance increment x cosine heading) into the $\triangle$E and $\triangle$N registers of the digital part of the bearing and range processor. The digital processor then adds the new $\triangle$E and $\triangle$N numbers to the contents of the East (E) and North (N) accumulators. The E and N accumulators, therefore, contain the east and north vector components of the range and bearing back to the LM. The digital vectoring process then does a vector conversion on the N and E numbers to obtain range and bearing, which are displayed on digital counters in the IPI. Each distance increment from the odometer logic initiates the entire sequence described, and results in the updating of bearing and range.

NOTE

The bearing digital display is "locked out" (i.e. does not display updated readings) until the vehicle is driven beyond a 50 meter radius of the nav initialization point.

LS006-002-2H
LUNAR ROVING VEHICLE
OPERATIONS HANDBOOK

1.7 CREW STATION

The crew station consists of seats, footrests, inboard handholds, outboard handholds, arm rest, floor panels, seat belts, fenders, and toeholds.

1.7.1 Seats

LRV seats are tubular aluminum frames spanned by nylon (figure 1-28). The seats are folded flat onto the center chassis for launch and erected to the operational position by the crew after LRV deployment on the lunar surface. The seat back is used to support and restrain the PLSS from lateral motion when the crew is positioned for LRV operation. Refer to Section 2 for seat erection sequence. The seat bottom contains a cutout to allow access to the PLSS flow control valves and includes provisions for vertical support of the PLSS.

1G Trainer Note

> The 1G Trainer is also equipped with removable seat pads which allow comfortable operation in a "shirt sleeve" training session.

1.7.2 Footrests

For launch, the footrests (figure 1-28) are stowed against the center chassis floor and secured by two velcro straps. The footrests are deployed by the crew on the lunar surface. The footrests are adjusted, before launch, to accommodate specific crewmen.

1.7.3 Inboard Handholds

Inboard handholds (figure 1-28) are constructed of 1 inch O. D. aluminum tubing and are used to aid the crew during ingress and egress. The handholds also contain payload attach receptacles for the 16 mm data acquisition camera and the LCRU low gain antenna.

1.7.4 Outboard Handholds

Outboard handholds are integral parts of the chassis (figure 1-28) and are used to provide crew confort and stability when seated on the LRV and for attachment of the seat belt.

1.7.5 Arm Rest

The arm rest (figure 1-28) is used to support the arm of crewmen during hand controller manipulation.

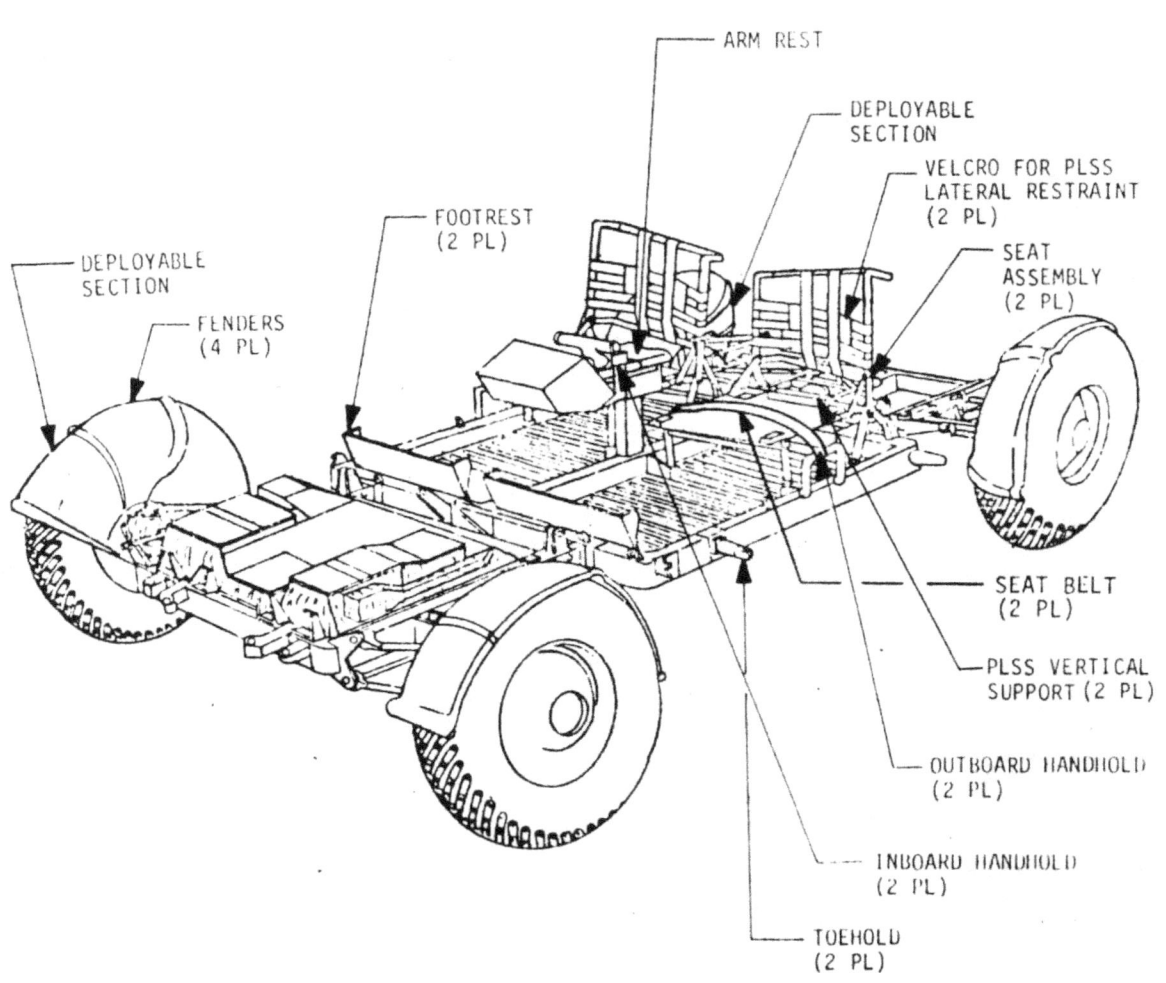

FIGURE 1-28 CREW STATION COMPONENTS

1.7.6   Seat Belts

> **NOTE**
>
> Before egress, the seat belt should be attached to the inboard handhold for accessibility upon ingress.

A seat belt is provided at each seat. The seat belts (figure 1-29) are constructed of nylon webbing. The belt end terminates in a hook which is secured to the outboard handhold. Belt length adjustment is provided by an adjustment buckle. A stretch section of the belt permits normal fastening and release.

1.7.7   Fenders

The deployable portion of each fender (figure 1-28) is positioned by the astronaut during LRV deployment on the lunar surface.

1.7.8   Toeholds

There are two toeholds, one on either side of the vehicle. The toehold is used to aid the crew in ingressing and egressing the LRV. The toehold is formed by dismantling the LRV/LM interface tripods and using the leg previously used as the tripod center member as the toehold. The tripod member is inserted into the chassis receptacle to form the operational position of the toehold.

> **NOTE**
>
> The toeholds are also used as tools to actuate the wheel-decoupling mechanism to release the telescoping tubes and saddle fitting on the forward chassis, and to free the steering decoupling rings from the stowed position. Either toehold may be used for decoupling.

1.7.9   Floor Panels

The floor panels in the crew station area are beaded aluminum panels (figure 1-30). The floor is structurally capable of supporting the full weight of standing astronauts in lunar gravity.

1G Trainer Note

> The 1G Trainer floor panels are flat plates in lieu of beaded panels.

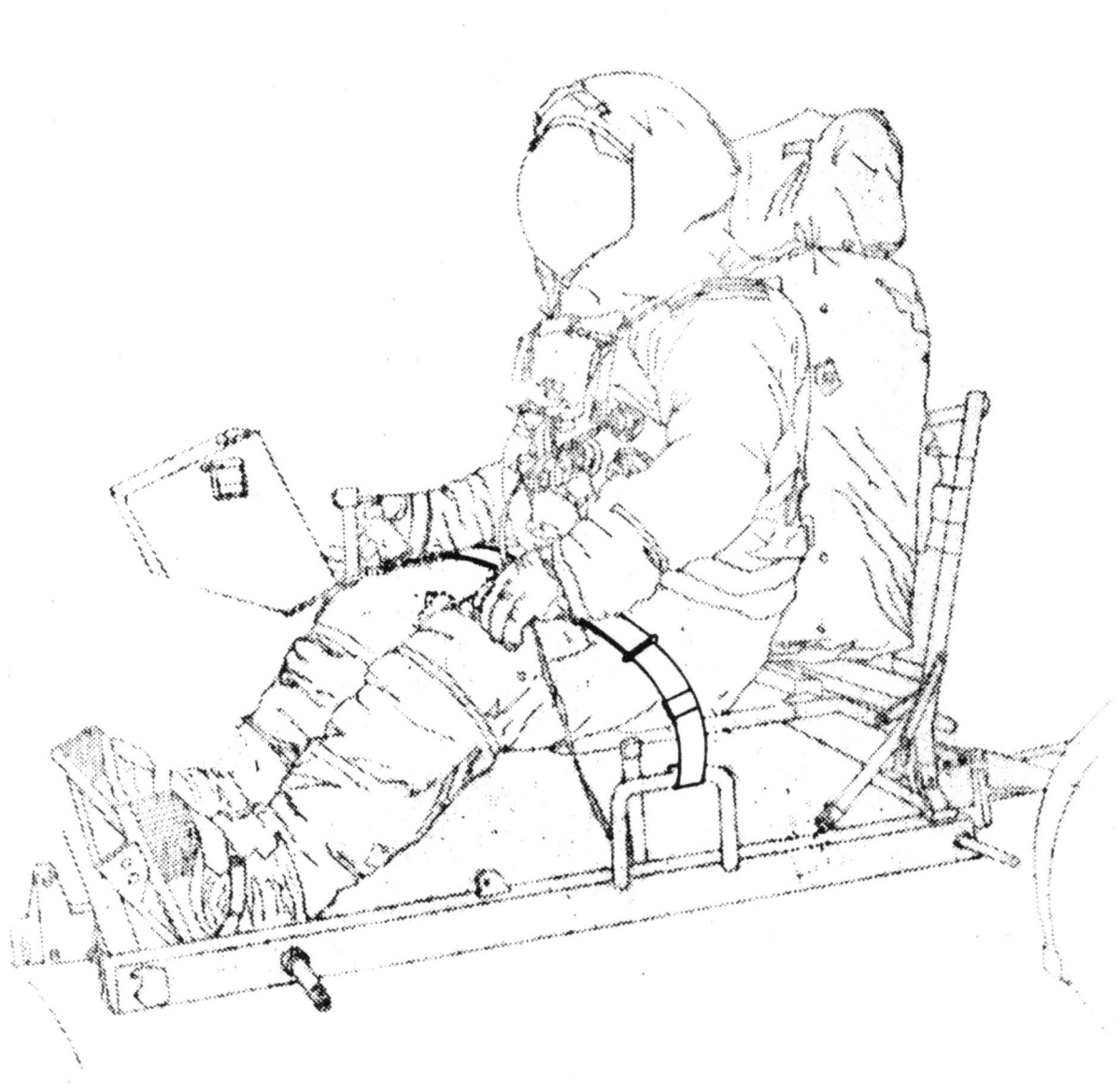

FIGURE 1-29   SEAT BELTS

LS006-002-2H
LUNAR ROVING VEHICLE
OPERATIONS HANDBOOK

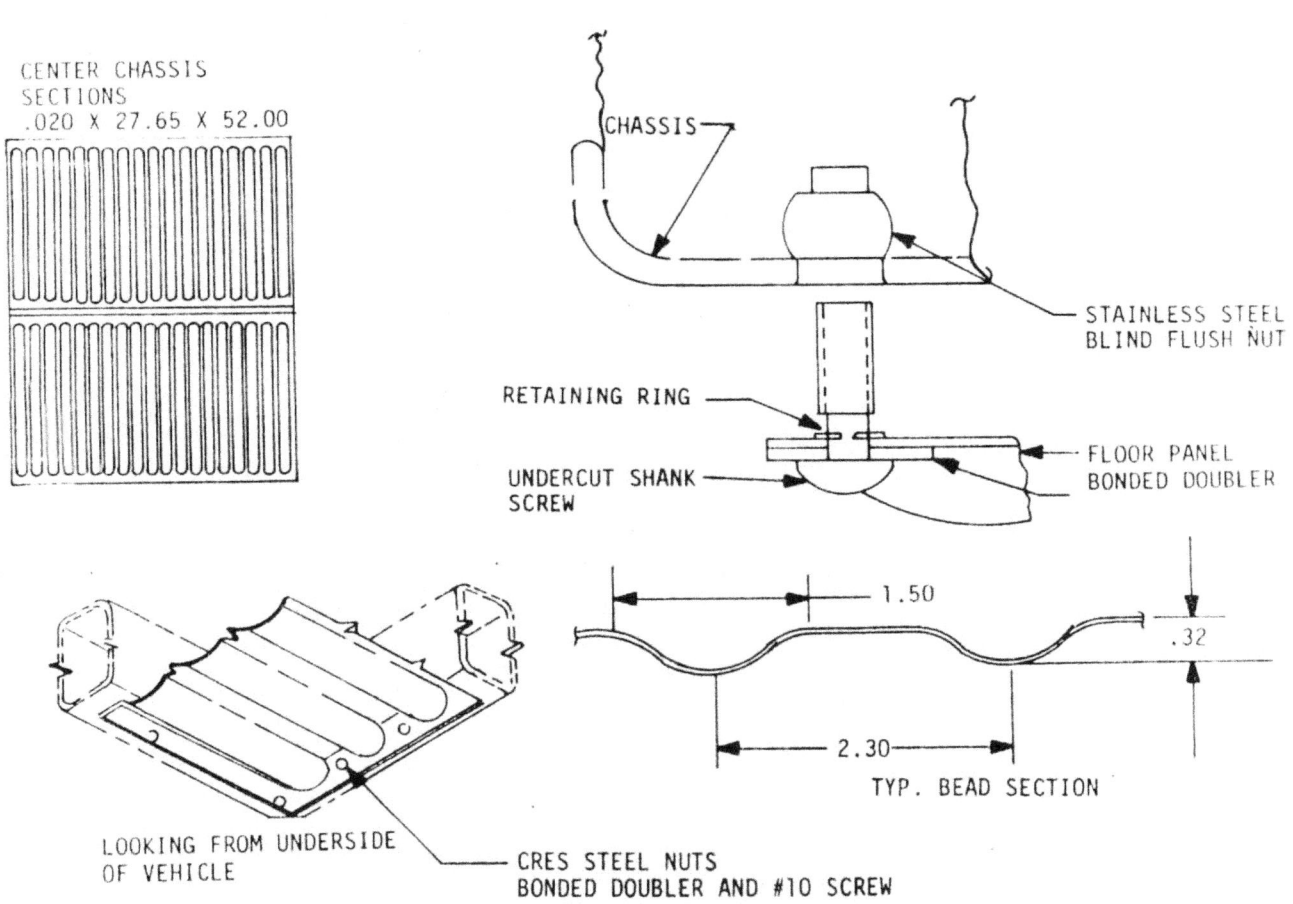

FIGURE 1-30 CREW STATION FLOOR PANELS

LS006-002-2H
LUNAR ROVING VEHICLE
OPERATIONS HANDBOOK

1.8   THERMAL CONTROL

1.8.1   LRV Thermal Control

Thermal control systems are incorporated into the LRV to maintain temperature sensitive components within the appropriate temperature limits during the translunar phase of a mission, during operation on the lunar surface, and during periods of inoperation on the lunar surface. Thermal control systems consist of special surface finishes, multilayer insulation, space radiators, second surface mirrors, thermal straps, and fusible mass heat sinks (figure 1-31).

1G Trainer Note

Thermal control for the 1G Trainer is described in paragraph 1.8.5.

1.8.2   Forward Chassis Thermal Control

The basic concept of thermal control for forward chassis components is energy storage during operation with subsequent energy transfer to deep space while the vehicle is parked between sorties. During operation heat energy released in the Drive Control Electronics (DCE) is stored in the DCE and the DCE thermal control unit (figure 1-32). Heat energy released in the Signal Processing Unit (SPU) is stored in the SPU, the SPU thermal control unit (a fusible mass device) and Battery No. 1. The SPU is thermally connected to Battery No. 1 by means of the SPU thermal strap (figure 1-33). Heat energy released in the Directional Gyro Unit (DGU) is stored in the DGU and Battery No. 2 by means of the DGU thermal strap. Space radiators are mounted on the top of the SPU, DCE, Battery 1 and Battery 2 (figure 1-34). Fused silica second surface mirrors are bonded to the radiators to minimize the solar energy absorbed by an exposed radiator, and to minimize the degradation of the radiating surface by the space and lunar environment. The space radiators are exposed only during the parking period between sorties. During sortie operation the space radiators are protected from the lunar surface dust by three dust covers (figure 1-31). The first dust cover protects the radiators on Battery No. 1 and the DCE. The second dust cover protects the SPU radiator. The third dust cover protects the radiator on Battery No. 2. These dust covers are opened manually at the end of the sortie, an over center latch (figure 1-35), holds the dust covers open until battery temperatures reach 45°F (+ 5°F), at which time a bimetallic device disengages the overcenter latch allowing the dust covers to close. The SPU dust cover is slaved to the Battery No. 1 dust cover.

In addition to the dust covers, a multi-layer insulation blanket (figure 1-36) is provided to protect the forward chassis components from the space environment and the lunar surface environment. The exterior, and certain portions of the interior, of the multi-layer insulation blanket are covered with a layer of Beta Cloth to protect against wear and tear and direct solar or hot gas heat loads.

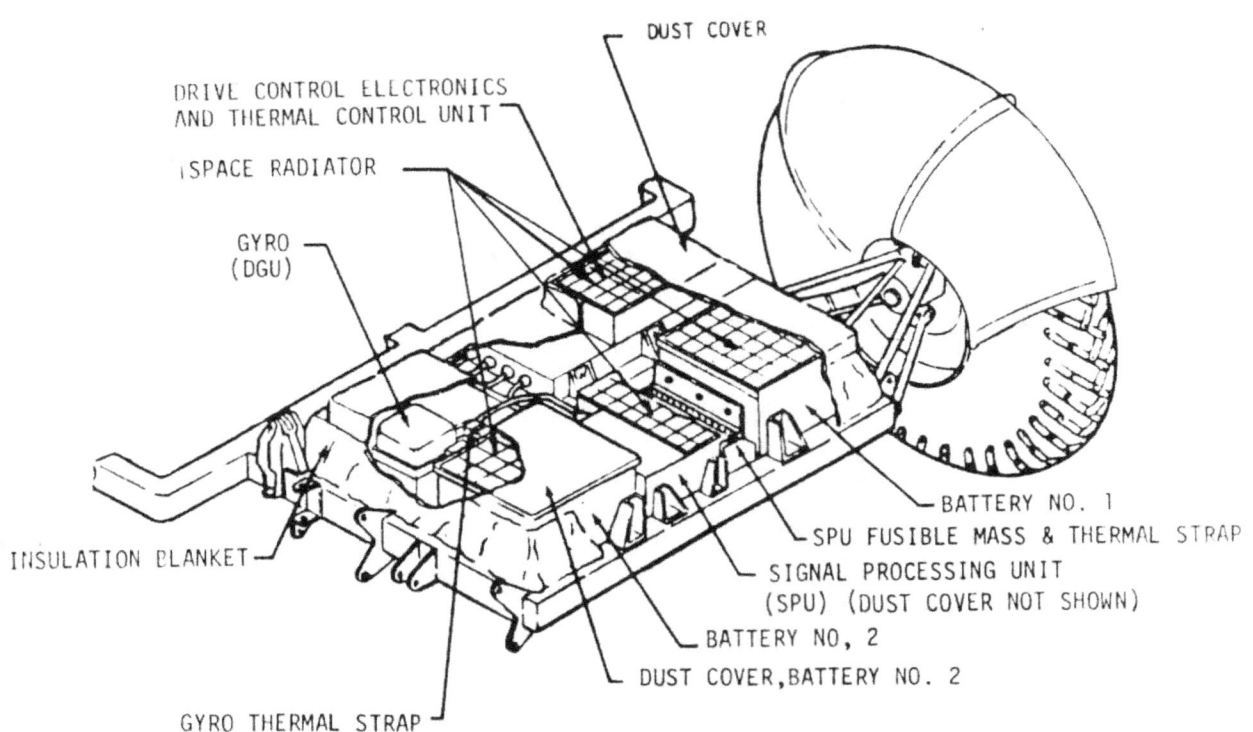

FIGURE 1-31   THERMAL CONTROL PROVISIONS (FORWARD CHASSIS)

LS006-002-2H
LUNAR ROVING VEHICLE
OPERATIONS HANDBOOK

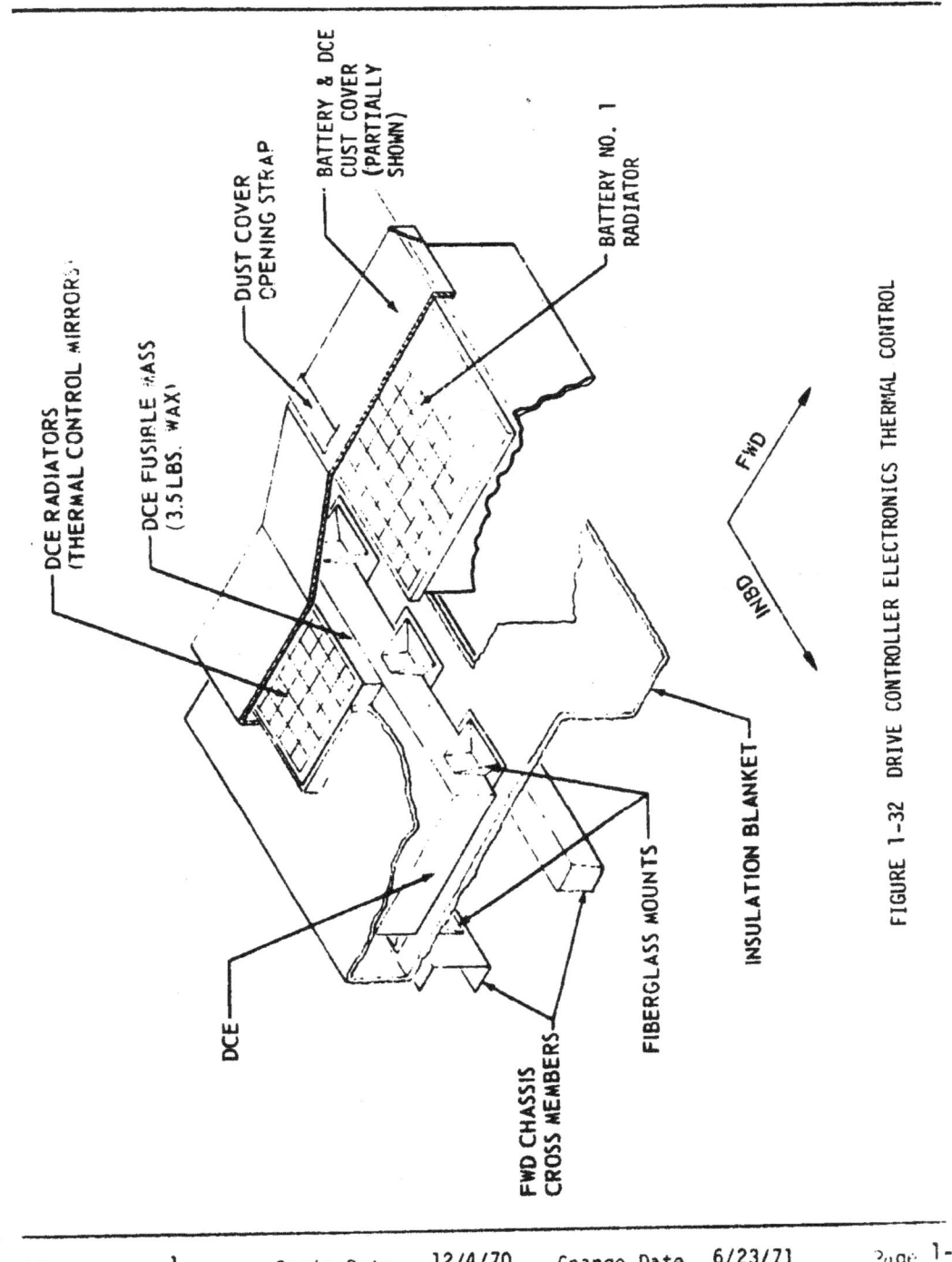

FIGURE 1-32 DRIVE CONTROLLER ELECTRONICS THERMAL CONTROL

LS006-002-2H
LUNAR ROVING VEHICLE
OPERATIONS HANDBOOK

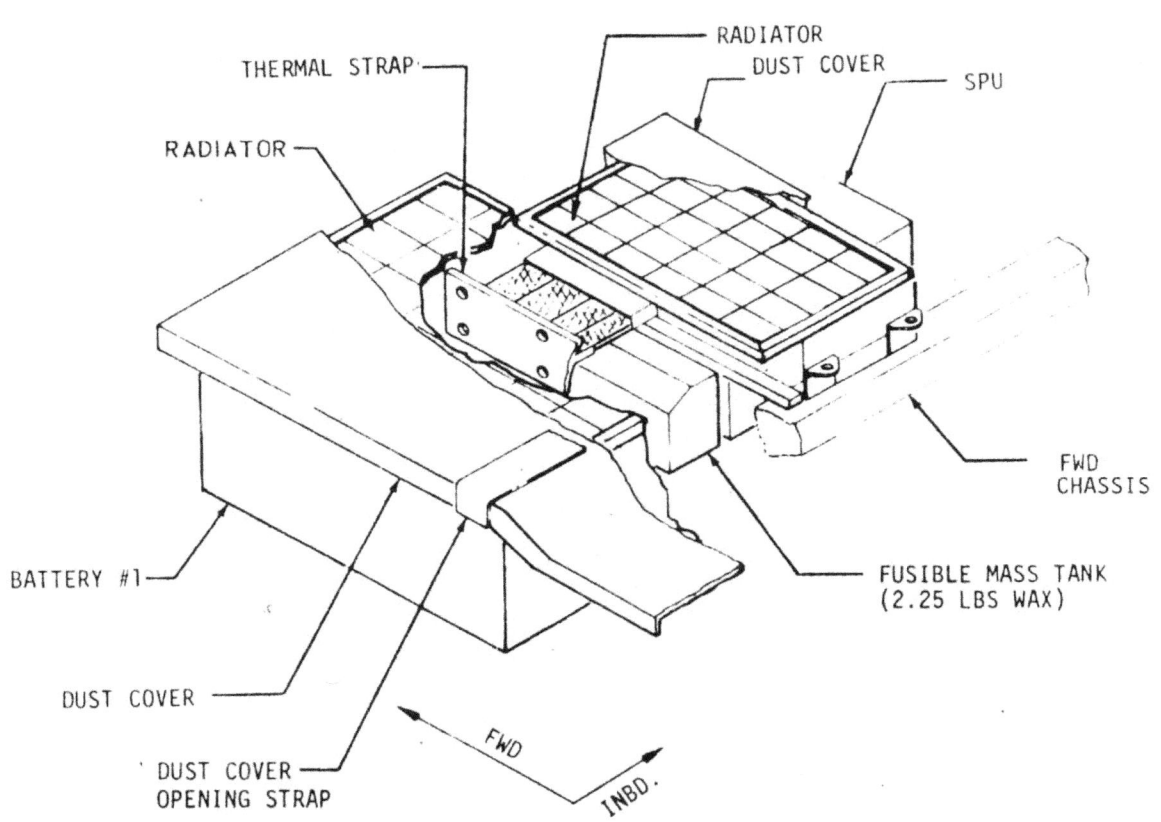

FIGURE 1-33 SPU ELECTRONICS AND BATTERY #1 THERMAL CONTROL

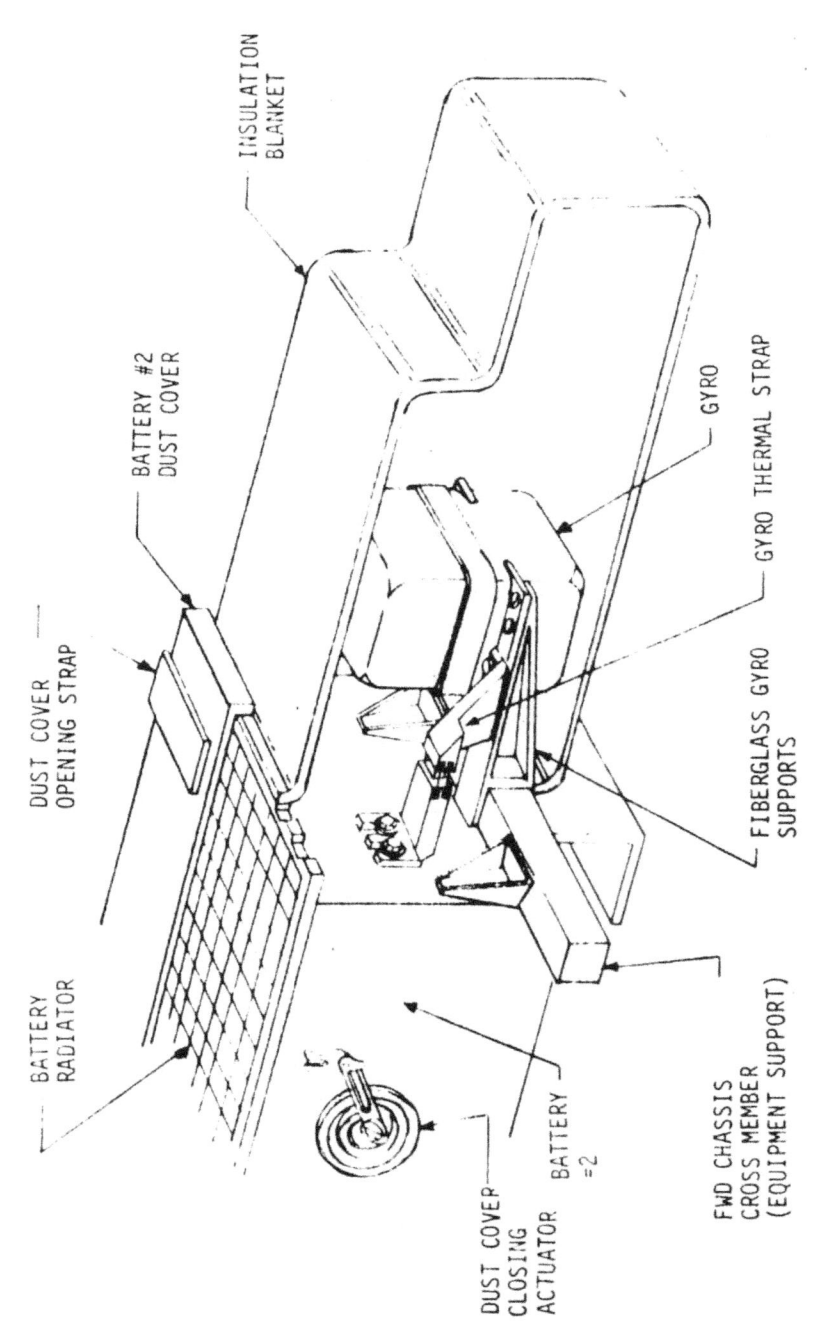

FIGURE 1-34 BATTERY NO. 2 AND DIRECTIONAL GYRO UNIT THERMAL CONTROL

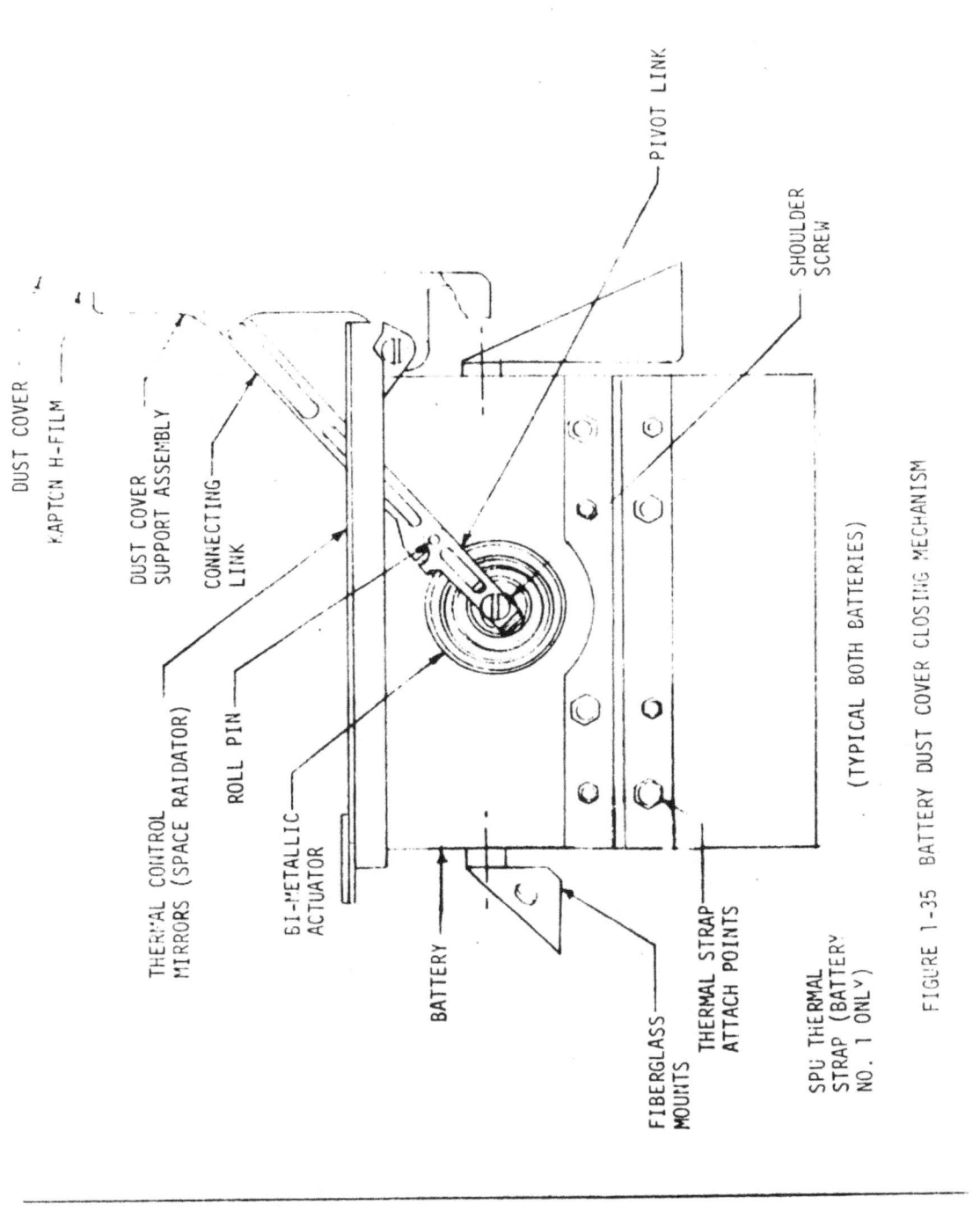

FIGURE 1-35 BATTERY DUST COVER CLOSING MECHANISM

LS006-002-2H
LUNAR ROVING VEHICLE
OPERATIONS HANDBOOK

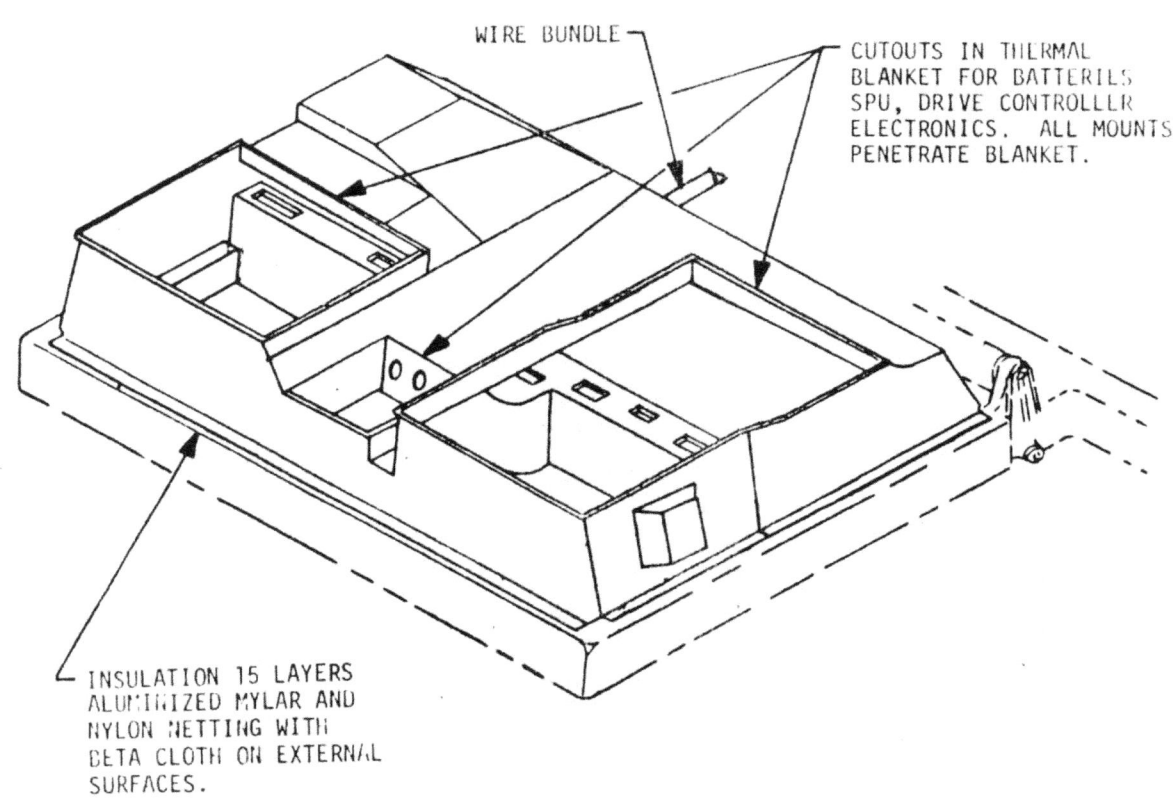

FIGURE 1-36 FORWARD CHASSIS INSULATION BLANKET

1.8.2 (Continued)

During a sortie, the astronauts can monitor the battery temperature by a meter on the Control and Display Console.

1.8.3 Control and Display Console

All instruments on the Control and Display Console are mounted to an aluminum plate which is isolated by radiation shields and fiberglass mounts. The external surfaces of the C&D console are coated with thermal control paint (Dow-Corning 92-007), and the face plate is black anodized and is isolated from the instrument mounting plate.

1.8.4 Center Chassis

Handholds, footrests, tubular sections of seats and center and aft floor panels are anodized.

1.8.5 1G Trainer Thermal Control

Whereas the LRV Flight Unit utilizes radiation to effect thermal control, the 1G Trainer relies primarily on convection to the atmospheric surroundings. The following 1G Trainer elements are cooled by a thermostatically controlled fan blower: traction drive (4 units) electronics assembly, battery (2 units), motor controller assembly (2 units), C&D console, DGU, and SPU. This makes a total of 12 separate blowers on the vehicle. Each fan motor thermostatic switch closes upon a rising temperature at a predetermined value, thereby applying battery voltage to a fan motor. When the temperature of the thermostat falls below a predetermined value, the switch opens, thereby shutting the fan motor off. Cooling fan motors and instrumentation warning thermostats are set as noted in Table 1-2.

NOTE

> The forward wheel traction drive fan blowers are enabled provided that there is power through at least one forward drive power switch. Similarly, the rear wheel traction drive fan blowers are enabled provided there is power through at least one rear drive power switch.

In addition to protective cooling, thermal instrumentation display at the C&D is provided for the traction drives and batteries. The display takes the form either of a discrete warning by a warning flag (activated by a thermostat) or an analog temperature display, (as sensed by a thermistor). Warning flag thermostats are located on the external motor case of each traction drive motor, on the external case of each traction drive gear reducer, and within each battery assembly. Thermostats are set as noted in Table 1-2. Note that a warning flag activation can be generated in a traction drive assembly by

COOLING FAN MOTOR THERMOSTATS SET AT:	Close On Rise Temp	Open On Fall Temp
Sub Assembly		
Traction Drive Motor (4 units)	160 ± 10°F	140 ± 10°F
Electronics Assembly (1 unit)	100 ± 5°F	85 ± 5°F
Battery (2 units)	100 ± 5 °F	85 ± 5°F
Motor Controller Assembly (2 units)	100 ± 5°F	85 ± 5°F
Display & Control Console (1 unit)	120 ± 3°F	114 ± 3°F
Navigation SPU	120 ± 3°F	114 ± 3°F
Navigation DGU	150 ± 5°F	140 ± 5°F

WARNING FLAG THERMOSTATS SET AT:	Close On Rise Temp	Open On Fall Temp
Sub Assembly		
Traction Drive Motor (4 units)	225 ± 5°F	175 ± 10°F
Traction Drive Gear Reducer (4 units)	200 ± 5°F	150 ± 10°F
Battery (2 units)	160 ± 5°F	120 ± 5°F

ANALOG DISPLAY THERMISTORS RESISTANCE VALUES ARE:	
Traction Drive Motor (4 units) Thermistor	3750 ohms ± 10% @ 80°F
	315 ohms ± 20% @ 225°F
Battery (2 units) Thermistor	553 ohms ± 5% @ 80°F
	640 ohms ± 5% @ 160°F

TABLE 1-2 1G TRAINER THERMAL CONTROL DEVICE SET POINTS

1.8.5 (Continued)

either the drive motor or the gear reducer assembly. Since only the drive motor has its temperature displayed on the console meters, a condition could arise where a "hot" gear reducer activated the warning flag but none of the temperature indicating meters show an abnormally high temperature, i.e., the drive motor associated with the hot gear reducer has a normal temperature. Under this condition it would be impossible for the operator to isolate the overtemperature to a specific subassembly. To alleviate this condition, the gear reducer warning thermostat is connected through a circuit such that whenever it closes it switches a low resistance across the associated drive motor thermistor; thereby, causing an abnormally high temperature reading to be displayed. Thus, the affected traction drive assembly can be isolated.

Analog temperature thermistors are located on the external case of each drive motor and on the main battery bus within each battery. Thermistors located in each assembly are monitored by a bridge circuit in the D&C console and the output of the bridge drives the display meters, which are calibrated in degrees Fahrenheit. By interrogating the temperature display meters, an overtemperature condition can be isolated to the specific subassembly and corrective action initiated. The nominal resistance values associated with both an ambient temperature and a "hot" temperature for these thermistors is noted in Table 1-2. It should be noted that only the battery readout meter is calibrated to indicate the thermistor temperature. The drive motor meters are calibrated to read the internal motor (rotor) temperature even though the thermistor is located external on the motor case and hence at much cooler location. Thus, the meter indicated temperature for each motor will be 450°F to 500°F when the thermistor is measuring an actual case temperature of 225°F. The cooling fan motors will activate when the meter indicated temperature is nominally 350°F.

LS006-002-2H
LUNAR ROVING VEHICLE
OPERATIONS HANDBOOK

1.9   SPACE SUPPORT EQUIPMENT (SSE)

The Space Support Equipment (SSE) consists of two basic subsystems of hardware, the structural support subsystem and the deployment hardware subsystem. The function of the structural support subsystem is to structurally support the LRV in the LM during launch boost, earth-lunar transit and landing. The function of the deployment hardware subsystem is to deploy the LRV from the LM to the lunar surface after landing.

1.9.1   Structural Support Description

The structural support subsystem by which the LRV is attached to the LM includes two steel support spools at the lower (left and right) sides of the LM quadrant. The spools are bolted to Grumman Aircraft Corporation (GAC) attach fittings. Aluminum tube tripod structures attached to the LRV center chassis terminate in apex fittings which steel are pinned together and clamped to the spools to support the LRV. The LRV is restrained against outboard rotation by an aluminum strut in the upper center of the LM, connecting the inboard quadrant corner structure to an LRV center chassis standoff with a pin.

1.9.2   Deployment Hardware Description

The deployment hardware system (figure 1-37) consists of bellcranks, linkages and pins to release the LRV from the structural support subsystem, thus allowing the LRV to deploy from the LM. It also consists of braked reels, braked reel operating tapes, braked reel cables, LRV rotation initiating push-off spring, deployment cable, telescopic tubes, chassis latches, release pin mechanisms, and LRV rotation support points.

1.9.3   Deployment Mechanism Operations

The LRV is attached to the SSE and deployment mechanism and held in position in the LM Quadrant as shown in figure 1-38. The deployment of the LRV from the LM to the lunar surface consists of five basic steps or phases:

Phase I   - Deployment from the stowed position of both braked reel separating tapes and the deployment cable (figure 1-39, inset Ⓐ)

Phase II   - Operating the D-Handle to disconnect the LRV from the structural support subsystem (figure 1-39, inset Ⓐ).

Phase III - Operating the double braked reel to unfold the LRV and lower the aft chassis wheels to the lunar surface (figure 1-39, insets Ⓑ Ⓒ and Ⓓ).

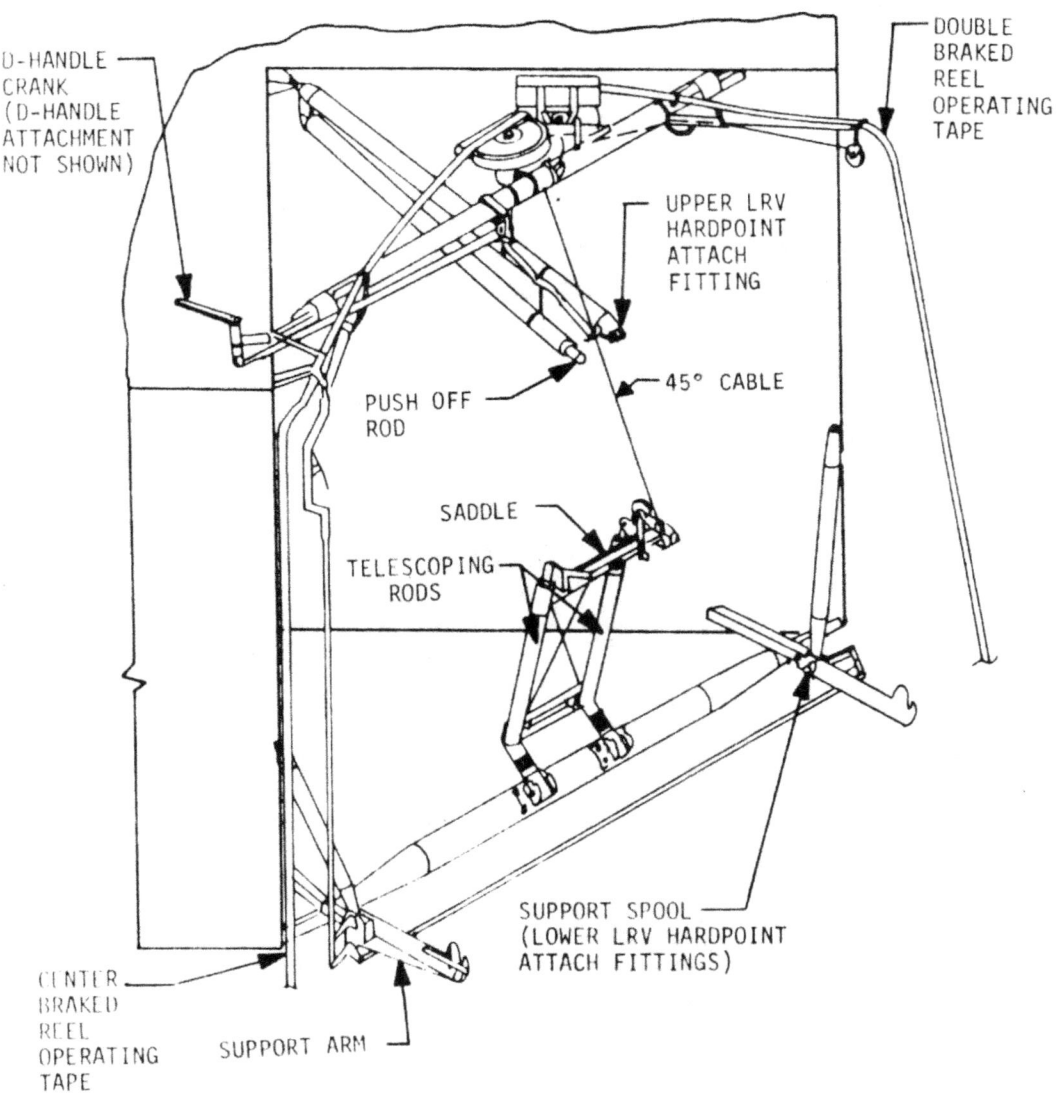

FIGURE 1-37   SPACE SUPPORT EQUIPMENT

LS006-002-2H
LUNAR ROVING VEHICLE
OPERATIONS HANDBOOK

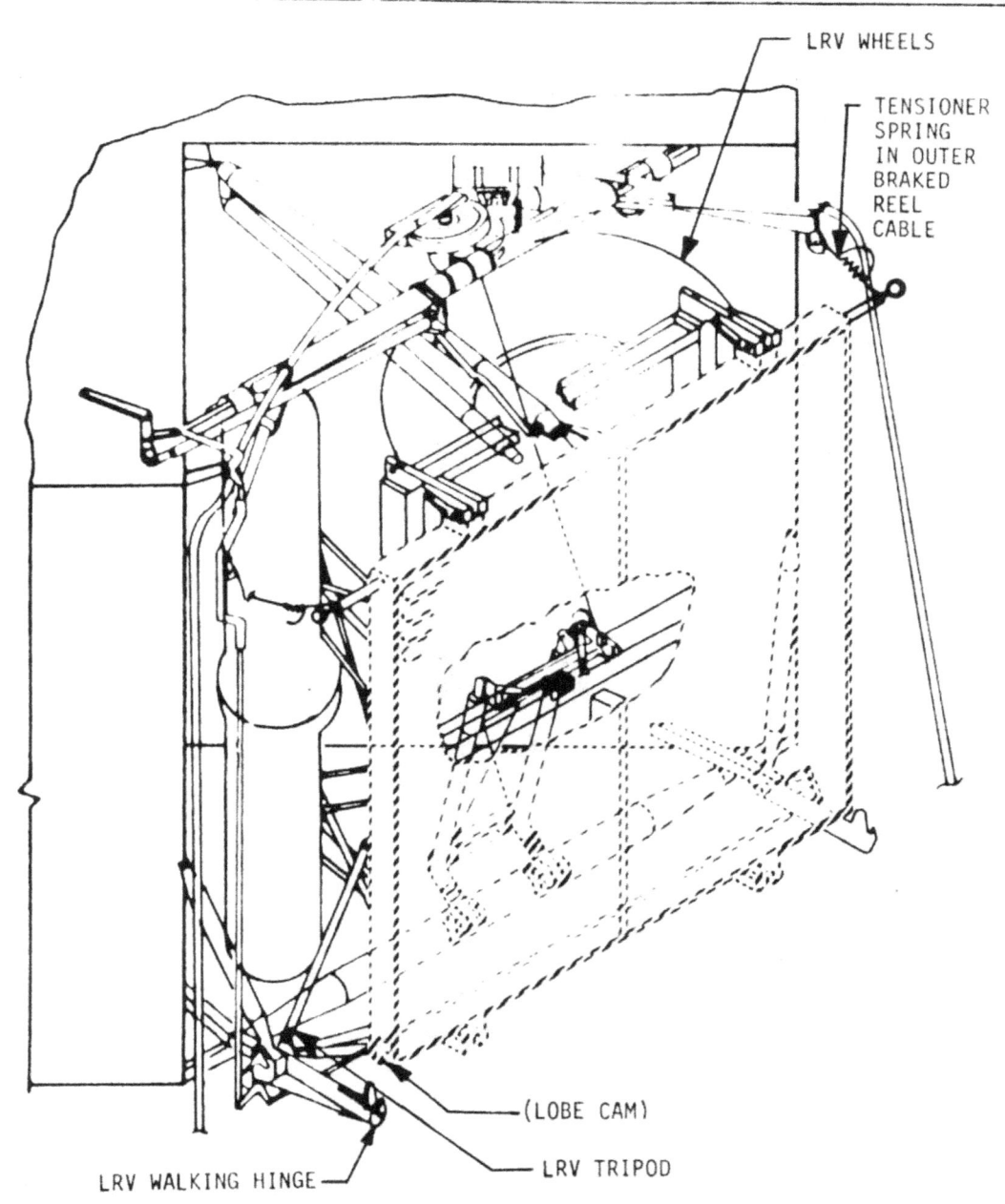

FIGURE 1-38   LM/SSE WITH LRV INSTALLED

LS006-002-2H
LUNAR ROVING VEHICLE
OPERATIONS HANDBOOK

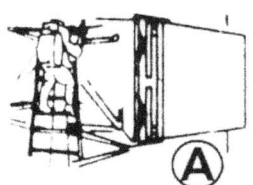

A
- LRV STOWED IN QUADRANT
- ASTRONAUT REMOVES INSULATION BLANKET, OPERATING TAPES
- ASTRONAUT REMOTELY INITIATES AND EXECUTES DEPLOYMENT

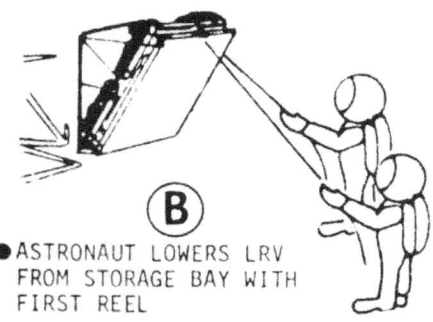

B
- ASTRONAUT LOWERS LRV FROM STORAGE BAY WITH FIRST REEL

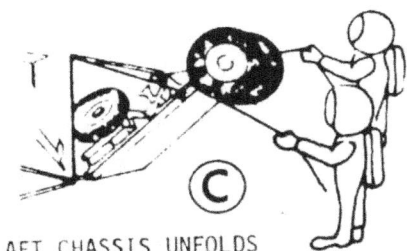

C
- AFT CHASSIS UNFOLDS
- REAR WHEELS UNFOLD
- AFT CHASSIS LOCKS IN POSITION

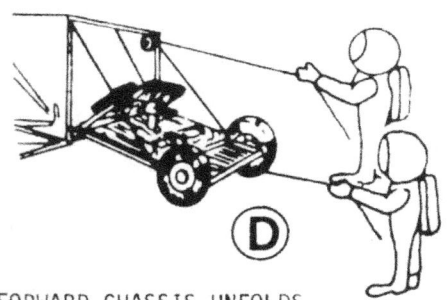

D
- FORWARD CHASSIS UNFOLDS
- FRONT WHEELS UNFOLD

E
- FORWARD CHASSIS LOCKS IN POSITION. ASTRONAUT LOWERS LRV TO SURFACE WITH SECOND LEVEL

F
- ASTRONAUT DISCONNECTS SSE
- ASTRONAUT UNFOLDS SEATS, FOOTRESTS, (FINAL STOP)

FIGURE 1-39 LRV DEPLOYMENT SEQUENCE

LS006-002-2H
LUNAR ROVING VEHICLE
OPERATIONS HANDBOOK

1.9.3 (Continued)

Phase IV - Operating the center braked reel to lower the forward chassis wheels to the lunar surface (figure 1-39, inset (E)).

Phase V - Disconnecting the SSE from the LRV after all four wheels are on the surface (figure 1-39, inset (F)).

1.9.3.1 Phase I Deployment Description

This phase consists of visual inspection, removal of insulation blanket (figure 1-40), deployment of the two braked reel operating tapes and the deployment cable from their stowed position. The double braked reel operating tape is stowed in a nylon bag attached to the lower, right support arm by velcro tape. The center braked reel operating tape is stowed on the left side of the LRV center chassis by velcro tape in a nylon bag attached to the lower left support arm by velcro tape (figure 1-41). The deployment cable is stowed on the left side of the LRV center chassis by teflon clips. The deployment cable is used to assist deployment of the LRV in the eventuality that the LRV stops during any phase of the deployment.

1.9.3.2 Phase II Deployment Description

At the completion of Phase I, the astronaut actuates the D-handle, which is located on the right side of the porch (figure 1-42). The first 5 to 6 inches of travel of the D-handle removes the two lower release pins (figure 1-43) out of the apex fittings, releasing the lower half of the apex fittings, allowing it to fall away immediately or during deployment rotation. The apex fittings are now configured to lift off of the spools when required. The last segment of travel of the D-handle removes the upper release pin. When the upper release pin is removed, the push-off spring rotates the LRV out from the LM approximately 4°, taking up the slack in the outer braked reel cables. The slack is due to cable tensioner springs in the cables.

1.9.3.3 Phase III Deployment Description

The LRV is now released from the LM and is ready to be deployed to the lunar surface. During the entire Phase III operations, the astronaut operates the double braked reel operating tape. The braked reel (figure 1-44) is a worm and worm gear arrangement. When the operating tape is pulled, the cable storage drum is rotated, thus releasing (feeding off) cable from the drum. The cable is attached to the LRV center chassis and, as the LRV rotates outboard due to gravity, supports the LRV. As the drum is rotated, feeding out cable, the LRV is allowed to rotate and deploy. For the first 15° of rotation, the LRV rotates on the apex fittings. At 15° rotation, the walking hinge is engaged by the LRV and the point of rotation shifts from the apex fittings to the walking hinge, at which point the apex fitting lifts off of the spools. The deployment cable may or may not be required at this time, depending on the landed attitude of the LM.

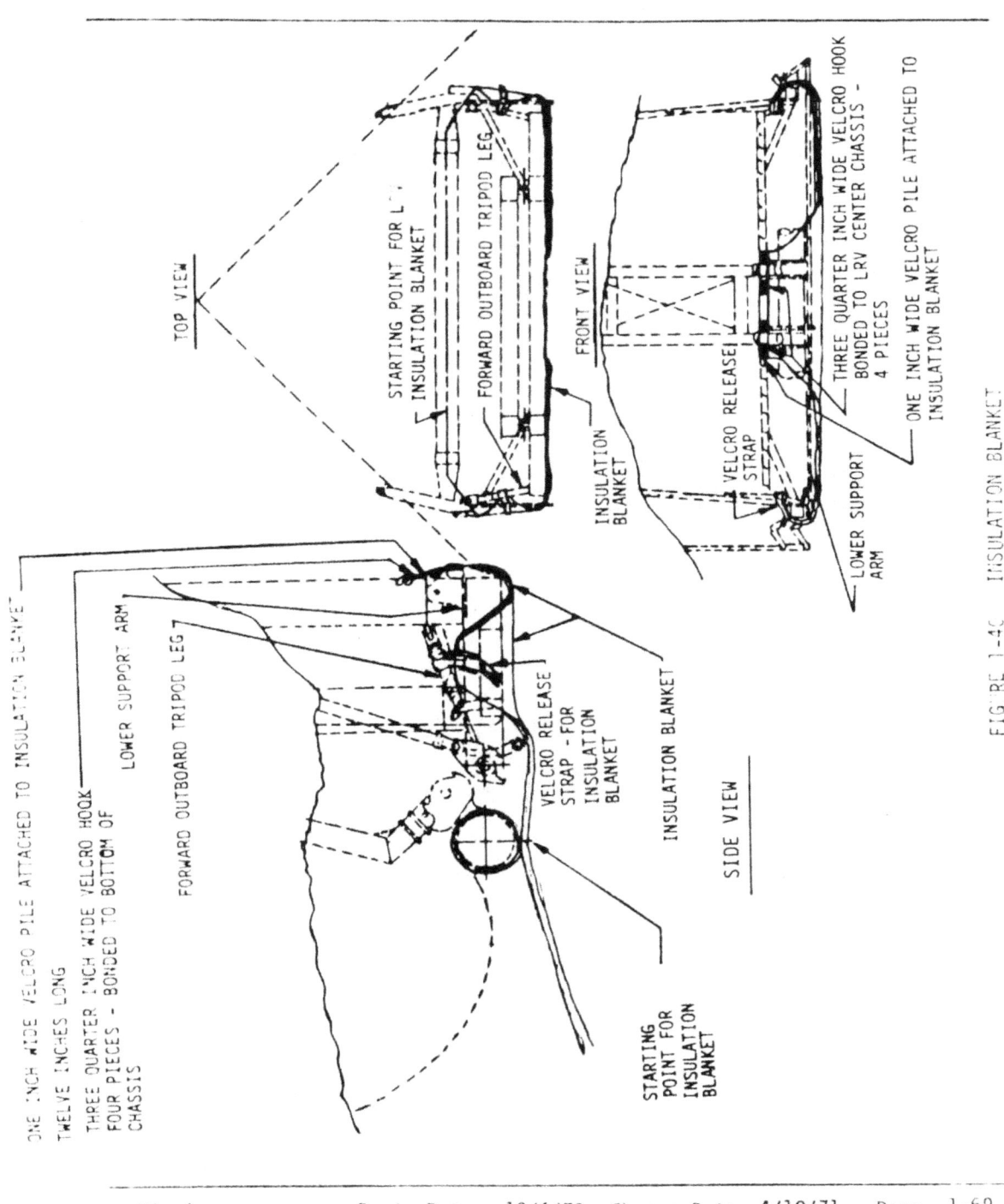

FIGURE 1-40 INSULATION BLANKET

LS006-002-2H
LUNAR ROVING VEHICLE
OPERATIONS HANDBOOK

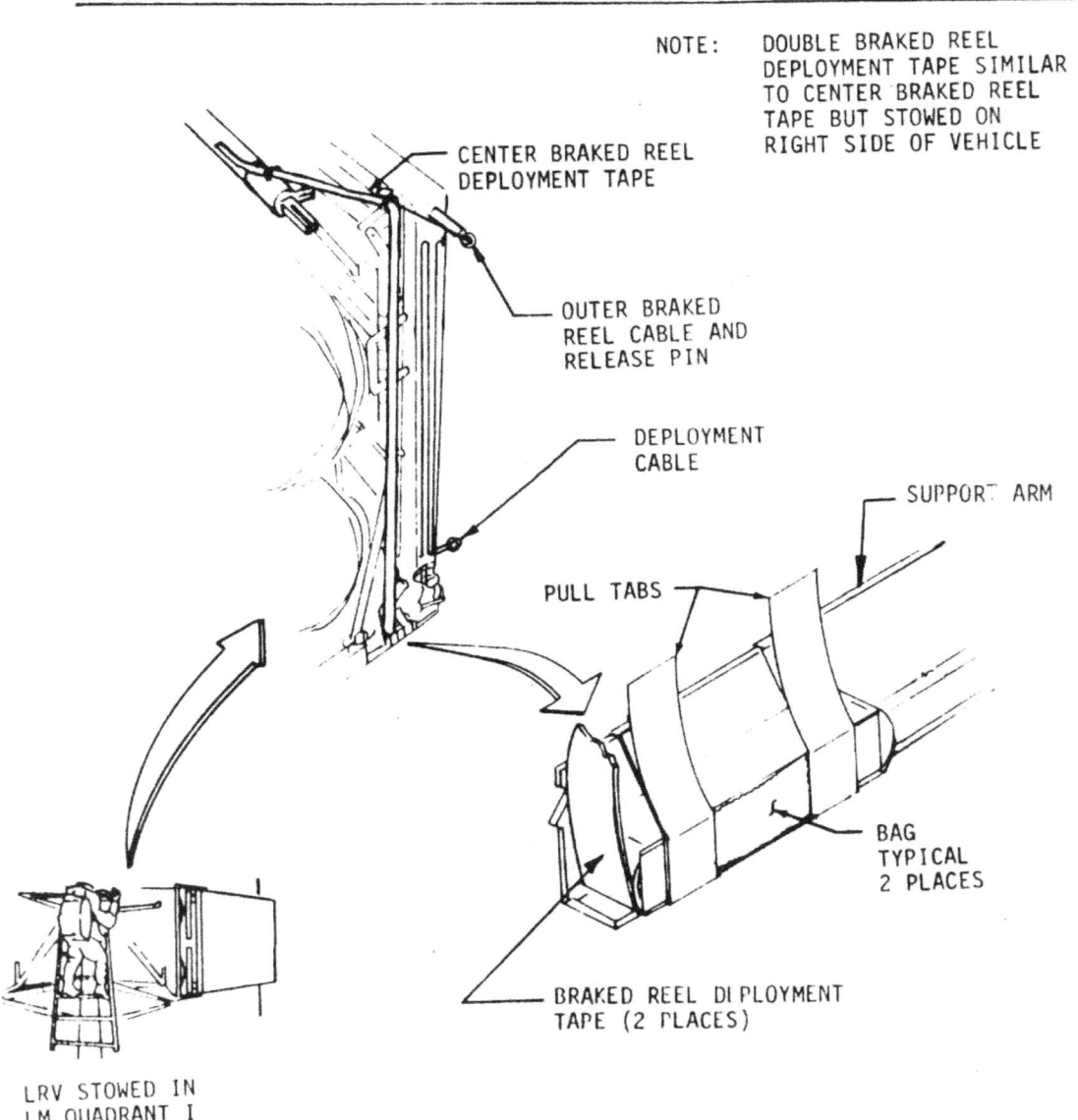

FIGURE 1-41   LRV DEPLOYMENT TAPES AND CABLES

Mission   J   Basic Date 12/4/70   Change Date 4/19/71   Page 1-70

LS006-002-2H
LUNAR ROVING VEHICLE
OPERATIONS HANDBOOK

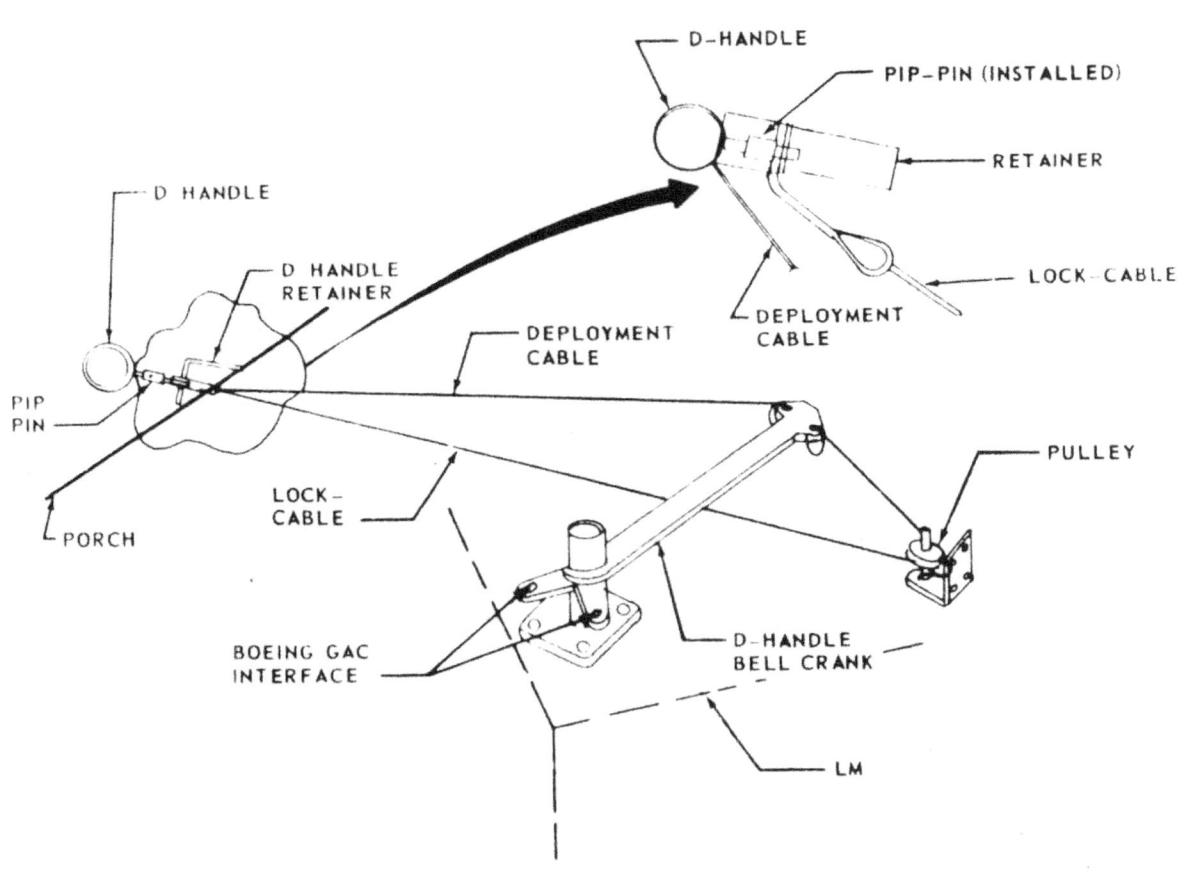

FIGURE 1-42 D-HANDLE RELEASE SYSTEM

LS006-002-2H
LUNAR ROVING VEHICLE
OPERATIONS HANDBOOK

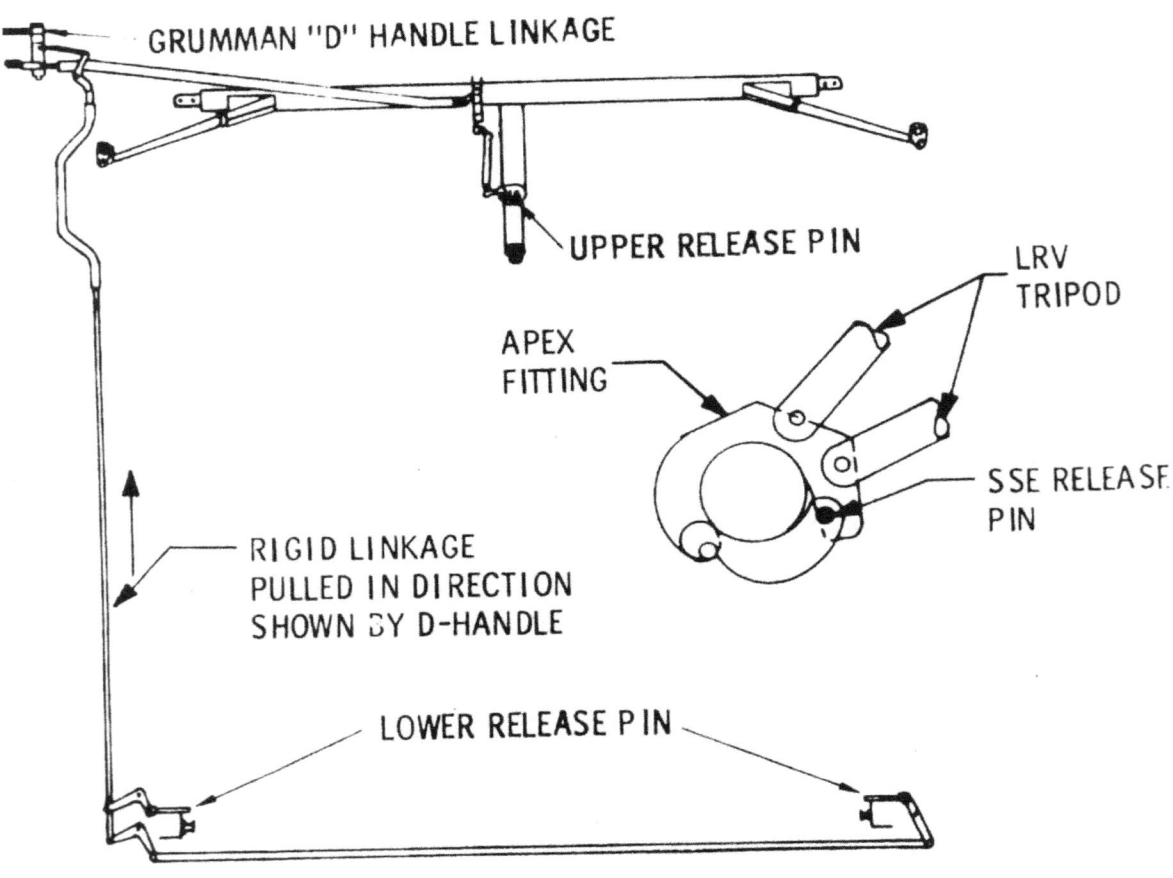

FIGURE 1-43   LRV/SSE SUPPORT STRUCTURE AND RELEASE SYSTEM

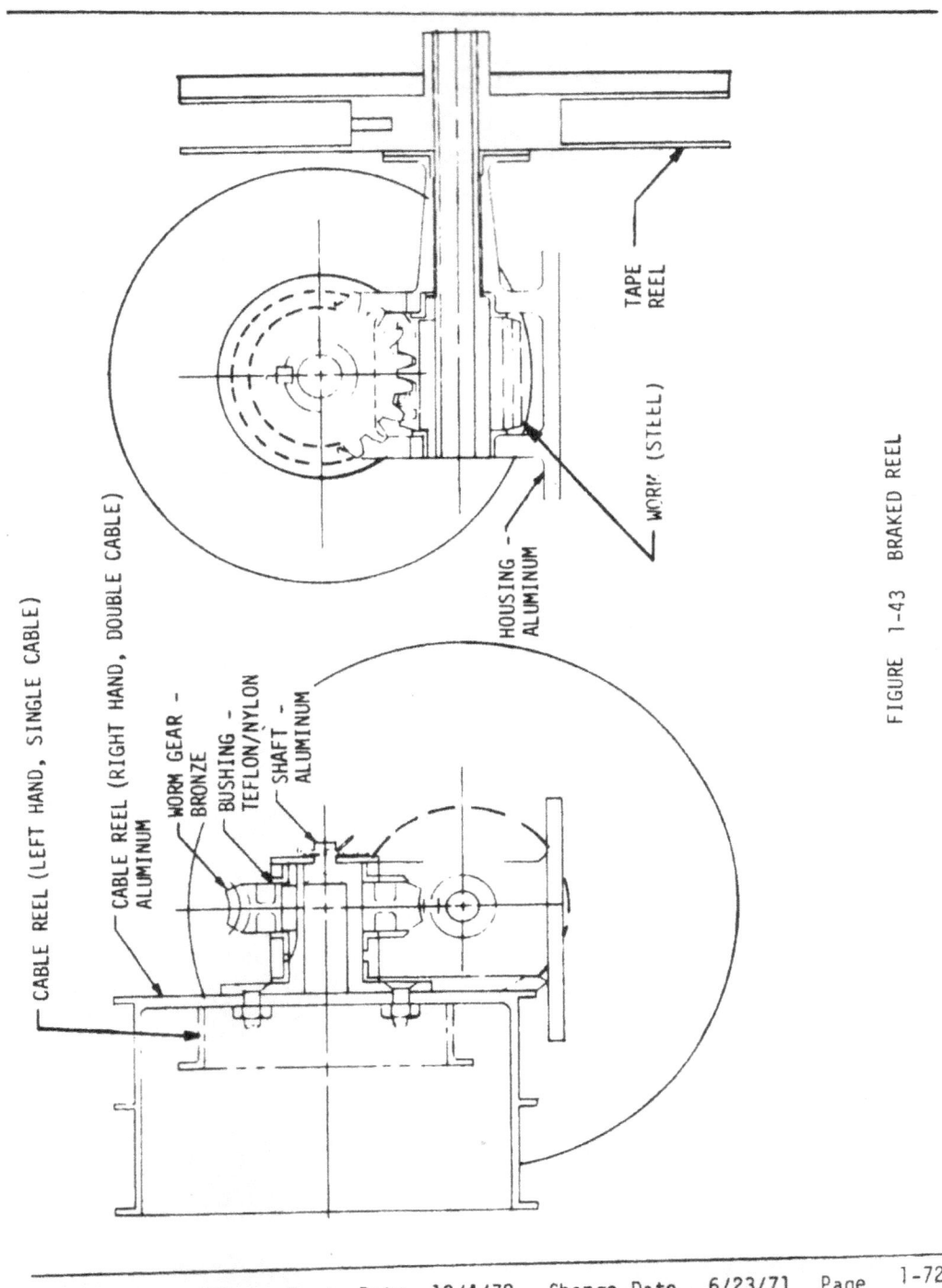

FIGURE 1-43 BRAKED REEL

1.9.3.3 (Continued)

The LRV continues to rotate about the walking hinge. At 35° rotation, the lower telescopic tubes ratchets are engaged, preventing any reverse rotation of the telescopic tube assembly about its lower pivot points. The telescopic tube assembly consists of a pair of three telescoped aluminum tubes, hinged to the GAC structure (lower center of the LM quadrant) and connected at the top by the saddle (figure 1-44). The aluminum saddle fits to the forward section of the forward chassis, held by two dowel pins and a ball-lock pin clevis joint. The saddle carries the pulleys, cables and pin mechanisms whereby the forward and aft LRV chassis are unlocked from the stowed (folded) LRV position. As the LRV moves outboard, the 45° cable tightens, and rotates each with a steel cable and ball-lock pin. The two ball-lock pins lock the connection between forward and aft chassis to the console post mounted on the center chassis (figure 1-45). If either the aft chassis latch pins or the forward chassis latch pin fails to pull, the deployment cable may be pulled to accomplish this action. (The mechanical advantage of the deployment cable to the pins themselves is 5 to 1).

The telescopic tubes and forward chassis stops at 45° due to the 45° cable (chassis latch actuating cable) becoming taut, then by counteracting forces of the LRV forward chassis hinge torsion spring, the telescopic tubes and forward chassis return to the 35° position (stop due to the telescopic tube ratchet). The center chassis and aft chassis continue to deploy. After it is unlocked, the aft chassis fully deploys (unfolds) due to the aft chassis hinge torque bars, until it latches with the center chassis.

The wire mesh LRV wheels are held in the stowed position by four aluminum tube struts. One end of each strut is held by a steel pin to the aft or forward chassis structure (figure 1-46). The other end of each strut is held to a wheel hub by a pin (in the hub). The pins in the chassis are pulled by a steel cable, so linked as to pull the pins as the chassis opens, approximately 170°. When the pins are pulled, the spring-loaded wheels move to deployed or operational position. As the wheels rotate forward to the deployed position, a mechanism within the wheel hub retracts the remaining pin retaining the wheel strut. The strut is thus freed at both ends, and falls free during wheel deployment movement. Each strut is retained by a 1/8 inch diameter mylar tether.

The LRV center/aft chassis continues outboard rotation, pivoting around the lower support arm latch. During LRV outboard rotation, the telescopic tubes extend (lengthen). Before 72° LRV rotation an anti-collapse telescopic tube latch in each tube engages to prevent shortening (but permit elongation) of the tubes.

At approximately 73° center chassis angle, the lobe (cam) on the forward sides of the center chassis strut (engaged in the lower support arm latch) strikes the steel latch lock arm. As the chassis rotates, the cam forces the latch lock arm down out of a safety retaining spring, and unlocks the latch. The center/aft chassis continues to rotate until the aft chassis wheels

LS006-002-2H
LUNAR ROVING VEHICLE
OPERATIONS HANDBOOK

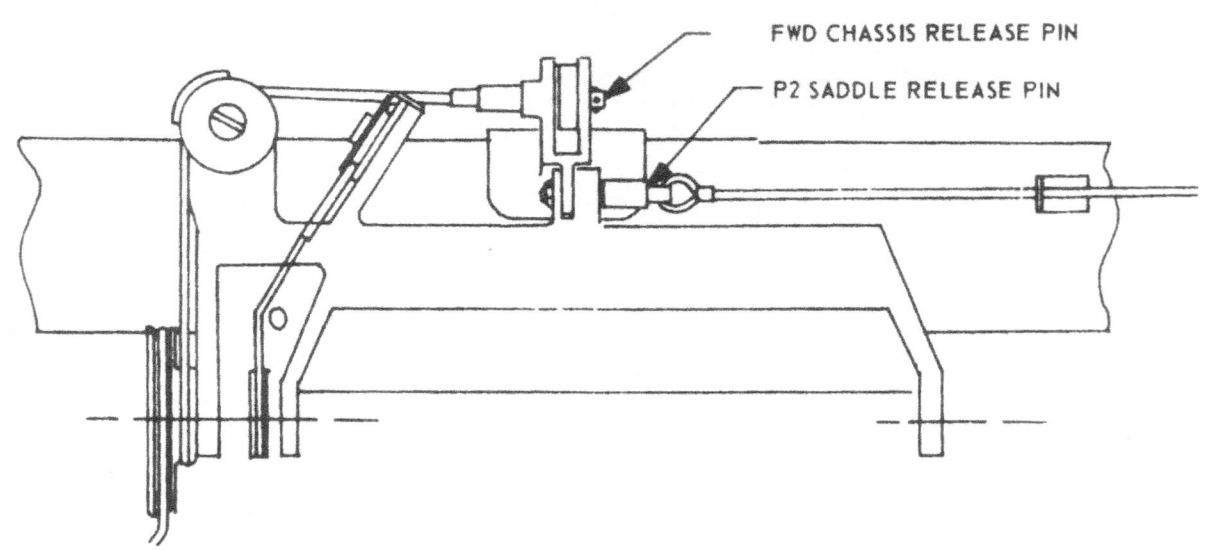

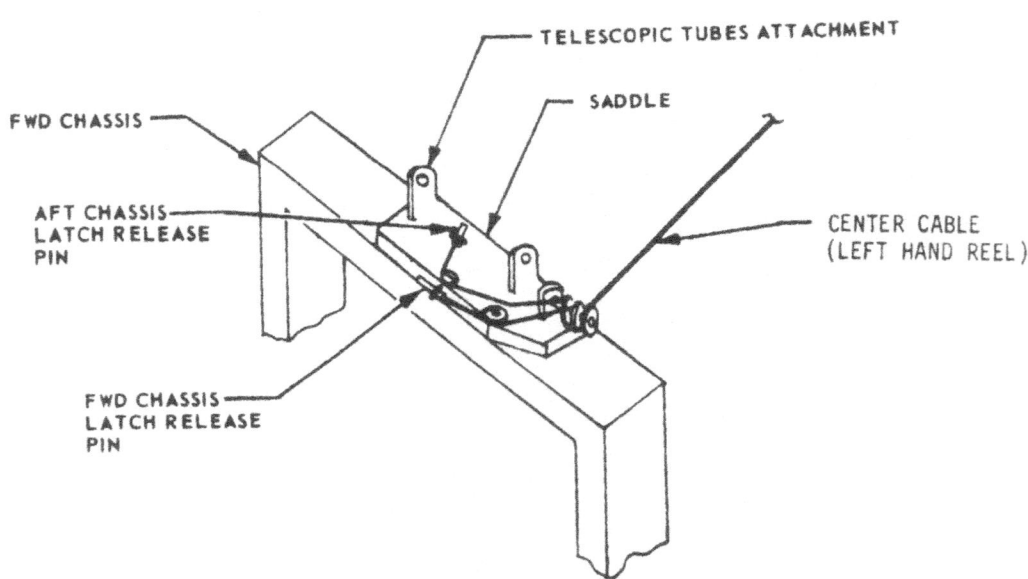

FIGURE 1-44   LRV SADDLE AND FORWARD CHASSIS LATCH RELEASE

LS006-002-2H
LUNAR ROVING VEHICLE
OPERATIONS HANDBOOK

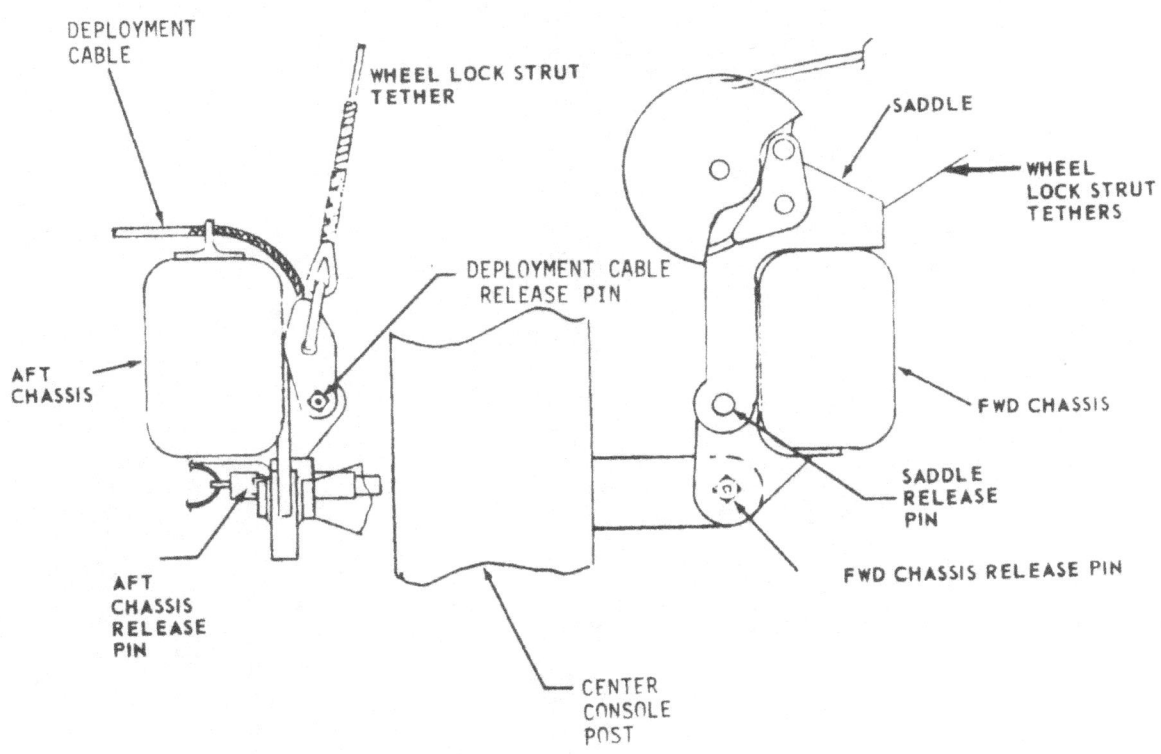

FIGURE 1-46 FORWARD AND REAR CHASSIS LATCH

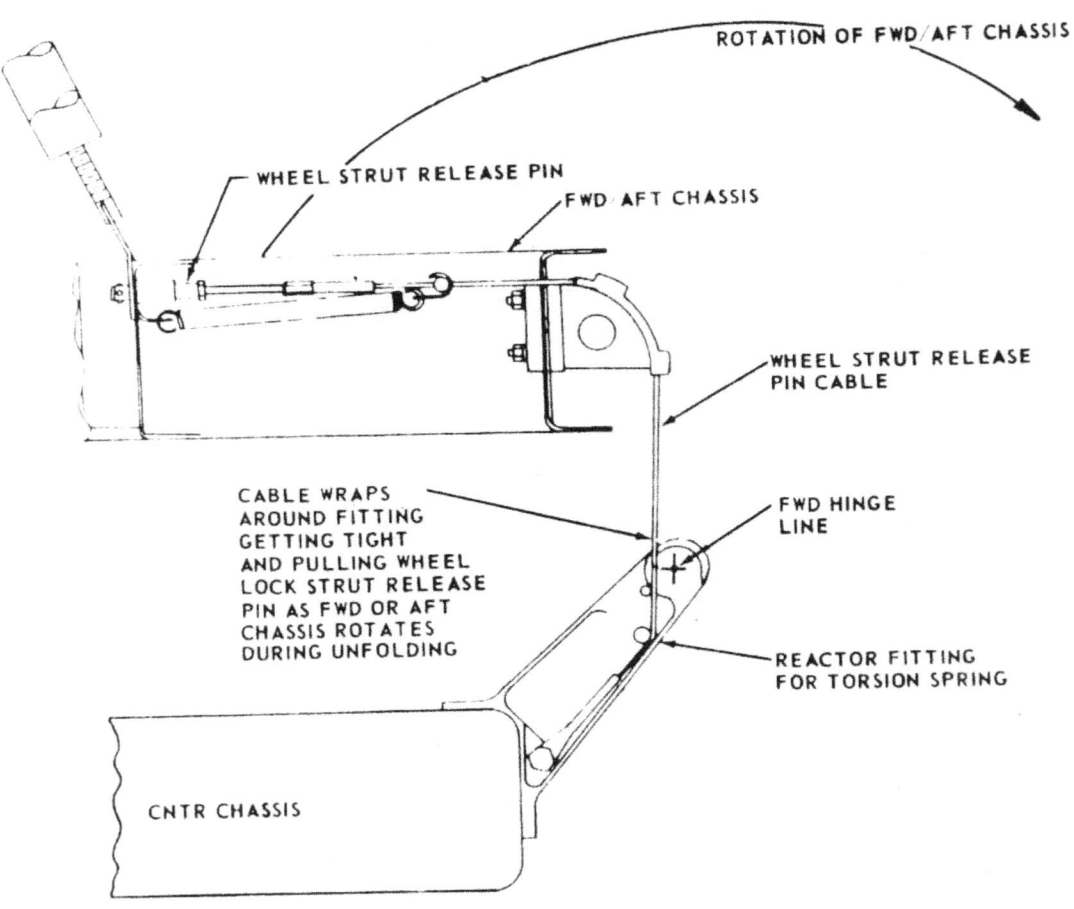

FIGURE 1-47 WHEEL LOCK STRUT RELEASE

LS006-002-2H
LUNAR ROVING VEHICLE
OPERATIONS HANDBOOK

1.9.3.3 (Continued)

are on the surface. The wheels are locked with the emergency hand brake, therefore must slide on the surface. Depending on the landing attitude of the LM and the condition of the surface, the wheels might not slide on the lunar surface, therefore use of the deployment cable by the astronaut would be required. The astronaut continues to actuate the double braked reel operating cable to allow the forward chassis hinge to deploy by virtue of the forward chassis hinge torsion spring. Concurrently, the center/aft chassis move outboard, away from the LM. At this point in the sequence, the 45° cable becomes taut due to the outboard movement of the entire LRV. The center chassis continues to move down and, driven by the forward chassis springs, outboard. As the angle between forward and center chassis approaches 170°, the forward wheel lock strut pins in the forward chassis release, and the forward wheels deploy like the previously described aft wheel deployment. The astronaut then pulls the pins that attach the two outer braked reel cables to the center chassis.

Phase III is complete (motion ceases) with the aft wheels on the lunar surface, with forward and aft chassis locked to the center chassis, all wheels deployed, all four wheel struts free and hanging from their tethers, the outer braked reel cables released, and with the forward chassis held up by the telescopic tube assembly and the 45° cable.

1.9.3.4 Phase IV Deployment Description

This phase of the deployment consists of the astronaut actuating the center braked reel operating tape, thus allowing the forward chassis to lower and the forward chassis wheels to lower to the surface. Again, the deployment cable may be required at this point if the aft wheels will not slide on the surface.

1.9.3.5 Phase V Deployment Description

This phase consists of releasing the deployment hardware from the LRV. The astronaut pulls up on the saddle release cable, located on the left rear side of the forward chassis. This operation releases a ball-lock pin which holds the saddle on the forward chassis. When the saddle is released, the following hardware goes with it:

a. Telescopic tube assembly.
b. Forward and aft chassis lock release pins.
c. Forward chassis wheel lock struts and tethers.

The astronaut then pulls a ball lock pin, located on the aft center of the aft chassis. This releases the deployment cable and the aft chassis wheel lock struts. At this point, the deployment from the LM to the lunar surface is complete.

Mission J   Basic Date 12/4/70   Change Date 4/19/71   Page 1-78

LS006-002-2H
LUNAR ROVING VEHICLE
OPERATIONS HANDBOOK

SECTION 2

NORMAL PROCEDURES

INTRODUCTION

This section defines the normal procedures to be followed by the astronauts for operating the LRV on the lunar surface and the 1G Trainer during earth training operations.

*** 1G TRAINER NOTE ***

The procedures contained in this section also apply to 1G Trainer operation with the exception of deployment operations, defined in 2.1. When performing training per paragraph 2.1 all steps can be performed except those dealing with pulling the LM D-rings, inspecting hinge pins, and deploying the inboard handholds.

Mission ___J___ Basic Date _12/4/70_ Change Date _4/19/71_ Page 2-1

LS006-002-2H
LUNAR ROVING VEHICLE
APOLLO OPERATIONS HANDBOOK

STA/STEP	PROCEDURE	REMARKS
2.1	UNLOADING AND CHASSIS DEPLOYMENT	
	Both crewmen are utilized for the unloading. Procedure steps subsequent to 2.1w are for one crewman unless otherwise noted.	Prior to initiation of the subsequent procedures, crew should position TV camera to monitor all LRV deployment operations.
a.	Perform an overall inspection of the LRV and SSE to verify proper configuration and no obvious damage.	To extent which lighting & landing angle permits inspect lower tube & bellcrank assys, left side tube assy, & upper tube & bellcrank assys & verify no obstructions preventing operation of these pin release mechanisms.
b.	Release LRV insulation blanket as follows: (1) At left side pull velcro straps & release blanket from tripod & peel away from floor panel velcro. (2) At right side pull velcro straps & release blanket from tripod & peel away from floor panel velcro. (3) Verify blanket detaches from LRV and falls back toward LM.	Figure 2-1. Perform overall inspection of those portions of LRV and SSE which were previously obscured by insulation blanket.
c.	Inspect each of two lower support arm latches to verify proper configuration. Latch should be in position and trip arm should be up, as determined by continuous white stripe on outer side of lower support arm. If either latch is in the "tripped" position, reset as follows: (1) Push trip arm down. (2) Rotate latch up into position. (3) Raise trip arm until locking dog on trip arm engages receptacle on latch. This will be indicated by continuous white stripe.	Figure 2-1.1.
d.	Release left hand deployment tape stowed in nylon bag attached to lower left support arm by velcro tape.	Figure 2-2.
e.	Stow left hand deployment tape by draping it over a LM landing strut for convenient future access.	Tape should be placed so that crewman is not required to move

Mission   J   Basic Date   12/4/70   Change Date   7/7/71   Page 2-2

LS006-002-2H
LUNAR ROVING VEHICLE
OPERATIONS HANDBOOK

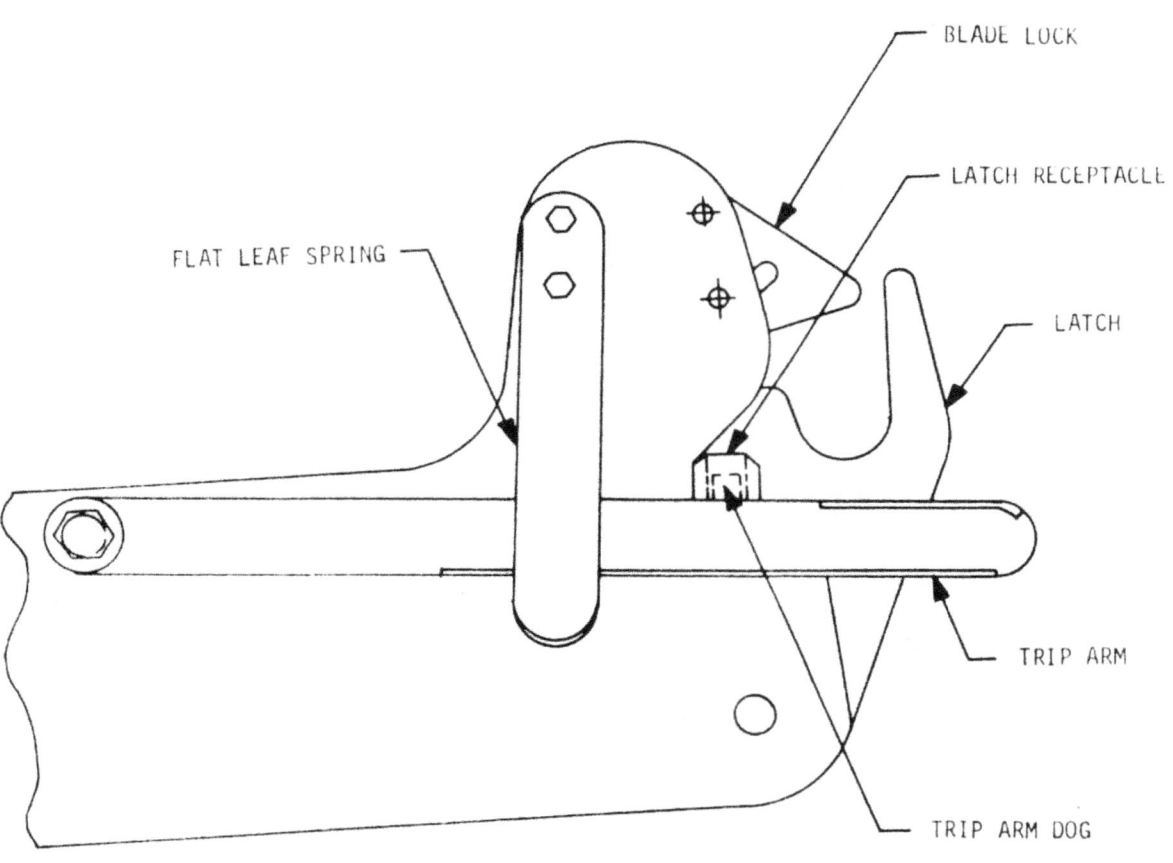

NOTE: LEFT SIDE SUPPORT ARM SHOWN. RIGHT SIDE SUPPORT ARM IS MIRROR IMAGE.

FIGURE 2-1 SUPPORT ARM LATCH MECHANISMS LATCHED CONFIGURATION

LS006-002-2H
LUNAR ROVING VEHICLE
OPERATIONS HANDBOOK

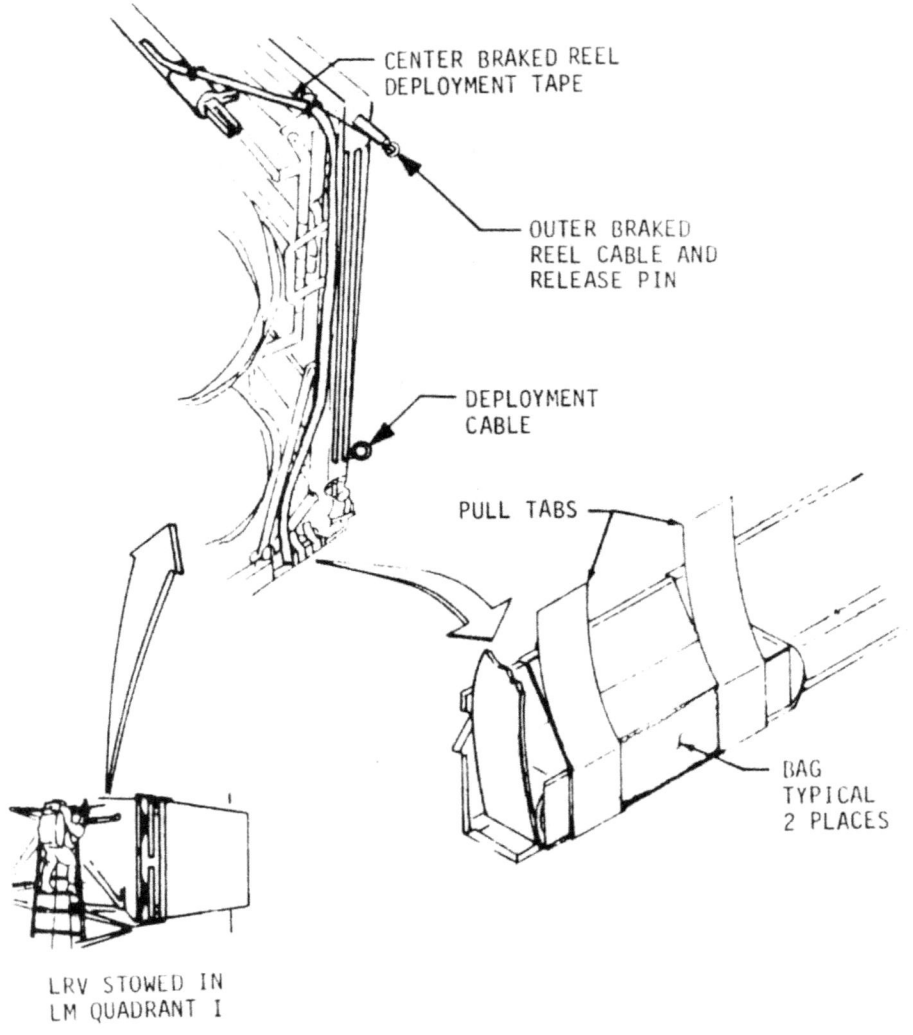

FIGURE 2-2 LRV DEPLOYMENT TAPES AND CABLES

LUNAR ROVING VEHICLE
OPERATIONS HANDBOOK

STA/STEP	PROCEDURE	REMARKS
2.1	(Continued)	into limited space between LRV and LM landing leg.
f.	Release deployment cable from teflon clips on left side of LRV center chassis and deploy cable to maximum length and at 45° angle from Quad I toward descent ladder.	
g.	Release right hand deployment tape stowed in nylon bag attached to lower right support arm by velcro tape. Place tape in convenient location for future access.	
h.	Ascend LM ladder.	
i.	Inspect D-handle and bellcranks to verify there are no obstructions preventing operation.	Figure 1-41.
j.	Inspect cable assembly connecting D-handle and bellcrank to ensure there is no fouling that would prevent operation.	Figure 2-3.
	CAUTION	
	During and subsequent to deployment D-handle operation, both crewmen should remain out of the LRV deployment envelope.	
k.	Pull LRV deployment D-handle. Verify LRV moves outward from LM about 4 degrees.	Figure 2-4.
	NOTE: If push-off rod fails to rotate LRV about 4° outward from LM quadrant, deployment cable may be pulled to initiate LRV movement.	First 5 to 6 inches of travel releases lower release pins, lower half of apex fittings may fall away immediately or during deployment rotation. The last segment of travel releases the upper pin. As the upper release pin is pulled, LRV rotates out of LM about 4 degrees.

Mission ___J___ Basic Date __12/4/70__ Change Date __7/7/71__ Page __2-5__

LS006-002-2H
LUNAR ROVING VEHICLE
OPERATIONS HANDBOOK

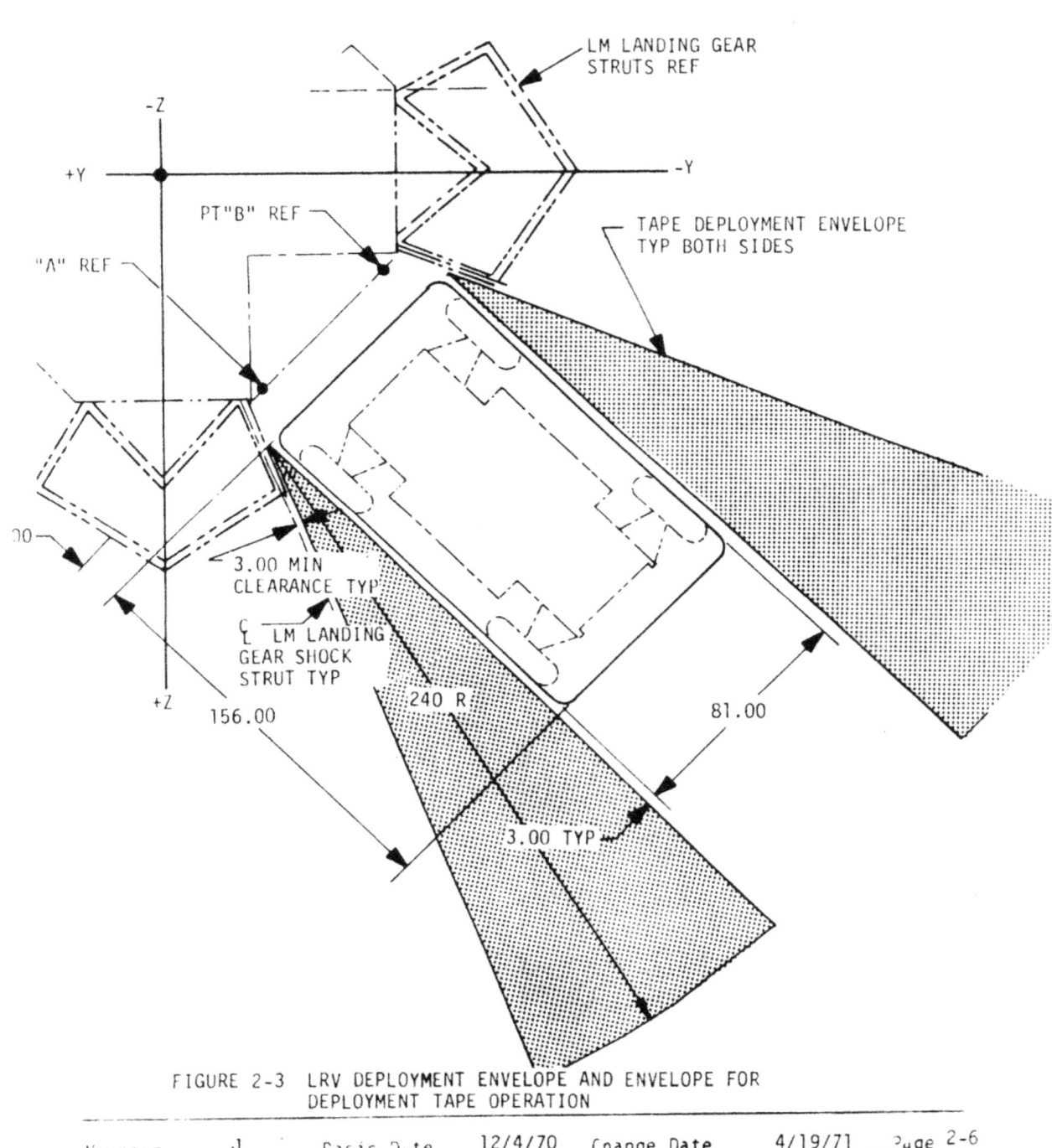

FIGURE 2-3 LRV DEPLOYMENT ENVELOPE AND ENVELOPE FOR DEPLOYMENT TAPE OPERATION

LS006-002-2H
LUNAR ROVING VEHICLE
OPERATIONS HANDBOOK

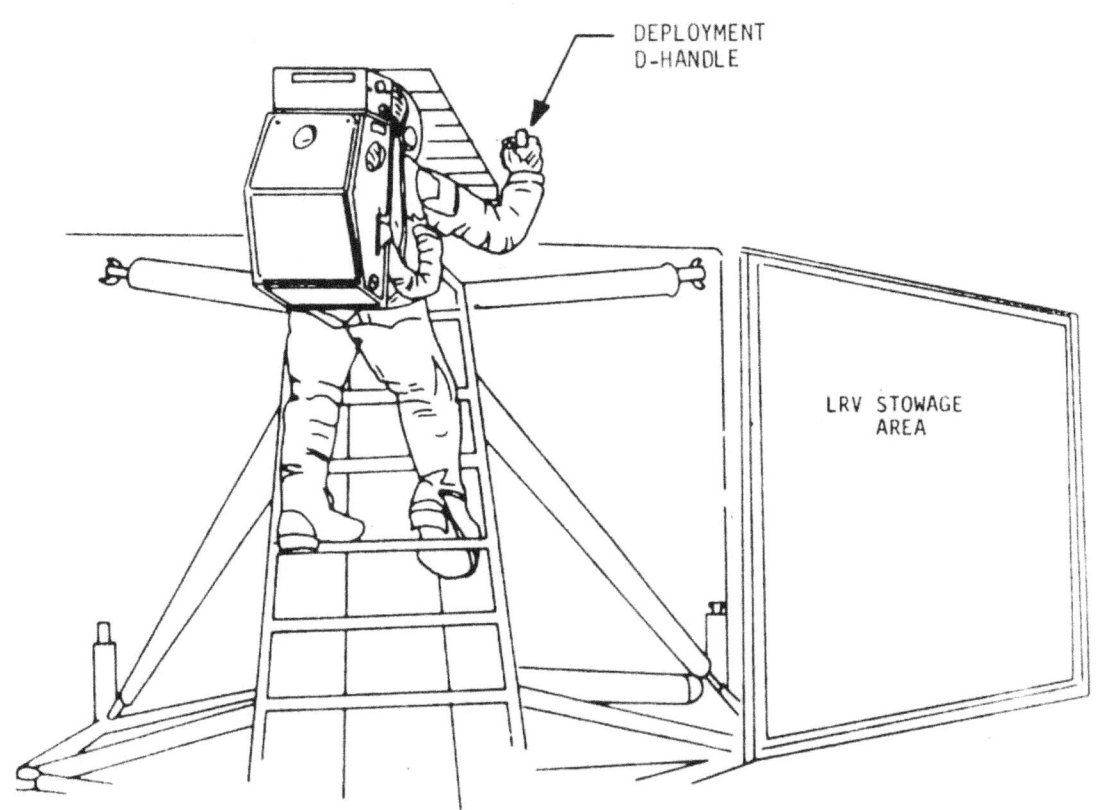

FIGURE 2-4 CREWMAN POSITIONED TO DEPLOY LRV

LUNAR ROVING VEHICLE
OPERATIONS HANDBOOK

STA/STEP	PROCEDURE	REMARKS
2.1	(Continued)	
l.	Descend LM ladder. Grasp deployment cable and monitor deployment activity.	Crewman operating deployment cable should keep slack out of double braked reel cables.
m.	Other crewman pulls double braked reel tape at right side of vehicle. Verify LRV rotates outward from LM (Figure 2-5, View B). NOTE: Crewman should remain within defined envelopes for deployment tape operation (Figure 2-3) to ensure that deployment tapes do not contact sensitive LM components.	For first 15 degrees of rotation LRV rotates on apex fittings, thereafter apex fittings lift off spools and rotation point shifts to walking hinge. Lower telescopic tubes ratchet engage at 35 degrees rotation.
n.	Continue to pull double braked reel tape (figure 2-5, View C). When vehicle rotates outboard to about 45 degrees, verify that: (1) Aft chassis unfolds and locks in position. (2) Rear wheels unfold and tethered rear wheel struts fall free. (3) Forward chassis is released from console post and returns to 35 degree position. NOTE: If either aft or forward chassis latch pins fail to pull automatically, deployment cable may be pulled to accomplish pin release.	At about 45 degrees the 45° cable tightens, pulling the forward and aft chassis latch pins at the console post mount on the center chassis. The aft chassis and wheels fully deploy and the forward chassis returns to the 35° position.
o.	Continue to pull double braked reel tape (Figure 2-5, View D). Verify that: (1) Center/aft chassis rotates until rear wheels contact lunar surface. (2) Rear wheels slide on surface permitting center/aft chassis to move away from LM.	At about 73 degrees, the cam on forward sides of center chassis strikes latch lock arm, forces arm down out of retaining spring and unlocks latch.

Mission J   Basic Date 12/4/70   Change Date 4/19/71

LS006-002-2H
LUNAR ROVING VEHICLE
OPERATIONS HANDBOOK

- LRV STOWED IN QUADRANT
- ASTRONAUT REMOVES INSULATION BLANKET, OPERATING TAPES
- ASTRONAUT REMOTELY INITIATES AND EXECUTES DEPLOYMENT

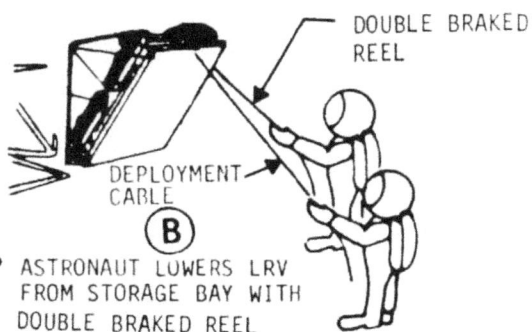

- ASTRONAUT LOWERS LRV FROM STORAGE BAY WITH DOUBLE BRAKED REEL

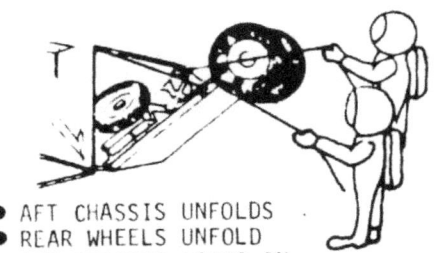

- AFT CHASSIS UNFOLDS
- REAR WHEELS UNFOLD
- AFT CHASSIS LOCKS IN POSITION

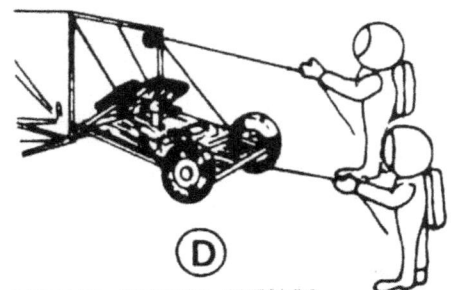

- FORWARD CHASSIS UNFOLDS
- FRONT WHEELS UNFOLD

- FORWARD CHASSIS LOCKS IN POSITION. ASTRONAUT LOWERS LRV TO SURFACE WITH CENTER BRAKED REEL

- ASTRONAUT DISCONNECTS SSE
- ASTRONAUT UNFOLDS SEATS, FOOTRESTS, (FINAL STOP)

FIGURE 2-5 LRV DEPLOYMENT SEQUENCE

Mission __J__ Basic Date __12/4/70__ Change Date __4/19/71__ Page __2-9__

LUNAR ROVING VEHICLE
OPERATIONS HANDBOOK

STA/STEP	PROCEDURE	REMARKS
2.1	(Continued)	
	NOTE: If wheels fail to slide, deployment cable may be pulled to permit center/aft chassis to move away from LM.	
p.	Continue to pull double braked reel tape (Figure 2-5, View D). Verify that: (1) Forward chassis continues to unfold and locks in position. (2) Forward wheels unfold. (3) Outer braked reel cables are slack. (4) 45° cable again becomes taut.	Forward wheel lock strut pins release and forward wheels deploy as the angle between the forward and center chassis approaches 170 degrees. (The 45 degree cable again becomes taut).
q.	Release double braked reel tape and at chassis RR grasp outer braked reel cable in right hand and remove cable pin P8 (Figure 2-6) with left hand.	At this time the forward and aft chassis sections are deployed and locked to the center chassis. All wheels are deployed. The forward chassis is held up by the telescopic tube assembly and the 45 degree cable.
r.	Discard cable and pin outside work area.	
s.	At chassis LR grasp outer braked reel cable in left hand and remove cable pin P1.	Figure 2-6.
t.	Discard cable and pin outside work area.	
u.	Pull center braked reel tape (Figure 2-5, View E). Verify that forward chassis lowers until all wheels contact lunar surface and support vehicle weight and 45° cable is slack.	This tape was previously stowed over a LM landing strut for convenient access.
	NOTE: If wheels fail to slide, deployment cable may be pulled to move LRV away from LM.	Using deployment cable to pull the LRV with parking brake

Mission   J   Basic Date   12/4/70   Change Date   4/19/71   Page 2-10

LUNAR ROVING VEHICLE
APOLLO OPERATIONS HANDBOOK

STA/STEP	PROCEDURE	REMARKS
2.1	(Continued)	
s.	At chassis LR grasp outer braked reel cable in left hand and remove cable pin P1.	Figure 2-6.
t.	Discard cable and pin outside work area.	
	CAUTION	
	To prevent any sudden surprise LRV motions, the center steel deployment cable should always be under tension when pulling on left hand deployment tape. This is accomplished by maintaining a force on the deployment cable.	
u.	Pull left hand deployment tape (Fig. 2-5, View E). Verify that forward chassis lowers until all wheels contact lunar surface and support vehicle weight and center cable is slack.	This tape was previously stowed over a LM landing strut for convenient access.
	NOTE: If wheels fail to slide, deployment cable may be pulled to move LRV away from LM.	Use of the deployment cable to drag the LRV with the brakes locked shall be kept to a minimum.

Mission _____J_____ Basic Date __12/4/70__ Change Date __7/7/71__ Page 2-10.1

LS006-002-2H
LUNAR ROVING VEHICLE
OPERATIONS HANDBOOK

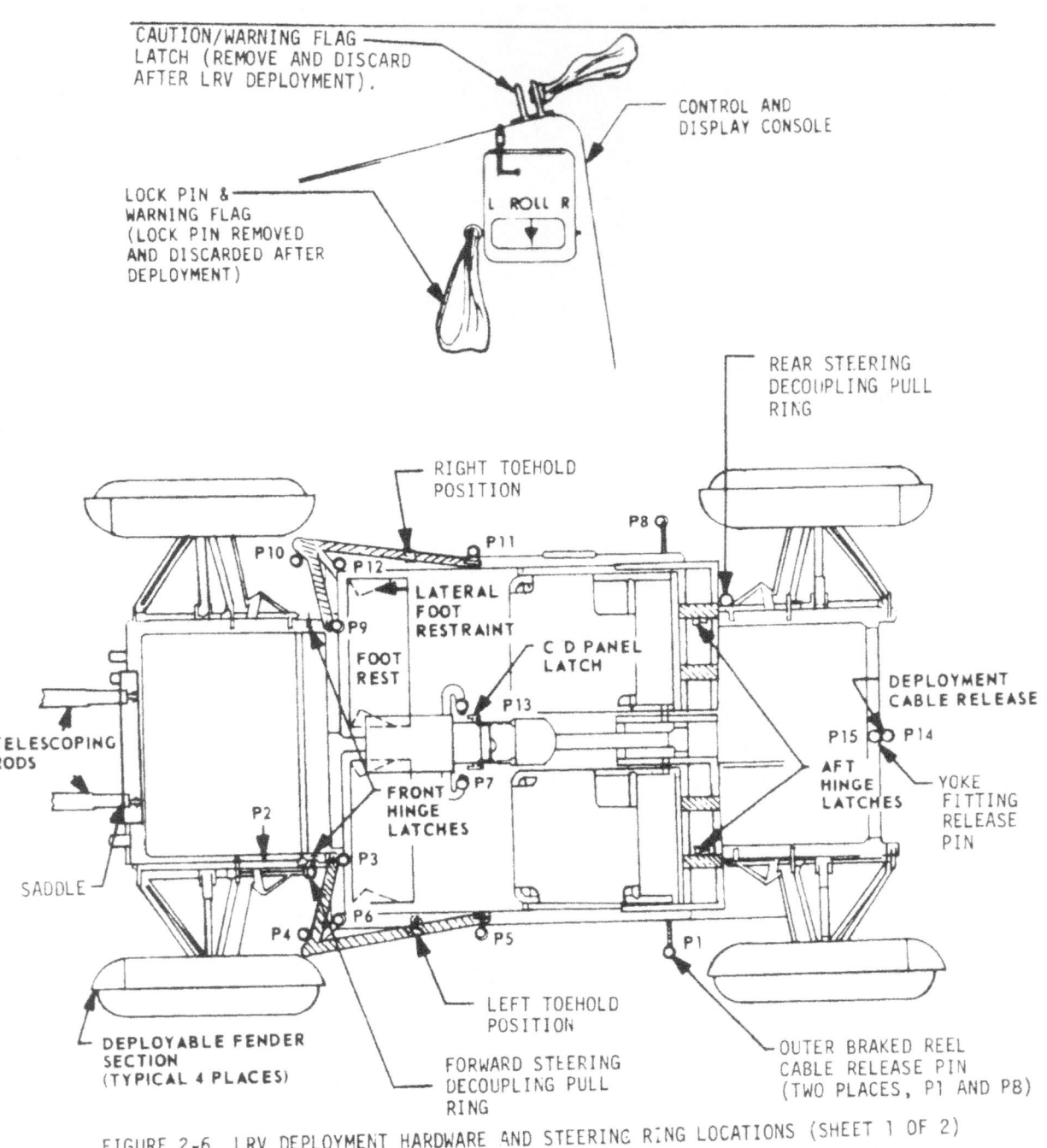

FIGURE 2-6 LRV DEPLOYMENT HARDWARE AND STEERING RING LOCATIONS (SHEET 1 OF 2)

LS006-002-2H
LUNAR ROVING VEHICLE
OPERATIONS HANDBOOK

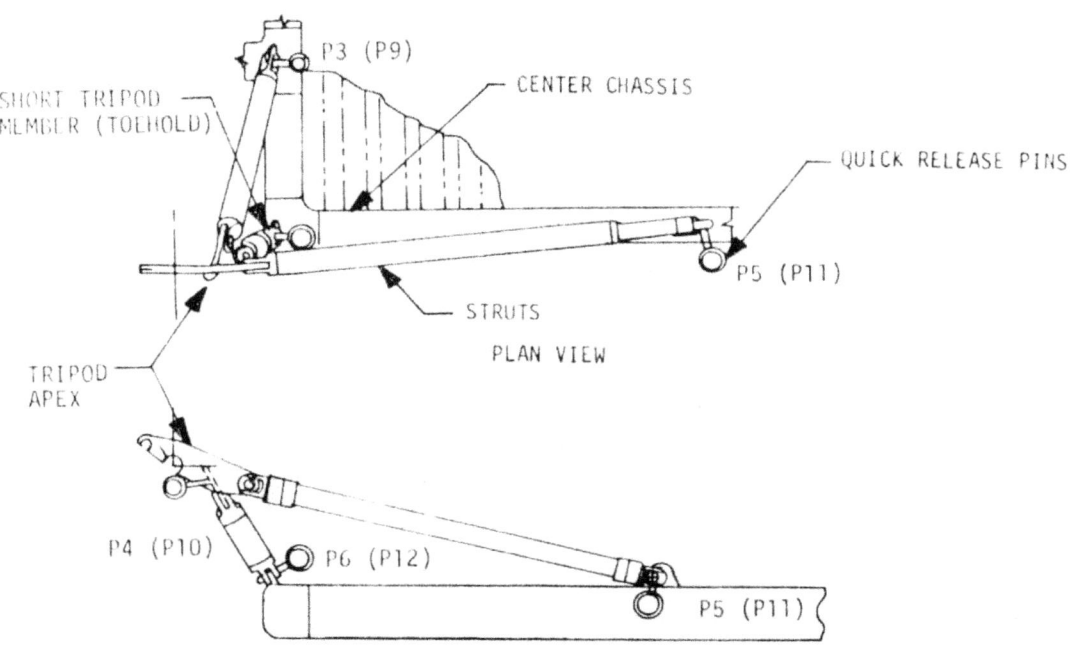

FIGURE 2-6 LRV DEPLOYMENT HARDWARE LOCATIONS (SHEET 2 OF 2)

LS006-002-2H
LUNAR ROVING VEHICLE
OPERATIONS HANDBOOK

STA/STEP	PROCEDURE	REMARKS
2.1	(Continued)	
v.	Release inboard handhold velcro tiedown strap (stand at left side of vehicle to effect release).	Figure 2-9, View 1
w.	Coil deployment cable and remove cable release pin P14 and chassis delatch fitting pin P15. Discard cable and deployment hardware outside of work area.	Figure 2-6. When pin P14 is pulled, deployment cable and rear wheel tethers fall free of vehicle. When pin P15 is pulled, yoke will either fall to surface or be retained in pinless fitting. If retained, pull yoke free at clevis and discard yoke.
x.	Deploy RF fender extension.	
y.	Verify both hinge pins flush at RF hinge.	If hinge pin is not flush, tap pin with toehold subsequently removed in step ac. Verify pin is latched by pressing down on chassis.
z.	Remove pins P9 and P10 from right tripod and discard clear of deployment area.	Figure 2-6.
aa.	Grasp tripod apex with right hand and remove pin P11 with left hand.	
ab.	Discard tripod main members and pin clear of deployment area.	
ac.	Grasp remaining short tripod member in right hand, remove pin P12 with left hand, and discard pin clear of deployment area.	

Mission J   Basic Date 12/4/70   Change Date 7/7/71   Page 2-13

LS006-002-2H
LUNAR ROVING VEHICLE
OPERATIONS HANDBOOK

STA/STEP	PROCEDURE	REMARKS
2.1	(Continued)	
ad.	Remove short tripod member and insert tripod member in right toehold position or stow in underseat stowage bag.	If short tripod member is installed in toehold position, end with hook should be outboard with hook pointing forward. This is also used as wheel decoupling tool.
ae.	Pull right footrest lift tabs.	Figure 2-7. Tabs pull free of footrests but remain attached to the floor panel.
af.	Rotate footrest upward and forward and lock into position.	Figure 2-8.
ag.	Release velcro tiedown strap (if necessary), pull out right C/D console "T" handle P13 with left hand and turn 90° CW.	
ah.	Rotate right seat to stable overcenter position.	
ai.	Rotate legs to full upright position.	
aj.	Attach forward seat legs velcro strap to outboard handhold.	
ak.	Verify underseat stowage bag erects.	
al.	Release right seat belt from underseat bag stowage position and stow in temporary location.	
am.	Pull seat pan frame forward to engage front legs.	
an.	Verify all seat latches latched.	
ao.	Verify both hinge pins flush at RR hinge.	If hinge pin is not flush, tap pin with toehold. Verify pin is latched by pressing down on chassis.

Mission   J   Basic Date 12/4/70   Change Date 7/7/71   Page 2-14

LS006-002-2H
LUNAR ROVING VEHICLE
OPERATIONS HANDBOOK

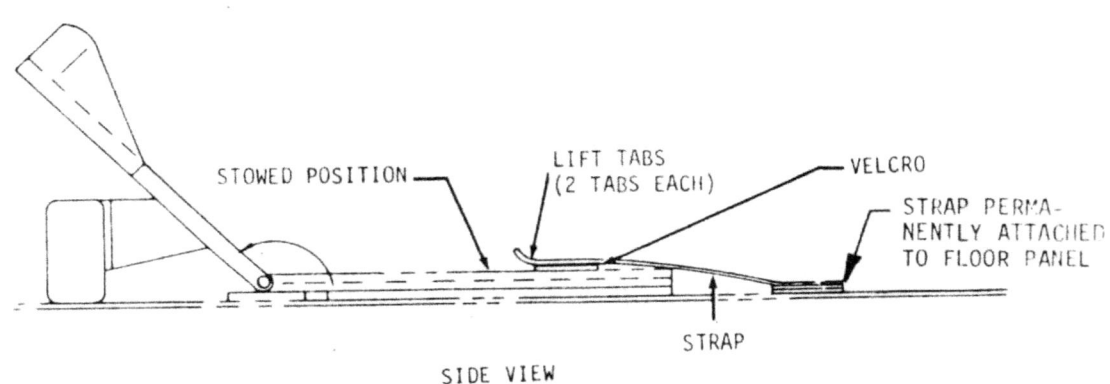

FIGURE 2-7 FOOT REST DEPLOYMENT

LS006-002-2H
LUNAR ROVING VEHICLE
OPERATIONS HANDBOOK

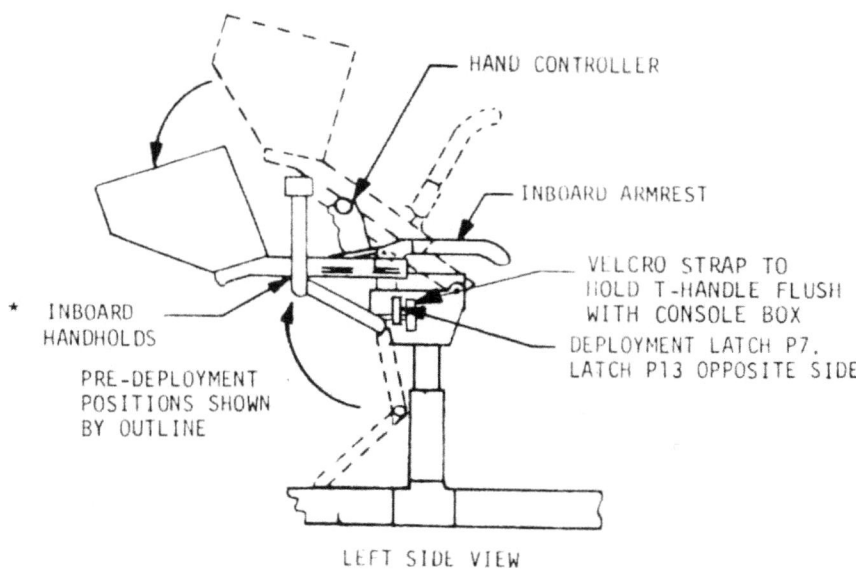

* 1G TRAINER INBOARD HANDHOLD CANNOT BE FOLDED DOWN TO
  SIMULATE PRE-DEPLOYMENT POSITION DUE TO CABLE RUNS
  FROM CONSOLE.

FIGURE 2-8 CONTROL AND DISPLAY CONSOLE DEPLOYMENT

LS006-002-2H
LUNAR ROVING VEHICLE
OPERATIONS HANDBOOK

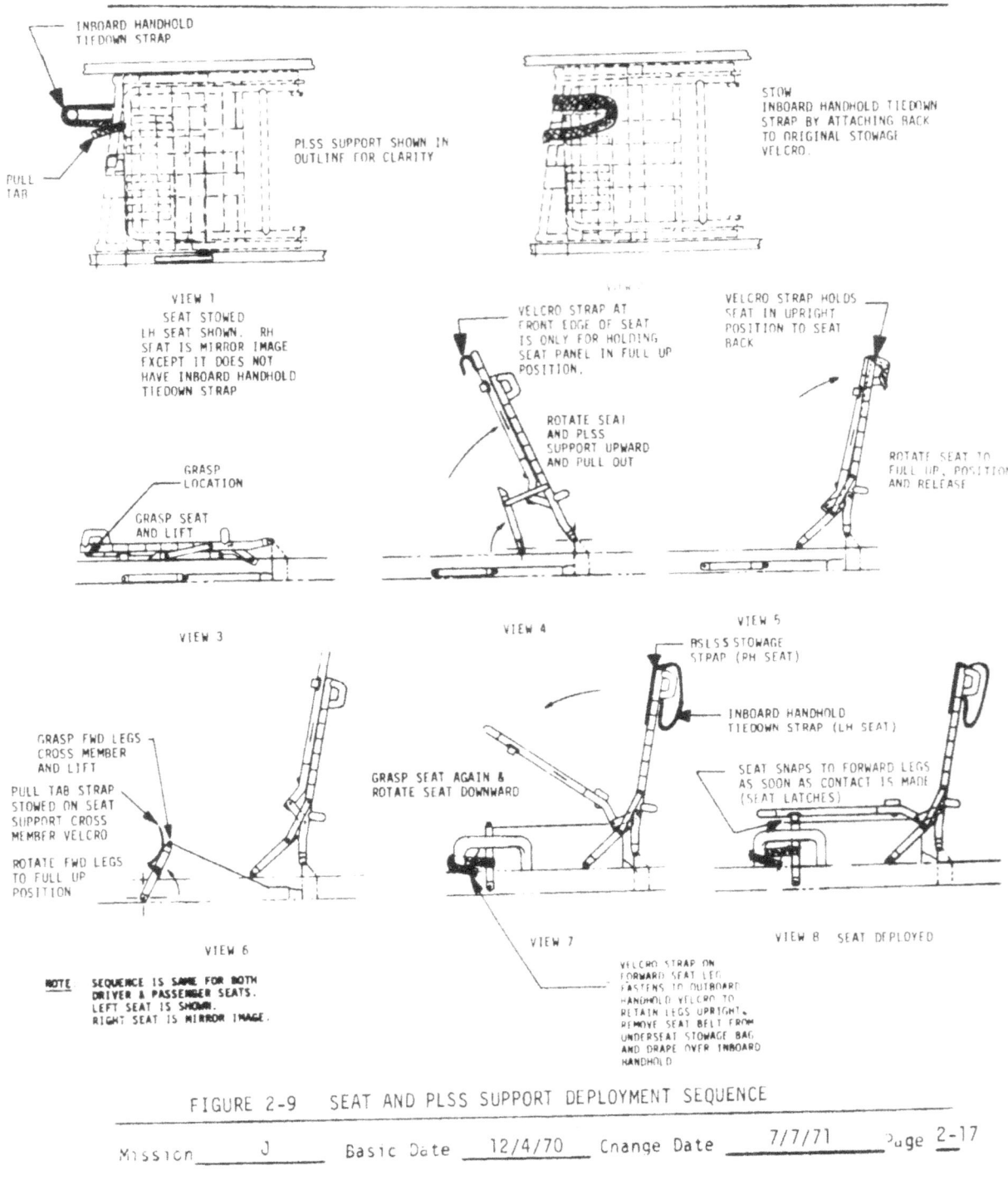

FIGURE 2-9 SEAT AND PLSS SUPPORT DEPLOYMENT SEQUENCE

LUNAR ROVING VEHICLE
APOLLO OPERATIONS HANDBOOK

STA/STEP	PROCEDURE	REMARKS
2.1	(Continued)	on chassis.
ap.	Visually verify the rear steering decoupling pull ring seal has not been broken.	Figure 2-6. If pull ring tie-down seal is broken and subsequent steering check using hand controller indicates steering is engaged, disregard broken seal. If hand controller is not engaged, recouple steering.
aq.	Deploy RR fender extension.	
ar.	Deploy LR fender extension.	
as.	Verify both hinge pins flush at LR hinge.	If hinge pin is not flush, tap pin with toehold. Verify pin is latched by pressing down on chassis.
at.	Rotate left seat to stable overcenter position.	
au.	Rotate legs to full upright position.	
av.	Attach forward seat legs velcro strap to outboard handhold.	
aw.	Verify underseat stowage bag erects.	
ax.	Release seat belt from underseat bag stowage position and place in temporary storage position.	
ay.	Pull seat pan frame forward to engage front legs.	

Mission J   Basic Date 12/4/70   Change Date 7/7/71   Page 2-18

LS006-002-2H
LUNAR ROVING VEHICLE
OPERATIONS HANDBOOK

STA/STEP	PROCEDURE	REMARKS
2.1	(Continued)	
az.	Verify all seat latches latched.	
ba.	Stow inboard handhold tiedown strap by making loop behind seat back and attaching end of strap to velcro patch on top of seat back.	
bb.	Fold inboard armrest down.	To prevent interference with hand controller armrest must be folded down to extent possible at this point.
bc.	Support console with left hand, with right hand release velcro tiedown strap (if required), pull out left C/D console "T" handle P7 and turn 90° CW.	
bd.	With right hand rotate inboard handhold to locked position while rotating console downward with left hand.	
be.	Rotate "T" handle P7 90° CW with right hand, fold "T" handle flush with console box and secure in position with velcro strap.	"T" handle should "snap-in", lock and fold down flush with console box.
bf.	Remove attitude indicator lock pin and discard.	Figure 2-6.
bg.	Remove C&W flag lock pin and discard.	
bh.	Pull pins P3 and P4 and discard clear of work area.	
bi.	Grasp tripod apex with left hand and pull pin P5.	
bj.	Discard pins and apex members clear of work area.	
bk.	Grasp short tripod member in left hand and pull pin P6 with right hand, and discard pin clear of deployment area.	

Mission J Basic Date 12/4/70 Change Date 7/7/71 Page 2-19

LS006-002-2H
LUNAR ROVING VEHICLE
OPERATIONS HANDBOOK

STA/STEP	PROCEDURE	REMARKS
bl.	Remove short tripod member and use hooked end to pull cable P2.	Figure 2-6. Tool hook interfaces with cable area color coded gold. Deflection of cable releases telescoping rods saddle and forward wheel strut tethers.
bm.	Visually verify that telescoping rods saddle falls away from LRV.	
bn.	Either insert short tripod member in left toehold position or stow in underseat stowage bag.	Figure 2-6. If short tripod member is installed in toehold position, end with hook should be outboard with hook pointing forward. This is also used as wheel decoupling tool.
bo.	Pull left footrest lift tabs.	Tabs pull free of footrests, but remain attached to floor panel.
bq.	Rotate footrest upward and forward and lock into position.	
br.	Verify both hinge pins flush at LF hinge.	
bs.	Deploy LF fender extension.	
bt.	Verify battery no. 1 and SPU dust covers closed and secured to velcro patch.	Verify by applying slight lift force on edge of cover.
bu.	Verify the forward steering decoupling pull ring seal has not been broken.	Figure 2-6. If seal is broken and subsequent steering check using hand controller indicates steering is engaged, disregard broken seal. If hand controller check indicates steering is

Mission J   Basic Date 12/4/70   Change Date 7/7/71   Page 2-20

LS006-002-2H
LUNAR ROVING VEHICLE
OPERATIONS HANDBOOK

STA/STEP	PROCEDURE	REMARKS
2.1	(Continued)	not engaged, center wheels in neutral steer, verify forward steering lock and continue mission using rear steering only.
bv.	Move to right side of vehicle and verify battery no. 2 dust cover closed and secured to velcro patch.	Verify by applying slight lift force on edge of cover.
bw.	At right side of LRV rotate right "T" handle P13 90° CW, fold "T" handle flush with console box and secure in position with velcro strap.	"T" handle should snap-in, lock and fold down flush with the console box.

Mission   J   Basic Date   12/4/70   Change Date   7/7/71   Page 2-21

LUNAR ROVING VEHICLE
APOLLO OPERATIONS HANDBOOK

STA/STEP	PROCEDURE	REMARKS
2.2	LRV POST DEPLOYMENT CHECKOUT AND DRIVE TO MESA	
a.	Verify Hand Controller in parking brake and neutral throttle position and reverse inhibit switch is on (pushed down).	Crewman stands along side the vehicle.
b.	Verify switches and circuit breakers in pre-launch positions as follows:	Figure 2-11. Crewman stands along side vehicle.
	NAV POWER Circuit Breaker - Open GYRO TORQUING Switch - OFF System RESET Switch - OFF AUX Circuit Breaker - Open BUS A, B, C, D, Circuit Breakers - Open + 15 VDC PRIM and SEC Circuit Breakers - Open ∓ 15 VDC Switch - OFF MOTOR TEMP Switch - FORWARD BATTERY Switch - AMPS PWM SELECT Switch - BOTH STEERING FORWARD and REAR Circuit Breakers - Open STEERING FORWARD and REAR Switches - OFF DRIVE POWER LF, RF, LR, RR Circuit Breakers - Open DRIVE POWER LF, RF, LR, RR Switches - OFF DRIVE ENABLE LF, RF, LR, RR Switches - OFF Verify all meters, indicators are at zero. Report off-zero indications.	

Mission J   Basic Date 12/4/70   Change Date 7/7/71   Page 2-22

LUNAR ROVING VEHICLE
APOLLO OPERATIONS HANDBOOK

STA/STEP	PROCEDURE	REMARKS
2.2	(Continued)	
	NOTE: The following step (2.2.c) may be performed at crew option. If this step is performed, do not perform steps 2.2.y, 2.2.z, or 2.2.aa.	
	c. Manually move the LRV away from the LM. (See remarks for LRV configuration for this operation).	Crew may manually move LRV away from LM prior to powerup; the hand controller should be placed in neutral throttle position and parking brake released. With a crewman standing on either side of vehicle outboard handholds may be used to lift, move, and tow LRV to and desired location.
	NOTE: After LRV is removed from area adjacent to LM, take precautions to prevent entaglement with wheel lock struts and cables remaining on lunar surface.	

Mission **J**   Basic Date **12/4/70**   Change Date **7/7/71**   Page **2-22.1**

LS006-002-2H
LUNAR ROVING VEHICLE
OPERATIONS HANDBOOK

STA/STEP	PROCEDURE	REMARKS
2.2	(Continued)	
d.	Set parking brake. (Or if step 2.2.c was not performed, verify brake is set).	Lunar weight of LRV at this point would be approximately 85 lbs. Hand controller is placed in neutral throttle position and brake disengaged to permit wheels to roll. Crewman stands along side vehicle, and should exercise care not to move vehicle while setting brake. Figure 2-10.
e.	Ingress left seat, fasten seat belt and initiate subsequent power up steps.	
f.	BUS A, BUS B, BUS C, BUS D Circuit Breakers - Close.	
g.	BATTERY Switch - VOLTS x 1/2.	
h.	Report BAT 1 and BAT 2 VOLTS indications.	
i.	BATTERY Switch - AMPS.	
j.	Report BAT 1 and BAT 2 temp (°F) indications.	
k.	Report BAT 1 and BAT 2 AMP-HR indications.	
l.	Report BAT 1 and BAT 2 AMPS indications.	
m.	± 15 VDC PRIM and SEC Circuit Breakers - Close.	
n.	STEERING FORWARD AND REAR Circuit Breakers - Close.	
o.	DRIVE POWER LF, RF, LR, RR Circuit Breakers - Close.	

Mission __J__ Basic Date __12/4/70__ Change Date __7/7/71__ Page 2-23

LS006-002-2H
LUNAR ROVING VEHICLE
OPERATIONS HANDBOOK

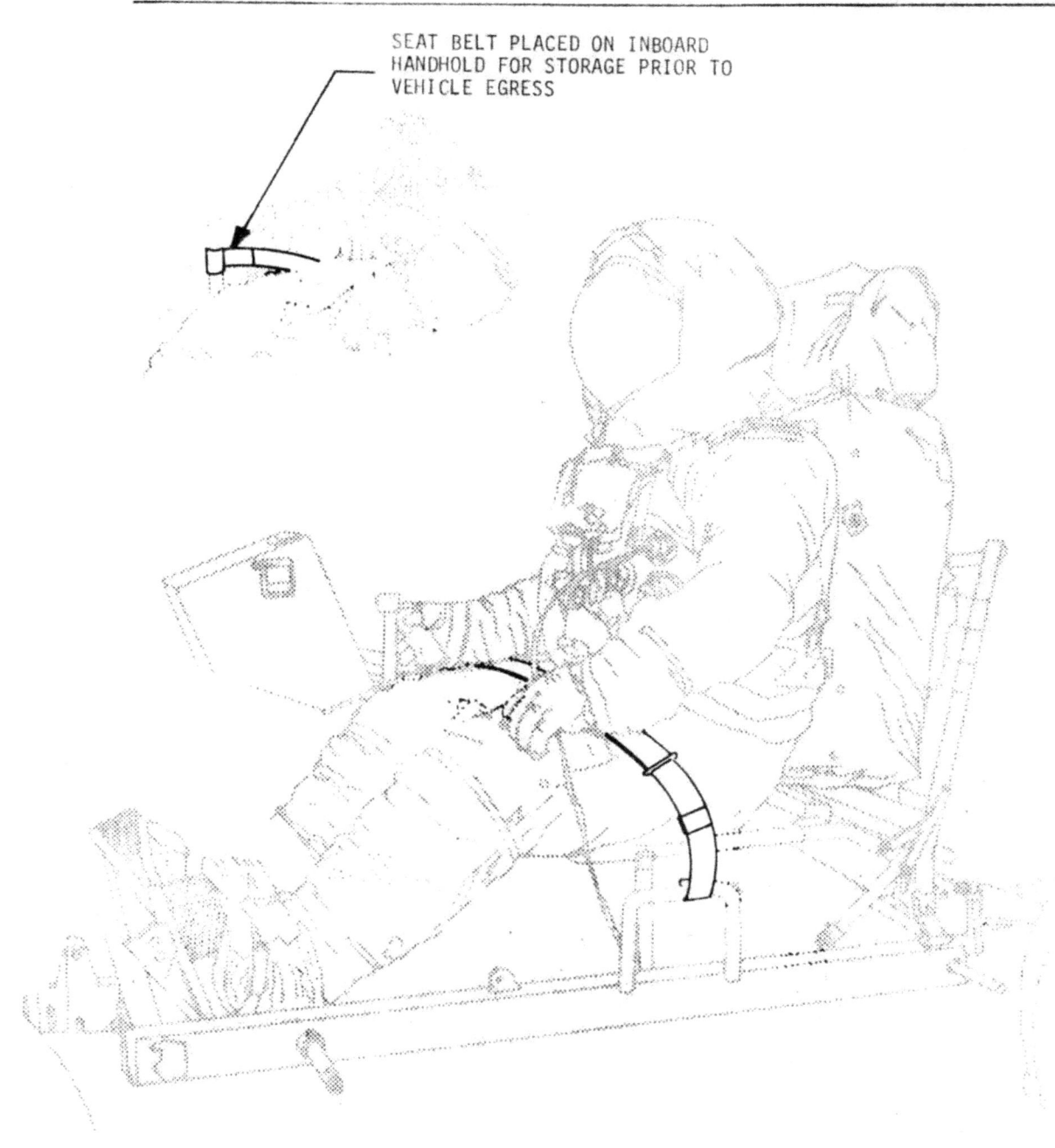

FIGURE 2-10 CREW POSITION

LS006-002-2H
LUNAR ROVING VEHICLE
OPERATIONS HANDBOOK

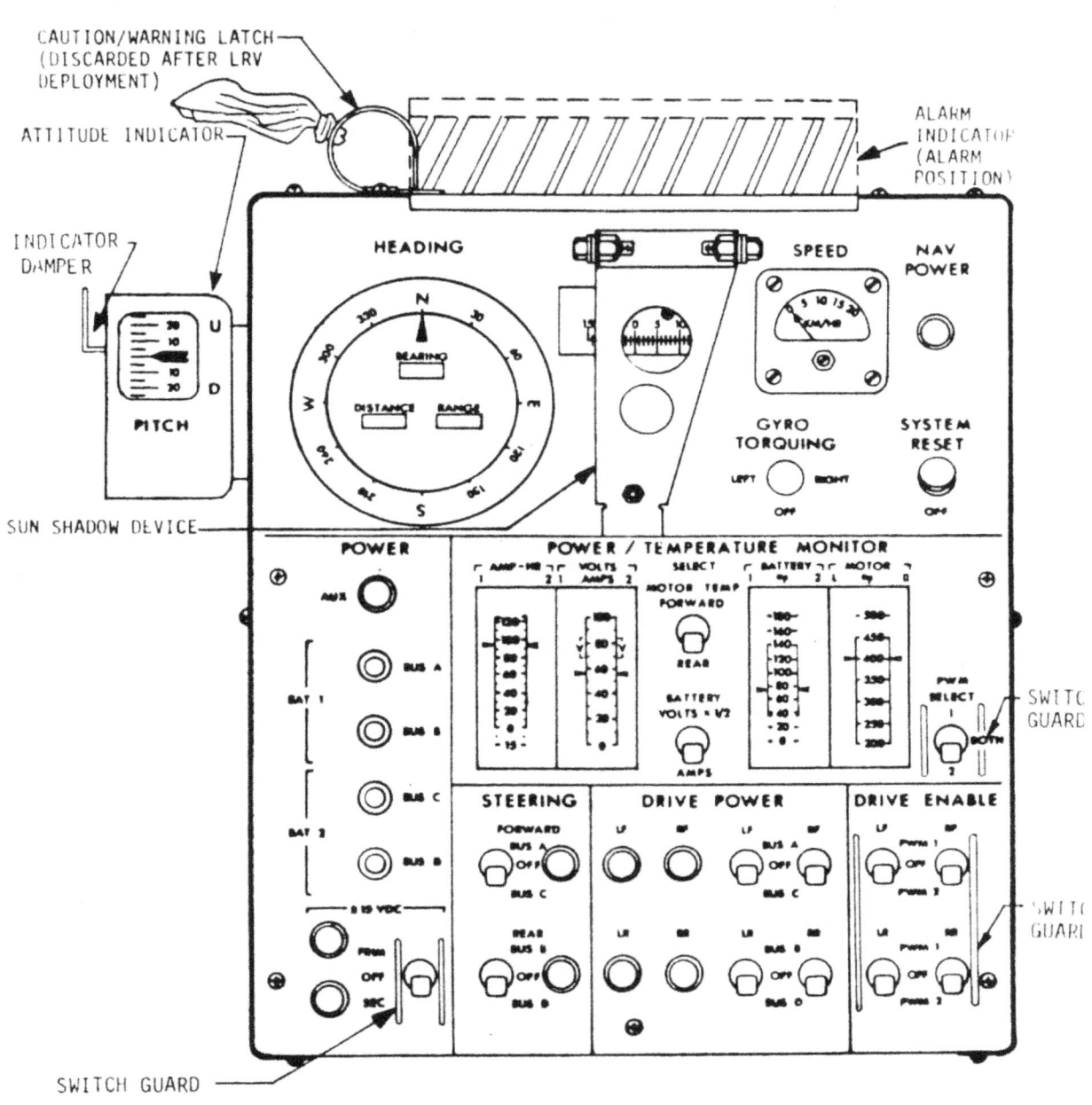

FIGURE 2-11 CONTROL AND DISPLAY CONSOLE

LS006-002-2H
LUNAR ROVING VEHICLE
OPERATIONS HANDBOOK

STA/STEP	PROCEDURE	REMARKS
2.2	(Continued)	
p.	DRIVE ENABLE LF and RF Switches - PWM 2.	
q.	DRIVE ENABLE LR and RR Switches - PWM 1.	
r.	±15 VDC Switch - SEC.	
s.	STEERING FORWARD Switch - BUS C.	Forward steering operates from Battery No. 2.
t.	STEERING REAR Switch - BUS B.	Rear steering operates from Battery No. 1.
	CAUTION	
	The hand controller should be in park brake position and the drive enable switches must be set to an active PWM prior to setting any drive power switch to an energized bus. If the drive power switch is turned on and the corresponding drive enable switch is not selected to an active PWM, then full power will be applied to the corresponding drive motor when the hand controller is released from brake position. Should this condition occur the hand controller should be immediately returned to park brake position.	The PWM select switch determines which PWM is active. The hand controller was verified set in park brake position in step 2.2.d. The PWM select switch was verified in "BOTH" position in step 2.2.b. The drive enable switches were set to active PWM positions in steps 2.2.p and 2.2.q.
u.	DRIVE POWER LF Switch - BUS C.	Front wheels operate from Battery No. 2.
v.	DRIVE POWER RF Switch - BUS C.	
w.	DRIVE POWER LR Switch - BUS B.	Rear wheels operate from Battery No. 1.

Mission __J__ Basic Date __12/4/70__ Change Date __4/19/71__ Page __2-26__

LS006-002-2H
LUNAR ROVING VEHICLE
OPERATIONS HANDBOOK

STA/STEP	PROCEDURE	REMARKS
2.2	(Continued)	
x.	DRIVE POWER RR Switch - BUS B.	
y.	Release parking brake.	
z.	Hand Controller reverse inhibit switch - UP position.	
	NOTE: The LRV driver may now back away from LM. LRV driver should request other crewman to direct and monitor any backing operations from an off-vehicle position.	
aa.	Stop LRV and set parking brake. Reset reverse inhibit switch (push switch down).	To the extent possible driver should verify steering, speed control and braking during this brief drive. The off-vehicle crewman should verify all four wheels rotating (not sliding).
ab.	Release parking brake and drive to MESA area for equipment loading.	
ac.	Stop LRV and set hand controller in parking brake position; Neutral throttle.	Parking brake should always be set prior to vehicle egress by either crewman.
ad.	Perform LRV partial power down as follows: DRIVE POWER Switches (2) - OFF. STEERING Switches (2) - OFF. + 15 VDC Switch - OFF.	Turning off drive power, steering, and + 15 VDC switches ensures that a failure in the DCE will not apply power to any vehicle motor thereby precluding any unnecessary power drain.

Mission J   Basic Date 12/4/70   Change Date 4/19/71   Page 2-27

LS006-002-2H
LUNAR ROVING VEHICLE
OPERATIONS HANDBOOK

STA/STEP	PROCEDURE	REMARKS
2.2	(Continued)	
	NOTE: The above step 2.2.ad assumes that payload loading and first LRV traverse will follow in order. Should crew rest period be scheduled subsequent to step 2.2.ae and prior to first LRV sortie, then Bus A, B, C, and D circuit breakers should be opened.	
	ae. Release and stow seat belt and egress vehicle.	

Mission J Basic Date 12/4/70 Change Date 4/19/71, Page 2-28

LS006-002-2H
LUNAR ROVING VEHICLE
OPERATIONS HANDBOOK

STA/STEP	PROCEDURE	REMARKS
2.3	PAYLOAD LOADING	
2.3.1	LCRU Installation	
a.	Place LCRU support post locks in the up position.	Figure 2-12. LRV arrives on lunar surface with LCRU support posts installed in LRV support tubes on forward chassis and with LRV/LCRU power cable connected to LRV auxiliary connector.
b.	Disconnect GCTA connector from LRV dummy connectors.	Figure 2-12.
	NOTES	
	1. Do not disconnect LCRU power cable from LRV auxiliary connector. Dust contamination could occur if this connector is disconnected.	
	2. Do not allow GCTA connector of cable to fall to lunar surface.	
	3. Do not place payload on battery cover.	
c.	Remove dummy connector from LRV GCTA receptacle and discard.	
d.	Remove LCRU from its LM stowage position and place onto LRV forward chassis LCRU support posts.	Figure 2-13.
e.	When LCRU is bottomed against support posts, position support post locks in horizontal position to secure LCRU.	
f.	Verify LRV AUX power circuit breaker - Open.	

Mission __J__ Basic Date __12/4/70__ Change Date __4/19/71__ Page 2-29

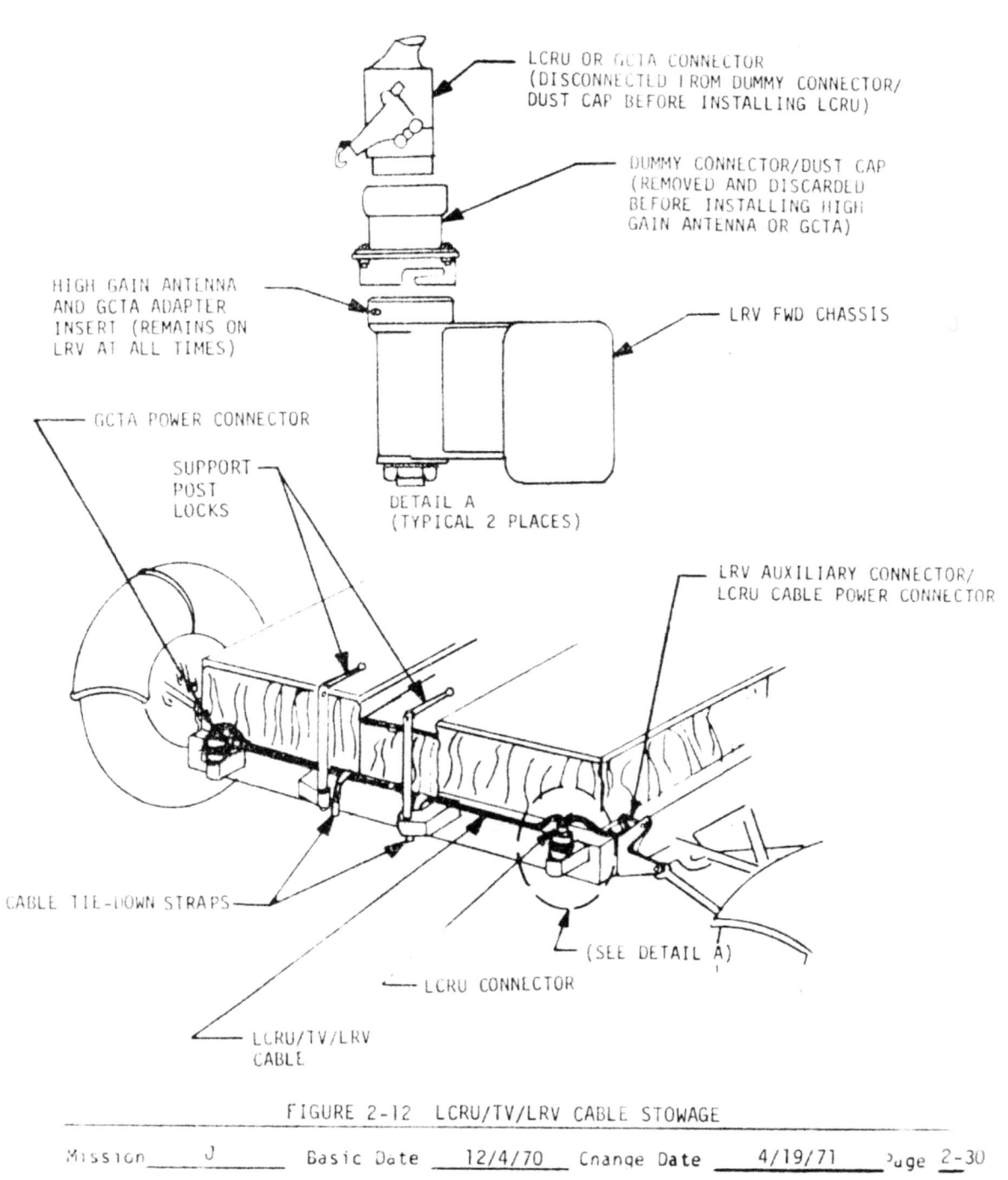

FIGURE 2-12 LCRU/TV/LRV CABLE STOWAGE

LS006-002-2H
LUNAR ROVING VEHICLE
OPERATIONS HANDBOOK

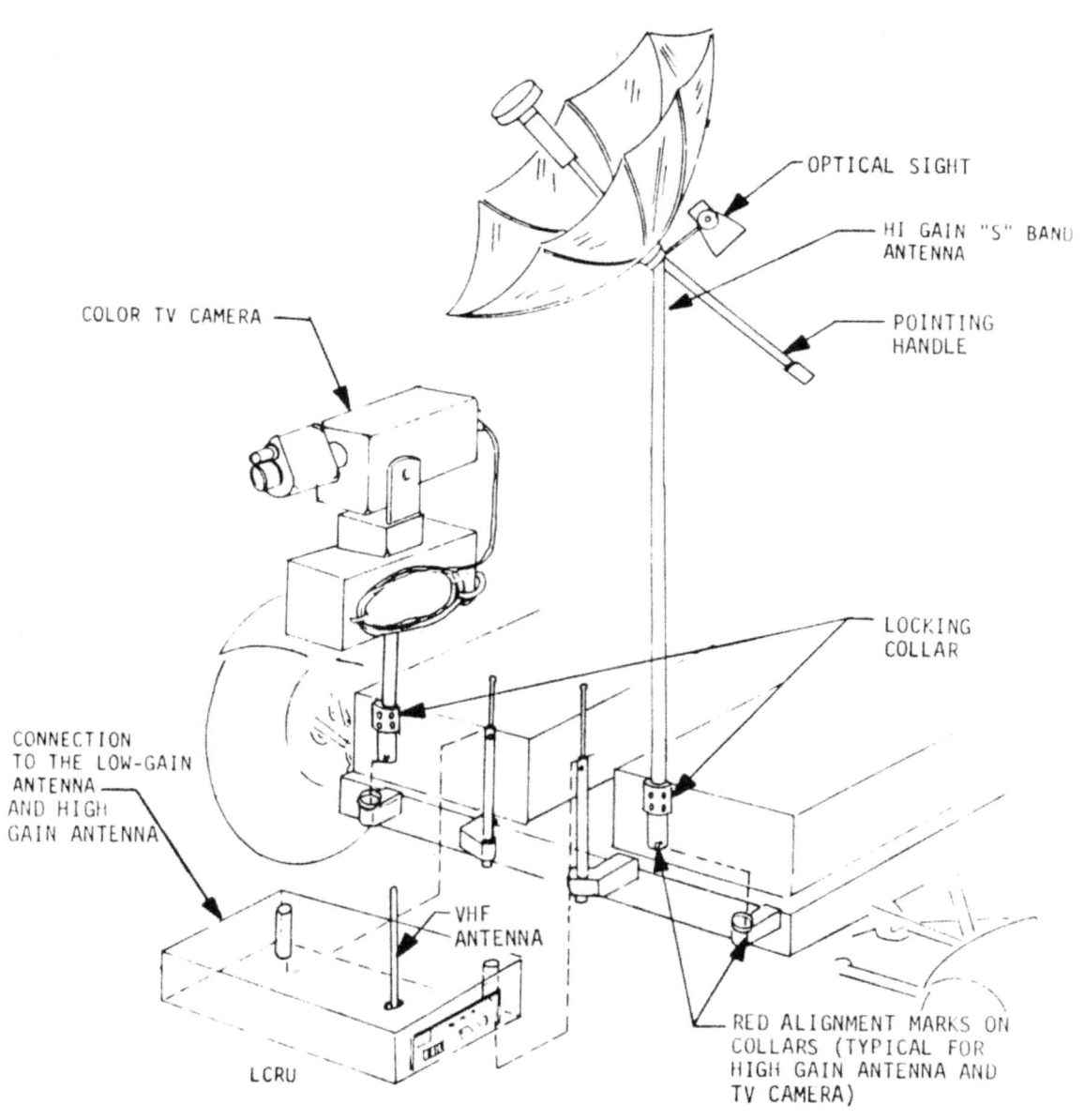

FIGURE 2-13 LCRU, HIGH GAIN ANTENNA, TV CAMERA INSTALLATION

LS006-002-2H
LUNAR ROVING VEHICLE
OPERATIONS HANDBOOK

STA/STEP	PROCEDURE	REMARKS
2.3.1	(Continued)	
g.	Disconnect LCRU power connector from LRV dummy connector and connect to LCRU.	
h.	Cover connector with thermal boot.	
i.	Remove dummy connector from LRV HGA receptacle and discard.	
2.3.2	GCTA Installation	
a.	At MESA, pull GCTA control unit pip pin release cable. **CAUTION** Do not strike GCTA control unit mirror surfaces on MESA.	Figure 2-13.
b.	Remove GCTA control unit and support staff from MESA.	
c.	Unfold GCTA support staff. Verify staff locked. **CAUTION** If GCTA staff is not properly locked, it could fall on LCRU and cause severe LCRU radiator damage.	
d.	With connector receptacles inboard, insert GCTA staff into mounting receptacle on right front corner of LRV.	
e.	Rotate staff to assure engagement of staff anti-rotational pins.	
f.	GCTA staff bayonet collar – Lock (CW).	Alignment marks are provided on GCTA staff locking collar.

Mission J   Basic Date 12/4/70   Change Date 7/7/71   Page 2-32

LS006-002-2H
LUNAR ROVING VEHICLE
OPERATIONS HANDBOOK

STA/STEP	PROCEDURE	REMARKS
2.3.2	(Continued)	
g.	Connect GCTA connector of LRV/LCRU cable to GCTA control unit.	
h.	On TV camera, LM PWR Switch - OFF.	
i.	Disconnect LM/TV cable from TV camera and rest connector on tripod handle.	
j.	Remove TV camera from LM tripod and install on GCTA azimuth/elevation unit.	
k.	Connect GCTA control unit connector to TV camera.	
2.3.3	16 mm Data Acquisition Camera Installation	Figure 2-14.
a.	Remove camera and staff from LM.	
b.	Assemble camera and staff into single unit.	
c.	Insert staff into receptacle on LRV right inboard handhold.	
d.	Verify staff locked in place by pulling up on the camera without depressing the push button on end of handhold. Camera staff should not move vertically.	
2.3.4	Low Gain Antenna Installation	Figure 2-14.
a.	Remove low gain antenna from LM stowage location.	
b.	Insert low gain antenna staff on LRV left inboard handhold.	
c.	Verify staff locked in vertically by pulling up on staff without depressing button on end of handhold. Low gain antenna staff should not move vertically.	

Mission J   Basic Date 12/4/70   Change Date 7/7/71   Page 2-33

LS006-002-2H
LUNAR ROVING VEHICLE
OPERATIONS HANDBOOK

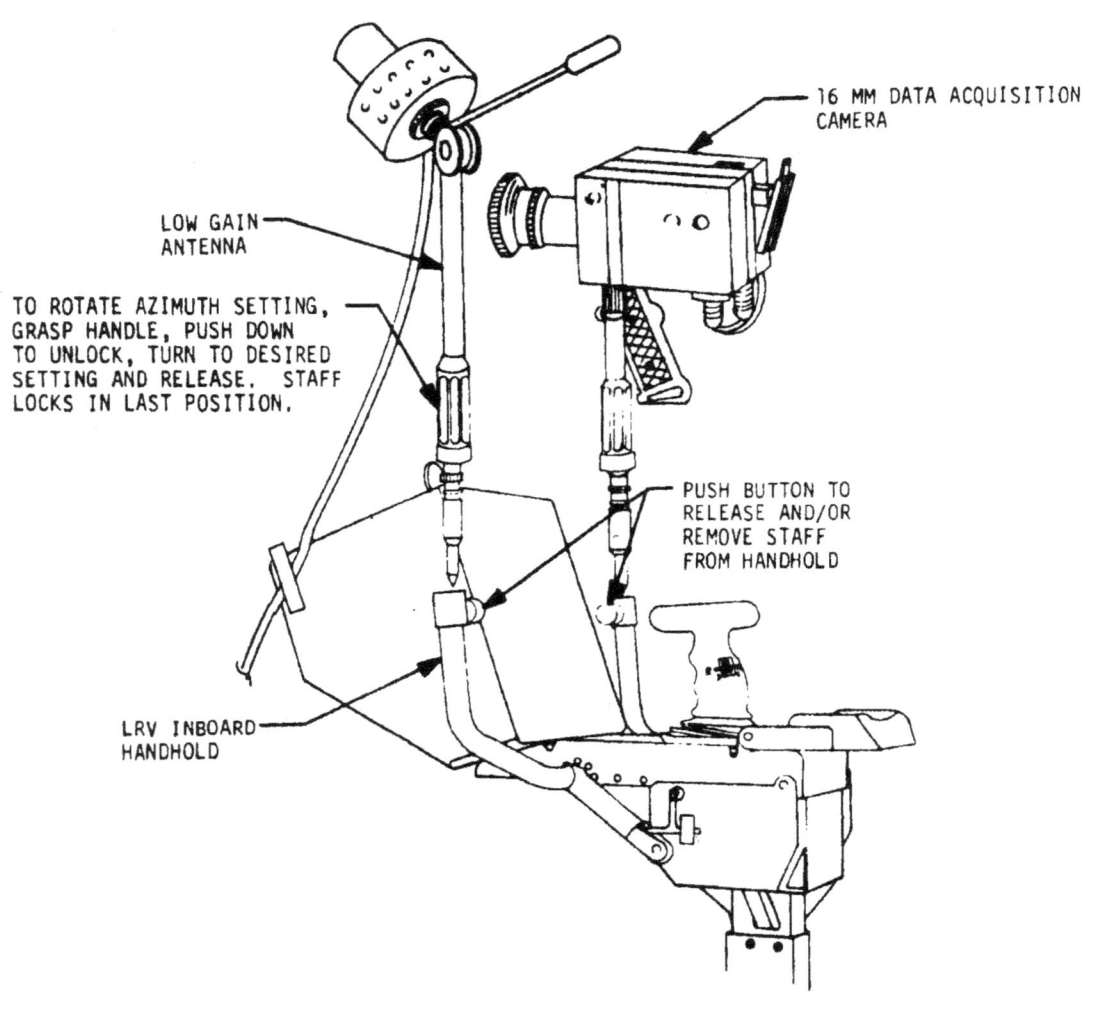

FIGURE 2-14 16 MM DAC AND LOW GAIN ANTENNA INSTALLATION

LS006-002-2H
LUNAR ROVING VEHICLE
OPERATIONS HANDBOOK

STA/STEP	PROCEDURE	REMARKS
2.3.4	(Continued)	
d.	Route low gain antenna cable to LCRU and secure to LRV with strap on console and clips on forward chassis.	Figure 2-15.
e.	Connect low gain antenna cable to LCRU.	
2.3.5	High Gain Antenna Installation.	Figure 2-13.
a.	Remove high gain antenna from LM stowage position.	
b.	Insert high gain antenna staff into the mounting receptacle on the left front corner of the LRV and lock.	Alignment marks are provided on HGA staff locking collar.
c.	Unfold and lock HGA staff.	
d.	Remove and discard optical sight retaining clamp.	
e.	Open and lock HGA dish.	
f.	Connect HGA cable to the LCRU.	
g.	Deploy LCRU VHF whip antenna.	
h.	Activate LCRU/GCTA and perform communication checks as required.	
i.	Deactivate LCRU/GCTA until needed.	
2.3.6	Aft Payload Pallet Installation	
a.	Release the pallet support post tiedown on LRV aft chassis.	Figure 2-16.
b.	Erect pallet support post.	
c.	Remove pallet from LM.	

Mission __J__   Basic Date __12/4/70__   Change Date __7/7/71__   Page 2-35

LS006-002-2H
LUNAR ROVING VEHICLE
OPERATIONS HANDBOOK

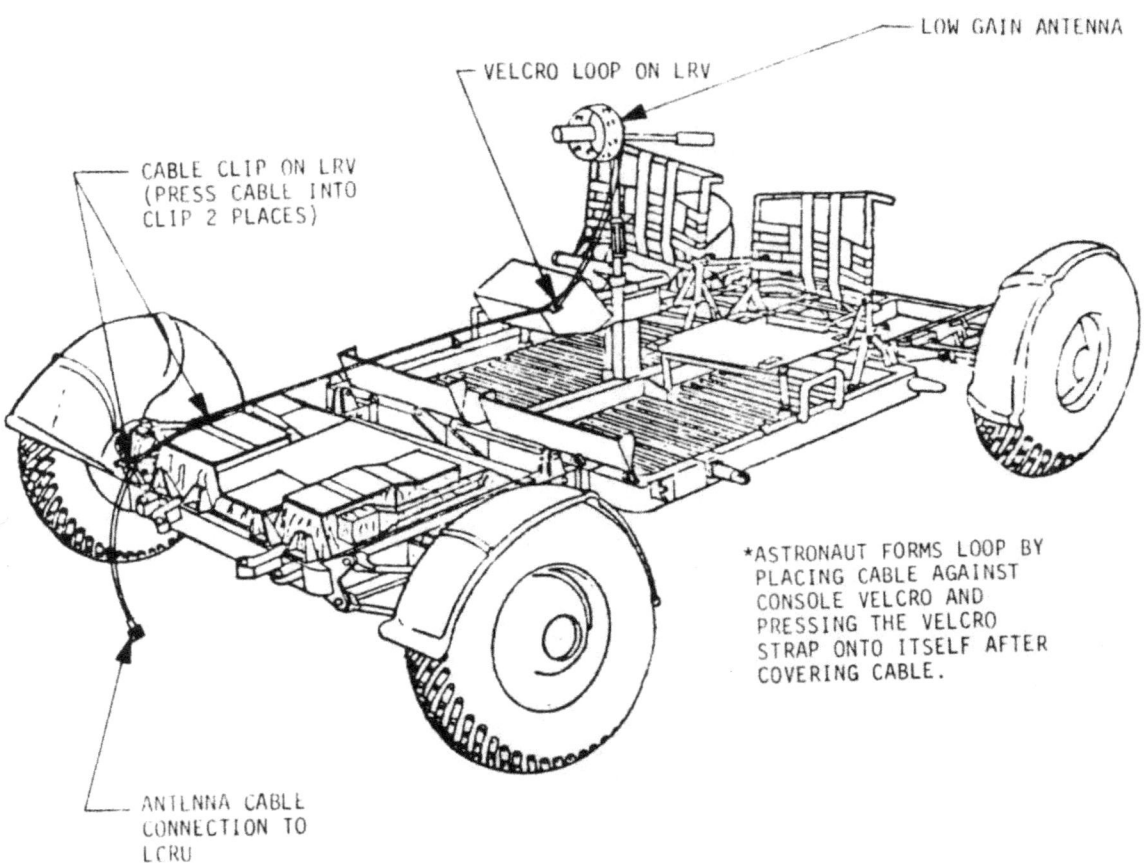

FIGURE 2-15 LCRU LOW GAIN ANTENNA CABLE INSTALLATION ON LUNAR SURFACE

LS006-002-2H
LUNAR ROVING VEHICLE
OPERATIONS HANDBOOK

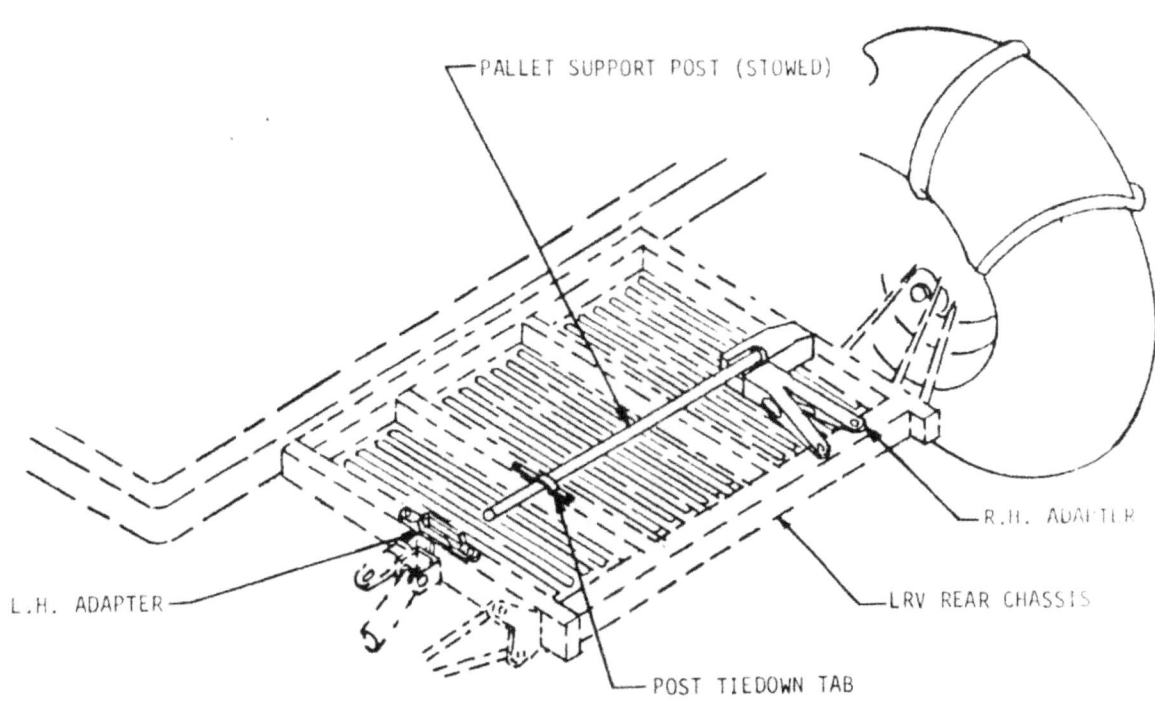

FIGURE 2-16 LRV REAR PAYLOAD PALLET ADAPTERS

LUNAR ROVING VEHICLE
APOLLO OPERATIONS HANDBOOK

STA/STEP	PROCEDURE	REMARKS
2.3.6	(Continued)	
	CAUTION	
	LRV steering recoupling tool must be cleared when installing pallet.	
d.	Connect pallet to pallet support post.	
e.	Rotate pallet about support post until pallet locks in pallet adapter on LRV LH aft chassis.	Figure 2-17.
2.3.7	Buddy SLSS Installation	
a.	Remove BSLSS bag from LM.	
b.	Release BSLSS support strap on back of right seat.	
c.	Feed strap through BSLSS bag handle and secure to PLSS support velcro on front of back seat.	Figure 2-18.
2.3.8	Map Holder Installation	
a.	Remove map holder from LM.	
b.	Place map holder clamp around inboard handhold vertical member.	Figure 2-18.1
c.	Tighten clamp by rotating clamp handle CW.	
2.3.9	Gnomon Bag Installation	
a.	Remove gnomon bag from LRV LH underseat stowage bag.	Figure 2-18.2
b.	Secure gnomon bag to back of LH seat per Figure 2-18.2	

Mission __J__   Basic Date __12/4/70__   Change Date __7/7/71__   Page 2-38

LUNAR ROVING VEHICLE
APOLLO OPERATIONS HANDBOOK

STA/STEP	PROCEDURE	REMARKS
2.3.10	Apollo Lunar Surface Drill (ALSD) Installation	
a.	Verify RH seat belt has been removed from RH underseat stowage bag and is stowed over inboard handhold.	Figure 2-18.3
b.	Leave LRV RH seat support in stowed-for-launch position and raise seat pan and PLSS support to configuration shown in View 5, Figure 2-9.	The ALSD is placed on the LRV only for transport from the LM to the ALSEP site with only one astronaut on the LRV.
c.	Remove ALSD from LM.	
d.	Place ALSD over stowed seat support frame and RH underseat bag. Orient ALSD per Figure 2-18.3	
e.	Slide ALSD rearward until ALSD butts against LRV center chassis rear cross-member.	
f.	Lower RH seat pan to operational position.	
	NOTE: Forward seat pan member will rest on ALSD structure. This is satisfactory configuration.	
2.3.11	Laser Ranging Retro Reflector (LR^3) Installation	
a.	Remove LR^3 from LM.	Figure 2-18.3
b.	Place LR^3 on RH seat of LRV per Figure 2-18.3	The LR^3 is placed on the LRV only for transport from the LM to the ALSEP site.
c.	Thread LRV RH seat belt through LR^3 handle and secure seat belt hook to LRV outboard handhold.	

Mission ___J___ Basic Date __12/4/70__ Change Date __7/7/71__ Page __2-38.1__

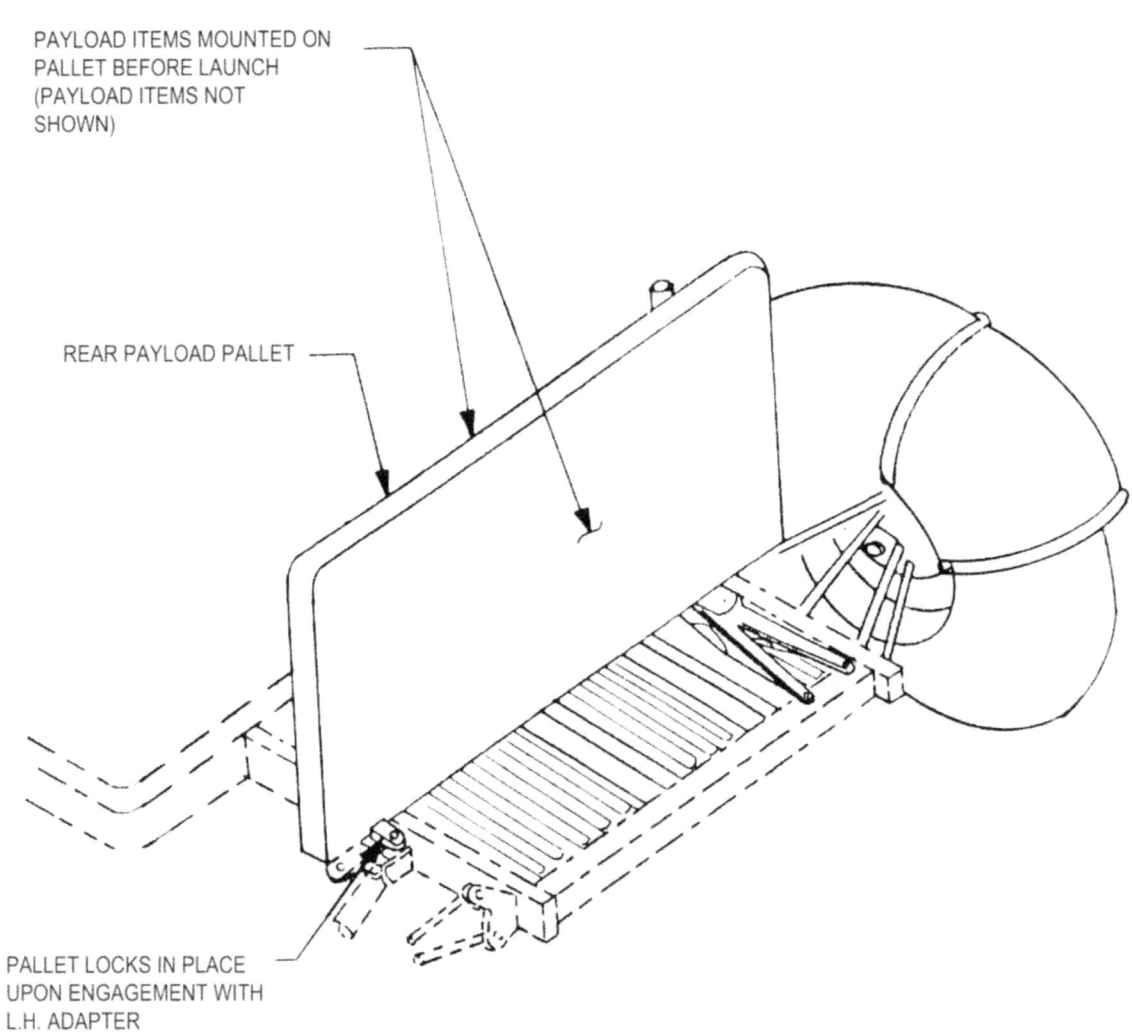

FIGURE 2-17 REAR PAYLOAD PALLET INSTALLED

LS006-002-2H
LUNAR ROVING VEHICLE
OPERATIONS HANDBOOK

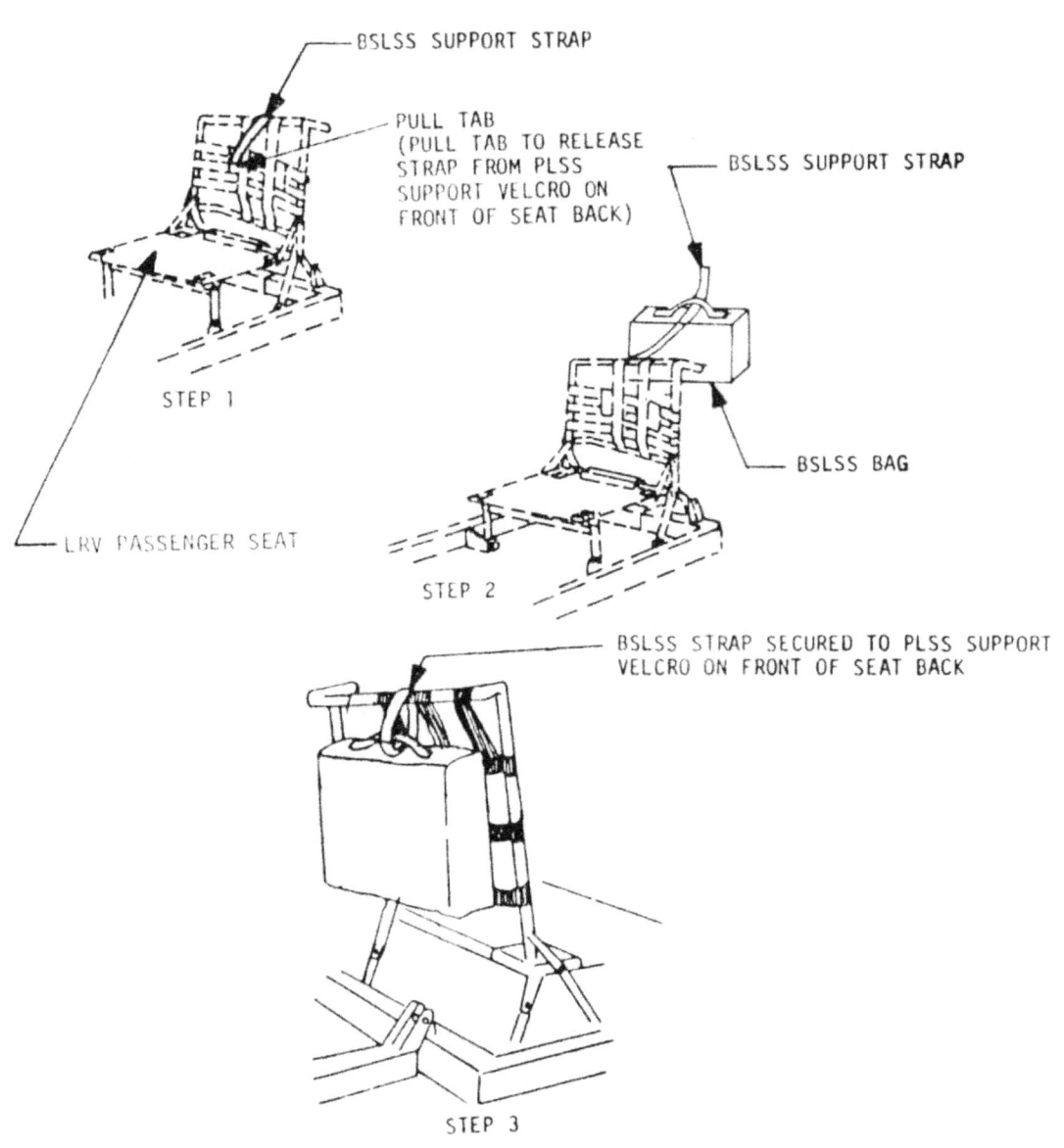

FIGURE 2-18 BUDDY SLSS INSTALLATION

LS006-002-2H
LUNAR ROVING VEHICLE
OPERATIONS HANDBOOK

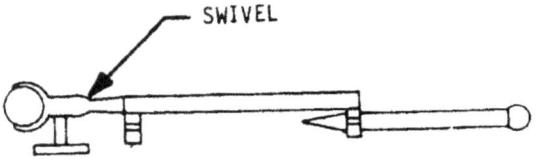

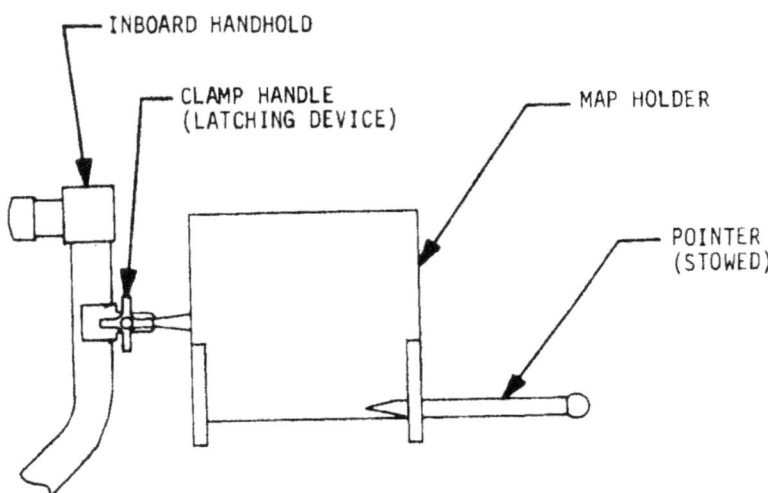

NOTE: MAP HOLDER CAN BE INSTALLED ON LEFT OR RIGHT HANDHOLD

FIGURE 2-18.1 MAP HOLDER INSTALLATION

LS006-002-2H
LUNAR ROVING VEHICLE
OPERATIONS HANDBOOK

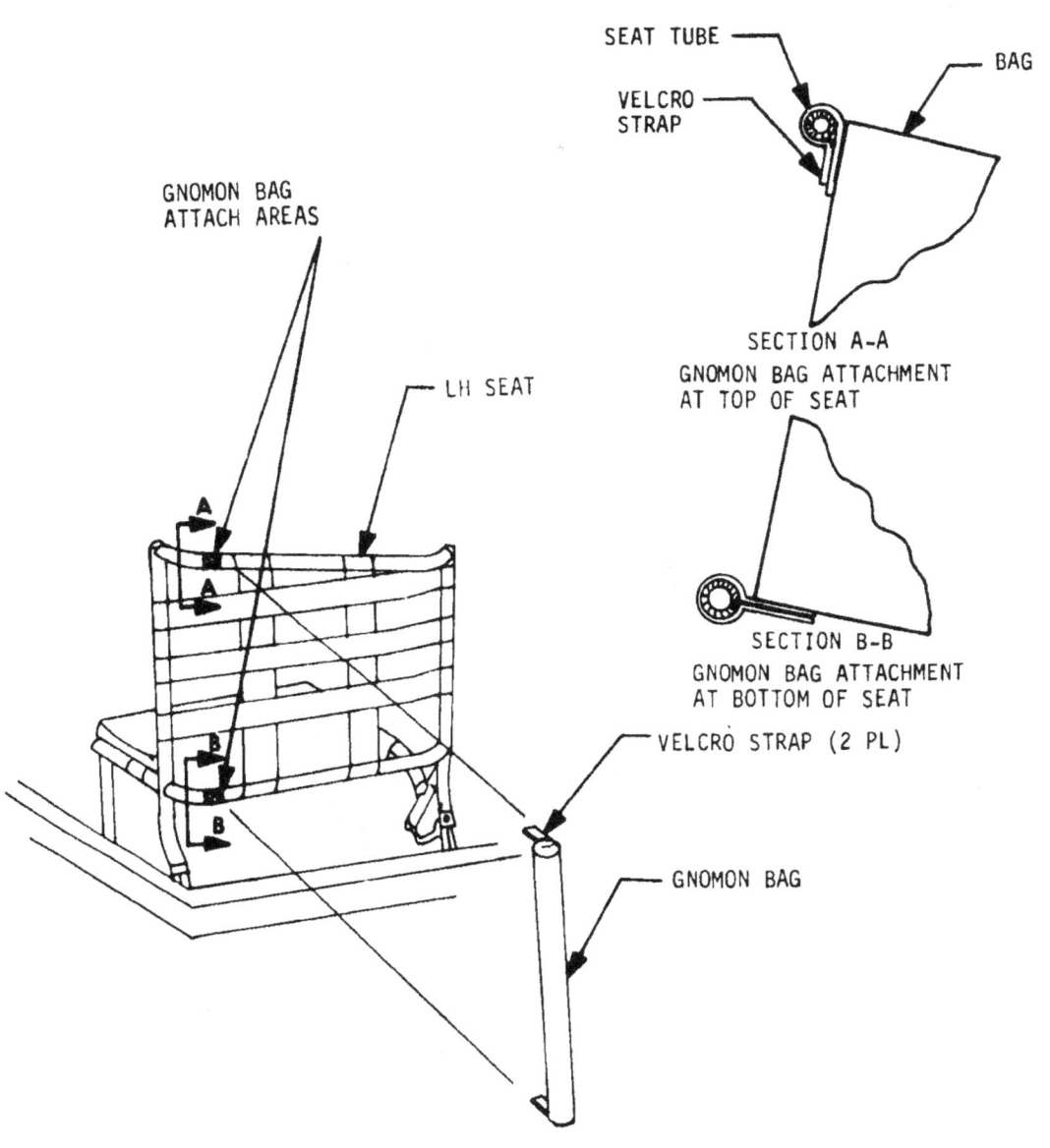

FIGURE 2-18.2 GNOMON BAG ATTACHMENT TO LRV

LS006-002-2H
LUNAR ROVING VEHICLE
OPERATIONS HANDBOOK
APPENDIX A

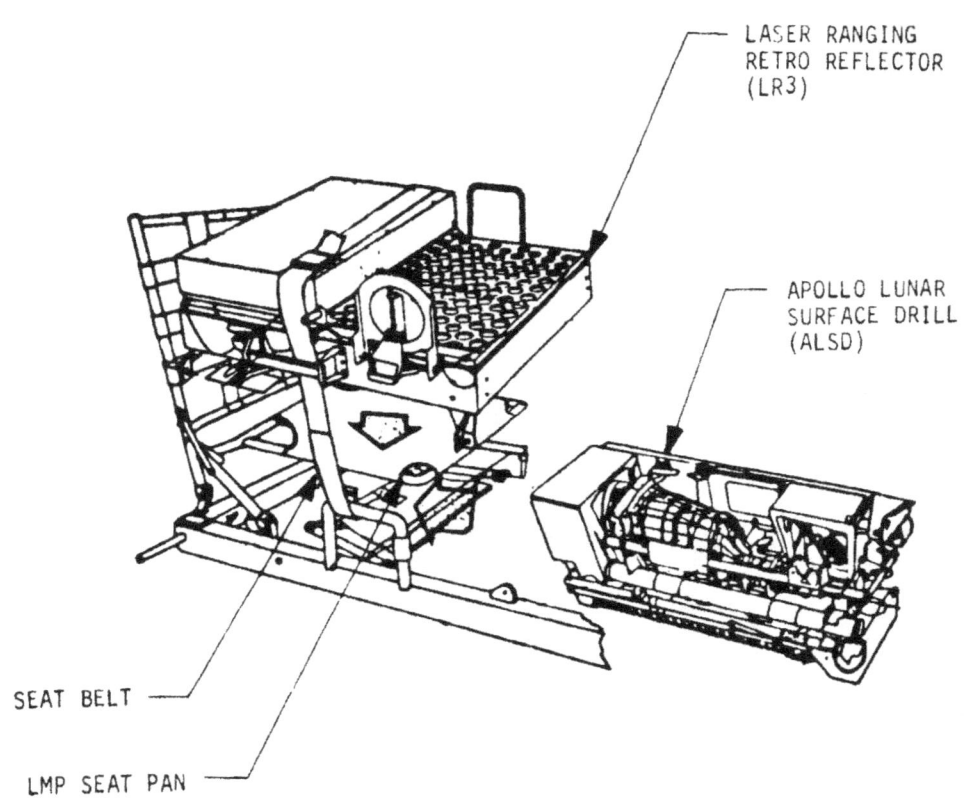

FIGURE 2-18.3 INTERIM STOWAGE OF LR^3 AND ALSD

LS006-002-2H
LUNAR ROVING VEHICLE
OPERATIONS HANDBOOK

STA/STEP	PROCEDURE	REMARKS
2.4	PRE SORTIE CHECKOUT AND PREPARATION	
a.	Verify battery and SPU dust covers closed.	
b.	Verify hand controller in parking brake and neutral throttle position and reverse inhibit switch is on (pushed down).	Board left seat first and verify parking brake set prior to other crewman boarding. Do not board both seats simultaneously.
	CAUTION	
	Do not grasp the 16 MM Data Acquisition camera staff or low gain antenna staff during ingress. The handholds are designed for ingress by grasping the handhold horizontal and vertical members below the payload staffs.	
c.	LRV driver ingress LRV left seat and fasten seat belt.	
d.	Other crewman ingress LRV, and fasten seat belt.	
	CAUTION	
	Do not grasp the 16 MM Data Acquisition camera staff or low gain antenna staff during ingress. The handholds are designed for ingress by grasping the handhold horizontal and vertical members below the payload staffs.	
	NOTE: If this is the first LRV sortie and this procedure sequence immediately follows initial payload loading (2.3) and LRV post deployment checkout (2.2), then at this point the C/D panel is in a power down configuration in accordance with step 2.2.ad. If this is the case then step 2.4 need only be a verification as it has been previously accomplished.	

Mission J   Basic Date 12/4/70   Change Date 4/19/71   Page 2-41

LS006-002-2H
LUNAR ROVING VEHICLE
OPERATIONS HANDBOOK

STA/STEP	PROCEDURE	REMARKS
2.4	(Continued)	
e.	BUS A, BUS B, BUS C, BUS D CB's - Close.	
f.	NAV POWER CB - Close.	Nav gyro is allowed to warm up and stabilize during payload loading.
	CAUTION	
	Do not torque nav gyro until nav power has been on for 3 minutes.	
g.	Report BAT 1 and BAT 2 AMPS.	
h.	BATTERY Switch - VOLTX x 1/2.	
i.	Report BAT 1 and BAT 2 VOLTS.	
j.	Report BAT 1 and BAT 2 AMP-HR. Insure reading is stabilized, indicating the amp-hr integrator has had adequate warmup time.	
k.	BATTERY Switch - AMPS.	
l.	DRIVE ENABLE LF and RF Switches - PWM 1.	
m.	DRIVE ENABLE LR and RR Switches - PWM 2.	
n.	+15 VDC Switch - PRIM.	
o.	STEERING FORWARD Switch - BUS A.	Forward steering operates from Battery No. 1.
p.	STEERING REAR Switch - BUS D.	Rear steering operates from Battery No. 2.
	CAUTION	The PWM select switch determines which PWM is active. The hand controller was verified in park brake position in step 2.2.ac.
	The hand controller should be in park brake position and the drive enable switches must be set to an active PWM prior to setting any drive power switch to an energized bus. If the drive power switch is turned on and	

Mission J   Basic Date 12/4/70   Change Date 4/19/71   Page 2-42

LS006-002-2H
LUNAR ROVING VEHICLE
OPERATIONS HANDBOOK

STA/STEP	PROCEDURE	REMARKS
2.4	(Continued)	
	CAUTION (Continued)	
	the corresponding drive enable switch is not selected to an active PWM, then full power will be applied to the corresponding drive motor when the hand controller is released from brake position. Should this condition occur the hand controller should be immediately returned to park brake position.	
q.	DRIVE POWER LF Switch - BUS A.	The PWM select switch was verified in "BOTH" position in step 2.2.b. The drive enable switches were set to active PWM positions in steps 2.4.1 and 2.4.m.
r.	DRIVE POWER RF Switch - BUS A.	Front wheels operate from Battery No. 1.
s.	DRIVE POWER LR Switch - BUS D.	Rear wheels operate from Battery No. 2.
t.	DRIVE POWER RR Switch - BUS D.	
u.	Drive LRV to level area ($\pm$ 6° pitch) near the LM.	
v.	Deploy SUN SHADOW DEVICE (SSD).	
w.	Deploy Vehicle Attitude Indicator to read roll.	
x.	Park down sun (within $\pm$ 3° per SSD) and level (within $\pm$ 6° roll) and set brake.	
y.	Report sun shadow device readings and LRV pitch and roll angles.	
z.	Fold (or reset) Sun Shadow Device (SSD).	Reset SSD to prevent it from obstructing drivers access to system reset switch.

Mission __J__ Basic Date __12/4/70__ Change Date __4/19/71__ Page 2-43

LS006-002-2H
LUNAR ROVING VEHICLE
OPERATIONS HANDBOOK

STA/STEP	PROCEDURE	REMARKS
2.4	(Continued)	
	aa. Fold Vehicle Attitude Indicator to drive position (pitch read position).	
	ab. Pull system reset switch from detent move momentarily to reset position and return to off.	
	ac. Verify BEARING, DISTANCE, RANGE Indicators - ZERO.	
	ad. Receive calculated heading from MCC.	
	CAUTION	
	Continuous torquing of nav gyro shall not exceed 2 minutes of any 7 minute period.	
	ae. Pull GYRO TORQUING Switch from detent and operate to LEFT or RIGHT for proper heading indication, then OFF.	Torque LEFT causes heading indication to move CCW. Torque RIGHT causes heading indication to move CW.
	af. Report battery 1 and 2 Amp-Hrs.	
	ag. Report battery and drive motor temperatures.	
	ah. Report battery current while vehicle is in motion one time between stops.	

Mission J   Basic Date 12/4/70   Change Date 4/19/71   Page 2-44

LS006-002-2H
LUNAR ROVING VEHICLE
APOLLO OPERATIONS HANDBOOK

STA/STEP	PROCEDURE	REMARKS
2.5	LRV CONFIGURATION FOR SCIENCE STOP	
a.	Stop LRV and set hand controller in parking brake position; Neutral throttle.	See remarks for step 2.2.ac.
b.	Perform LRV partial power down as follows:	
	DRIVE POWER Switches (4) - OFF. STEERING Switches (2) - OFF. + 15 VDC Switch - OFF.	
c.	Read and report displays in accordance with paragraph 2.8.	
d.	Crewman in right seat release and stow seat belt and egress vehicle.	
e.	Crewman in left seat release and stow seat belt and egress vehicle.	
f.	Align HGA.	
g.	LCRU mode Switch - FM/TV or TV RMT.	TV RMT will provide improved TV performance when the LM is available for voice relay.
h.	Open LCRU thermal blanket per ground request.	

Mission __J__   Basic Date __12/4/70__   Change Date __7/7/71__   Page __2-45__

LS006-002-2H
LUNAR ROVING VEHICLE
OPERATIONS HANDBOOK

STA/STEP	PROCEDURE	REMARKS
2.6	LRV CONFIGURATION PRIOR TO LEAVING SCIENCE STOP	
a.	Align LGA.	
b.	LCRU Mode Switch – PM1/WB.	
c.	Board LRV left seat and fasten seatbelt.	
	CAUTION	
	Do not grasp the 16 MM Data Acquisition camera staff or low gain antenna staff during ingress. The handholds are designed for ingress by grasping the handhold horizontal and vertical members below the payload staffs.	
d.	Verify hand controller in parking brake and neutral throttle position and reverse inhibit switch is on (pushed down).	
e.	Other crewman ingress LRV right seat and fasten seat belt.	
	CAUTION	
	Do not grasp the 16 MM Data Acquisition camera staff or low gain antenna staff during ingress. The handholds are designed for ingress by grasping the handhold horizontal and vertical members below the payload staffs.	
f.	Read and report displays in accordance with paragraph 2.8.	
g.	Update Nav System to correct for drift, if required by MCC.	
h.	+15 VDC Switch – PRIM.	

Mission ___J___ Basic Date __12/4/70__ Change Date __7/7/71__ Page __2-46__

LS006-002-2H
LUNAR ROVING VEHICLE
OPERATIONS HANDBOOK

STA/STEP	PROCEDURE	REMARKS
2.6	(Continued)	
i.	STEERING FORWARD Switch - BUS A.	The PWM select switch determines which PWM is active. The hand controller was verified set in park brake position in step 2.6.d.
j.	STEERING REAR Switch - BUS D.	The PWM select switch was verified in "BOTH" position in step 2.2.b. The drive enable switches were set to active PWM positions in steps 2.2.i and 2.2.j.
	CAUTION	
	The hand controller should be in park brake position and the drive enable switches must be set to an active PWM prior to setting any drive power switch to an energized bus. If the drive power switch is turned on and the corresponding drive enable switch is not selected to an active PWM, then full power will be applied to the corresponding drive motor when the hand controller is released from brake position. Should this condition occur the hand controller should be immediately returned to park brake position.	
k.	DRIVE POWER LF Switch - BUS A.	
l.	DRIVE POWER RF Switch - BUS A.	
m.	DRIVE POWER LR Switch - BUS D.	
n.	DRIVE POWER RR Switch - BUS D.	

Mission   J   Basic Date   12/4/70   Change Date   7/7/71   Page 2-47

LS006-002-2H
LUNAR ROVING VEHICLE
APOLLO OPERATIONS HANDBOOK

STA/STEP	PROCEDURE	REMARKS
2.7	POST SORTIE CHECKOUT	
a.	Park LRV cross sun from the right in view of LM windows (except if last sortie park LRV at selected LM ascent TV site.)	Figure 5-2. The LRV will be parked 300 feet due east of LM with a heading of 255 degrees for LM ascent.
b.	Hand controller in parking brake position, throttle in neutral - SET BRAKE.	
c.	Read and report displays in accordance with paragraph 2.8.	
d.	BATTERY Switch - AMPS.	
e.	DRIVE POWER Switches (4) - OFF.	
f.	STEERING Switches (2) - OFF.	
g.	± 15 VDC Switch - OFF.	
h.	Nav Power Circuit Breaker - OPEN.	
i.	BUS A, BUS B, BUS C, BUS D Circuit Breakers - OPEN. (Except if last sortie, BUS B and D circuit breakers - OPEN).	
j.	Crewman in right seat release and stow seat belts and egress LRV.	
k.	Crewman in left seat release and stow seat belts and egress LRV.	
l.	Align HGA.	
m.	LCRU mode Switch - TV RMT.	

Mission __J__ Basic Date __12/4/70__ Change Date __7/7/71__ Page __2-48__

STA/STEP	PROCEDURE	REMARKS
2.7	(Continued)	
q.	Open LRV Battery and SPU Dust Covers.	
r.	Prior to LM Ingress - perform the following:	
	AUX Power CB - Open (except if last sortie AUX PWR Circuit Breaker-Close)	
	LCRU Power Switch - OFF (except if last sortie LCRU Power Switch - EXT)	
	Adjust LCRU Thermal Blankets per ground request.	

LS006-002-2H
LUNAR ROVING VEHICLE
OPERATIONS HANDBOOK

2.8 DISPLAY READING SEQUENCE AND TIME INTERVALS

a. Report the following displays in the order shown at the time intervals specified below:

DISPLAY	TIME INTERVAL				
	Beginning of Each EVA	Beginning of Each Major Science Stop Separated by at Least 1 KM	End of a Long Stop of 45 to 60 Minutes on EVA 1 and EVA 2	Beginning of Each Stop During Traverses When Navigation Up-Date is Required	End of Each Major Traverse When LRV Has Returned to LM
Amp-Hr 1	X	X			X
Amp-Hr 2	X	X			X
Battery °F 1	X	X	X		X*
Battery °F 2	X	X	X		X*
Motor Temp:					
(Fwd) °F L	X	X	X		X
(Fwd) °F R	X	X	X		X
(Rear) °F L	X	X	X		X
(Rear) °F R	X	X	X		X
Heading (Deg)	X	X			X
Bearing (Deg)		X			X
Distance (KM)	X	X			X
Range (KM)		X			X
Pitch U or D (Deg)	X			X	
Roll L or R (Deg)	X			X	
Sun Shadow (Deg)	X			X	

NOTE: Make simultaneous Reading of volts, amps, slope, and speed one time between each science stop.

*These battery temperature measurements will be made both before opening the battery dust covers and as late as possible before ingress to the LM.

Mission J Basic Date 12/4/70 Change Date 7/7/71 Page 2-50

LUNAR ROVING VEHICLE
APOLLO OPERATIONS HANDBOOK

STA/STEP	PROCEDURE	REMARKS
2.9	SPECIAL PROCEDURES	
	The following procedures are to be followed when the occasion arises as a result of malfunctions or crew preference.	
2.9.1	Wheel Decoupling	
a.	Hand Controller - parking brake position with throttle control in neutral.	If right rear wheel is decoupled SPEED meter will not function. If two wheels are decoupled navigation system distance and range will be disabled. Brake on decoupled wheel will not be functional.
b.	DRIVE POWER switches (4) - OFF	
c.	STEERING Switches (2) - OFF	
d.	+15 VDC Switch - OFF	
e.	Egress LRV	
f.	Remove decoupling tool from stowage location (toehold position).	
g.	At affected wheel, engage hook end of decoupling tool with wheel decoupling mechanism (figure 2-19).	
h.	Pull out and rotate decoupling mechanism CCW. Release mechanism when rotated far enough toward center of wheel to be within wheel hub screw lock wire (figure 2-20).	If decoupling mechanism is not rotated inboard of the lockwire, mechanism could be damaged by wheel rotation, making recoupling impossible.
i.	Disengage decoupling tool from decoupling mechanism and repeat steps g and h for other decoupling mechanism on the affected wheel.	
j.	Disengage decoupling tool and replace in stowage location.	

Mission J   Basic Date 6/23/71   Change Date   Page 2-51

LS006-002-2H
LUNAR ROVING VEHICLE
OPERATIONS HANDBOOK

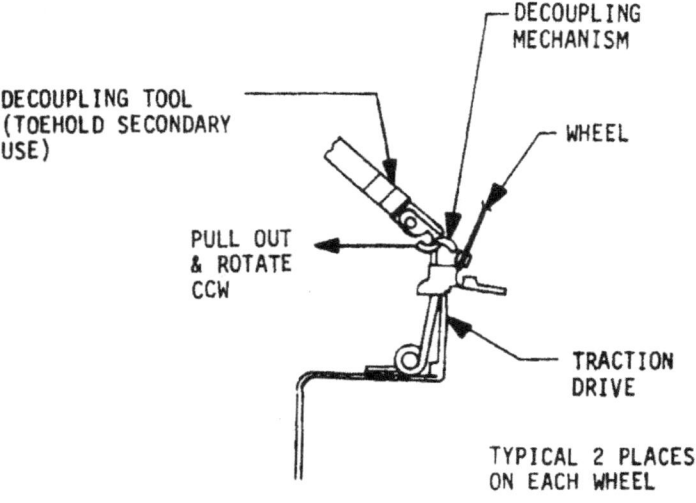

FIGURE 2-19 WHEEL DECOUPLING MECHANISM/DECOUPLING TOOL ATTACHMENT

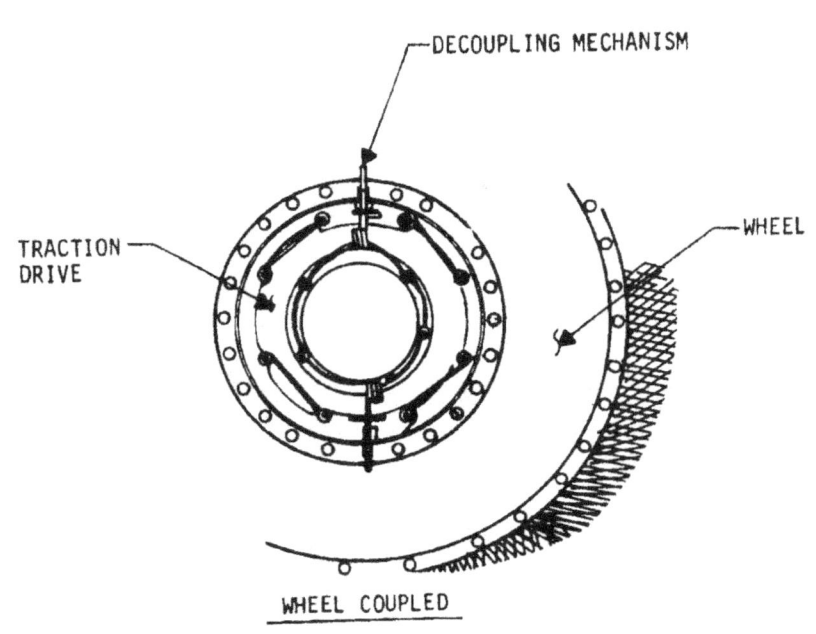
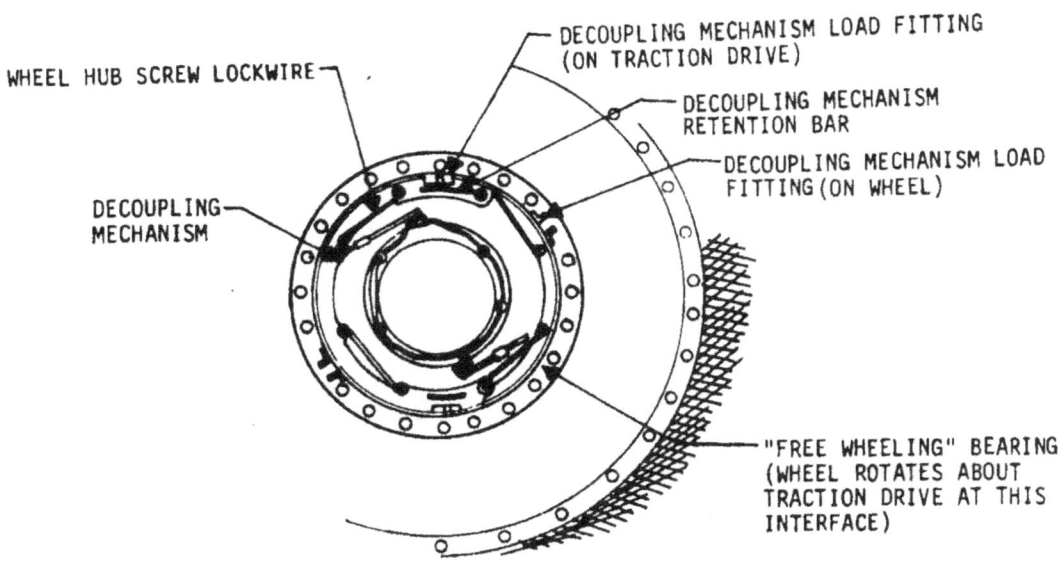

FIGURE 2-20 WHEEL DECOUPLING MECHANISM POSITIONS

LUNAR ROVING VEHICLE
APOLLO OPERATIONS HANDBOOK

STA/STEP	PROCEDURE	REMARKS
2.9.1	(Continued)	
k.	Ingress LRV.	
l.	Reset caution and warning flag if tripped.	
m.	± 15 VDC Switch – PRIM	
n.	STEERING Forward Switch – BUS A.	
o.	STEERING REAR Switch – BUS D.	
p.	DRIVE Power LF Switch – BUS A (If LF wheel was decoupled, leave in OFF position).	
q.	DRIVE Power RF Switch – BUS A (If RF wheel was decoupled, leave in OFF position).	
r.	DRIVE Power LR Switch – BUS D (If LR wheel was decoupled, leave in OFF position).	
s.	DRIVE Power RR Switch – BUS D (If RR wheel was decoupled, leave in OFF position).	
t.	Release parking brake.	
u.	Resume normal operations.	
2.9.2	Wheel Recoupling	
a.	Hand controller – parking brake position with throttle control in neutral.	
b.	DRIVE Power Switches (4) – OFF.	

Mission _____ Basic Date 6/23/71 Change Date _____ Page 2-54

LUNAR ROVING VEHICLE
APOLLO OPERATIONS HANDBOOK

STA/STEP	PROCEDURE	REMARKS
2.9.2	(Continued)	
c.	STEERING Switches (2) - OFF.	
d.	± 15 VDC Switch - OFF.	
e.	Egress LRV.	
f.	Remove decoupling tool from stowage location (outboard toehold position).	
g.	At affected wheel, engage hook end of decoupling tool with wheel decoupling mechanism (figure 2-19).	
h.	Pull out and rotate decoupling mechanism CW toward traction drive load fitting, sliding mechanism under mechanism retention bar, until mechanism is in load fitting on traction drive.	It is not necessary to align mechanism retention groove on wheel hub with wheel disc drive fitting. Later rotation of wheel will cause automatic engagement of wheel disc with mechanism.
i.	Repeat steps g and h for other decoupling mechanism.	
j.	Disengage decoupling tool from decoupling mechanism and replace tool in stowage location.	
k.	Ingress vehicle.	
l.	± 15 VDC Switch - PRIM.	
m.	STEERING Forward Switch - BUS A.	
n.	STEERING Rear Switch - BUS D.	

Mission J Basic Date 6/23/71 Change Date _____ Page 2-55

LUNAR ROVING VEHICLE
APOLLO OPERATIONS HANDBOOK

STA/STEP	PROCEDURE	REMARKS
2.9.2	(Continued)	
	CAUTION	
	If DRIVE POWER Switch of recoupled wheel is not in OFF position, upon application of power, the wheel hub will rotate rapidly inside the free-wheeling bearing, possibly shearing the decoupling mechanism.	
	o. DRIVE POWER LF Switch-BUS A. (If LF was recoupled, leave in OFF pos).	
	p. DRIVE POWER RF Switch-BUS A. (If RF was recoupled, leave in OFF pos).	
	q. DRIVE POWER LR Switch-BUS D. (If LR was recoupled, leave in OFF pos).	
	r. DRIVE POWER RR Switch-BUS D. (If RR was recoupled, leave in OFF pos).	
	s. Release parking brake.	
	t. Drive forward at very slow speed (less than 1 KPH) for at least 3 meters to allow engagement of wheel disc with decoupling mechanism.	
	u. DRIVE POWER Switch of recoupled wheel - BUS A if LF or RF, BUS D if LR or RR.	
	v. Resume normal operations.	
2.9.3	Steering Decoupling	
	NOTE	
	Forward steering decoupling is irreversible. Rear steering recoupling can be accomplished per Section 2.9.4.	
	a. Hand Controller - parking brake position with throttle control in neutral.	
	b. STEERING Switches (2) - OFF.	

Mission J   Basic Date 6/23/71   Change Date _____   Page 2-56

LS006-002-2H
LUNAR ROVING VEHICLE
OPERATIONS HANDBOOK

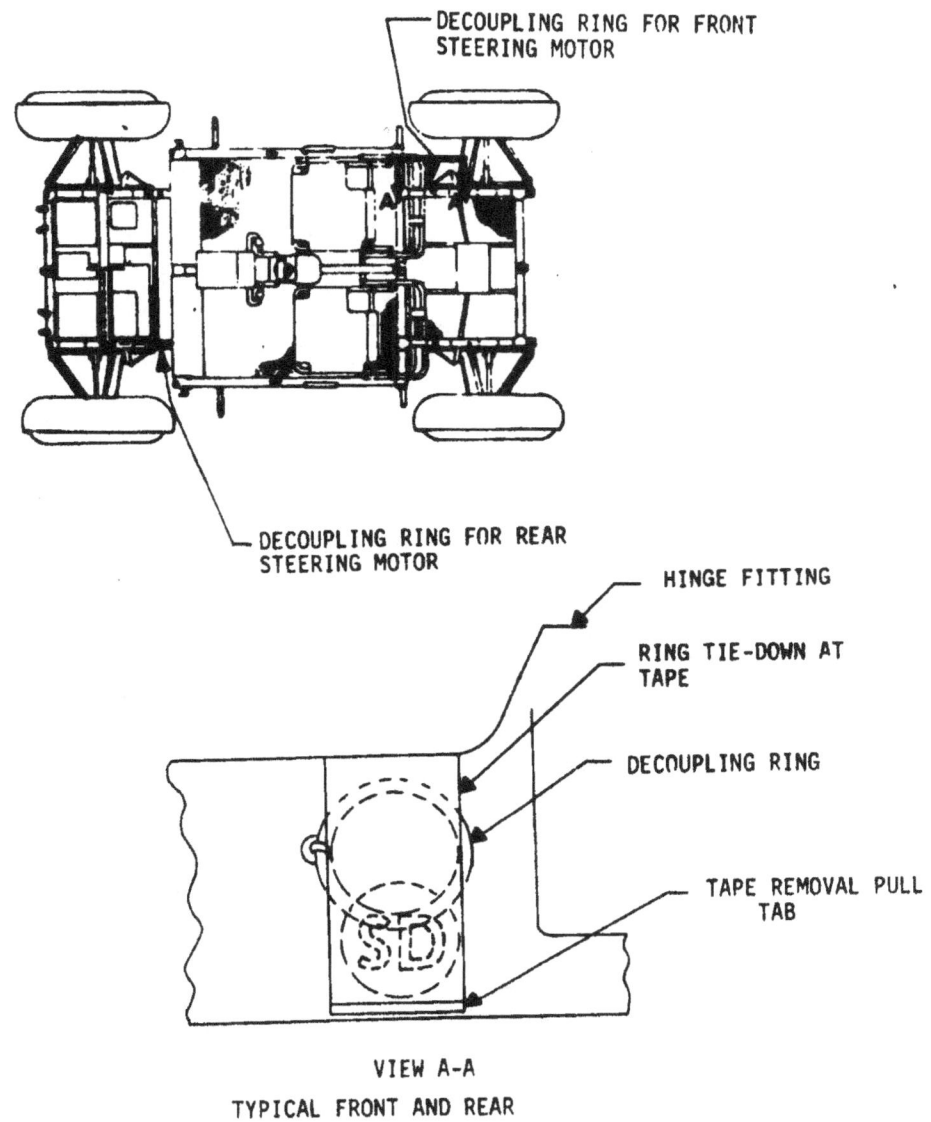

FIGURE 2-21 STEERING DECOUPLING RING LOCATIONS

LUNAR ROVING VEHICLE
APOLLO OPERATIONS HANDBOOK

STA/STEP	PROCEDURE	REMARKS
2.9.3	(Continued)	
c.	DRIVE POWER Switches (4) - OFF.	
d.	±15 VDC Switch - OFF.	
e.	Egress LRV.	
f.	At appropriate decoupling ring location, remove tape to free ring (figure 2-21).	
g.	Pull decoupling ring to maximum stroke possible and release.	
	CAUTION	
	Do not contact wheel with EVA glove. Use sample bag between wheel and glove to prevent possible damage to glove from possible broken wires in LRV wheel.	
h.	Manually straighten wheels to straight ahead position by pushing against wheel until steering is locked in straight ahead position.	To reduce required force push on wheel with larger turning angle, i.e. if wheels are turned to right push on right wheel to straighten.
i.	Ingress LRV.	
j.	±15 VDC Switch - PRIM.	
k.	STEERING FORWARD Switch - BUS A (if forward steering was decoupled, leave in OFF position).	
l.	STEERING REAR Switch - BUS D (If rear steering was decoupled, leave in OFF position).	
m.	DRIVE POWER LF Switch - BUS A.	
n.	DRIVE POWER RF Switch - BUS A.	
o.	DRIVE POWER LR Switch - BUS D.	

Mission J   Basic Date 6/23/71   Change Date _____   Page 2-58

LUNAR ROVING VEHICLE
APOLLO OPERATIONS HANDBOOK

STA/STEP	PROCEDURE	REMARKS
2.9.3	(Continued)	
p.	DRIVE POWER RR Switch - BUS D.	
q.	Release parking brake.	
r.	Resume operations.	
2.9.4	Rear Steering Recoupling	
a.	STEERING FORWARD Switch - BUS A.	
b.	STEERING REAR Switch - BUS D.	
c.	Allow steering electronics to null out by leaving hand controller in neutral steering position for 3 seconds.	Turning radius of LRV will now be 6 meters.
d.	STEERING Switches (2) - OFF.	
e.	Hand Controller - Parking brake position with throttle control in neutral.	
f.	DRIVE POWER Switches (4) - OFF.	
g.	±15 VDC Switch - OFF.	
h.	Egress LRV.	
i.	Fold LH seat to stowed-for-launch position.	
	CAUTION	
	Exercise caution while using recoupling tool as it consists of long, slender shaft which could prove hazardous if broken or fallen on.	
j.	Obtain recoupling tool from stowage location (figure 2-22).	

Mission __J__ Basic Date __6/23/71__ Change Date _____ Page 2-59

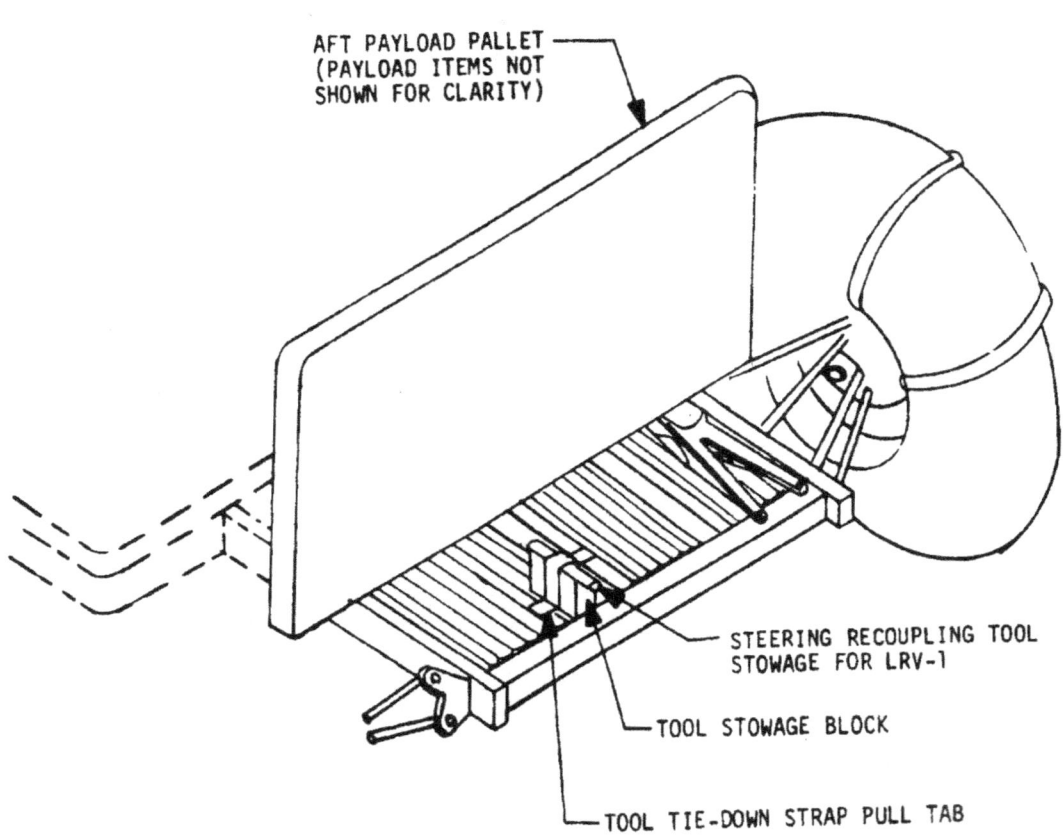

FIGURE 2-22 STEERING RECOUPLING TOOL STOWAGE

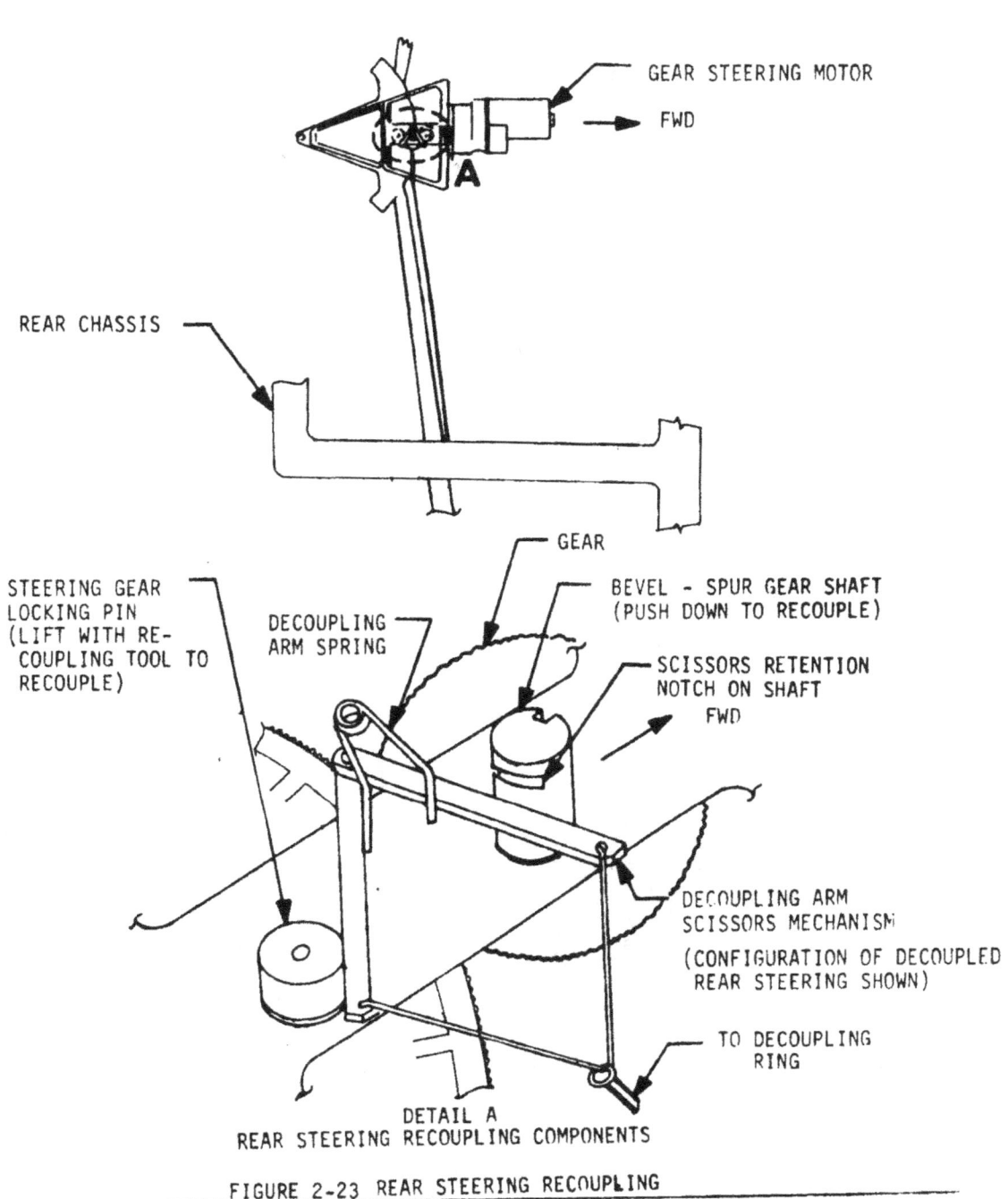

FIGURE 2-23 REAR STEERING RECOUPLING

LUNAR ROVING VEHICLE
APOLLO OPERATIONS HANDBOOK

STA/STEP	PROCEDURE	REMARKS
2.9.4	(Continued)	
k.	Standing beside LH seat, peel back beta-cloth dust cover from over rear steering transmission.	
l.	Push down gear shaft until scissors mechanism retains shaft in depressed position (figure 2-23). If gear shaft will not fully depress, manually rotate shaft gear by one to three teeth. If scissors arm does not retain shaft, manually assist scissors arm into notch.	This sequence must be followed or sector gear could be engaged to motor, which is at electrical zero, while wheels are not aligned with chassis.
m.	Insert recoupling tool in hole in steering gear lock pin until hook on tool is engaged with bottom of lock pin.	
n.	Lift tool and pin until scissors mechanism retains lock pin in raised position.	
o.	Remove tool from lock pin.	
p.	Replace dust cover over steering transmission.	
q.	Restow recoupling tool in stowage block.	
r.	Raise LH seat to operational position.	
s.	Ingress LRV.	
t.	±15 VDC Switch - PRIM.	
u.	STEERING FORWARD Switch - BUS A.	
v.	STEERING REAR Switch - BUS D.	
w.	Release parking brake and have other crewman verify wheels steer properly.	

Mission J Basic Date 6/23/71 Change Date _____ Page 2-62

LUNAR ROVING VEHICLE
APOLLO OPERATIONS HANDBOOK

STA/STEP	PROCEDURE	REMARKS
2.9.4	(Continued)	
	x. Set parking brake with throttle in neutral.	
	y. DRIVE POWER LF Switch - BUS A.	
	z. DRIVE POWER RF Switch - BUS A.	
	aa. DRIVE POWER LR Switch - BUS D.	
	ab. DRIVE POWER RR Switch - BUS D.	
	ac. Release parking brake.	
	ad. Resume normal operations.	

Mission __J__   Basic Date __6/23/71__   Change Date _____   Page __2-63__

LS006-002-2H
LUNAR ROVING VEHICLE
OPERATIONS HANDBOOK

SECTION 3

MALFUNCTION PROCEDURES

3.0 INTRODUCTION

Malfunction procedures encompass the recognition, diagnosis, and corrective action for system malfunctions. In most cases, the crew is alerted to a malfunction condition by Control and Display Panel meters or indicators. The crew will then locate, correct, or isolate the malfunction and determine its effect on the scheduled mission. In general, the procedures cover significant single failures. Double unrelated failures are not covered to prevent procedures from becoming complex and unmanageable. Malfunctions of a minor nature not requiring detailed procedures are covered in Section 2.

The malfunction procedures are arranged in logic flow diagram format and arranged by symptom routine. A three column format is used for symptom routine logic flow diagrams. A description and use of each of these columns is as follows.

3.1 SYMPTOM COLUMN

The primary purpose of the symptom column is to allow entry into the malfunction procedures. This block explains and qualifies the situation so that the reader understands the symptom or condition that exists. All symptoms are numbered in sequence starting with the number 1.

3.2 PROCEDURE COLUMN

The procedure column presents a step-by-step logic flow diagram of actions and decisions used to isolate or correct a malfunction symptom. This information is presented with several types of logic blocks. These blocks contain the procedures, decisions, and actions to locate and isolate the failure. Remote event symbols are used to reference items in the remarks column or to refer to other procedural steps.

3.3 REMARKS COLUMN

This column will include the following information:

- Amplifying additional remarks related to the symptom.

- Amplifying remarks which relate to a decision and/or action items (e.g., why a step is taken, etc.)

- Explain resultant system status or operational capability after a failure has been identified, i.e., how subsystem is degraded, can degraded subsystem support primary mission, early termination of mission, etc.

- Cautions or warnings, as necessary, to cover conditions that may exist because of a failure.

LS006-002-2H
LUNAR ROVING VEHICLE
OPERATIONS HANDBOOK

MALFUNCTION PROCEDURES	NO.	PAGE
CAUTION AND WARNING FLAG ACTUATES	1	3-3
EITHER BATTERY TEMP > 125°F	2	3-3
ONE DRIVE MOTOR TEMP > 400°F	3	3-4
ABNORMAL IMBALANCE BETWEEN BAT 1 AND BAT 2 AMPS (VEHICLE ACCELERATION NORMAL OR LOW)	4	3-4
FRONT (REAR) WHEELS DO NOT RESPOND TO HAND CONTROLLER STEERING COMMANDS	5	3-5
ONE OR MORE WHEELS DRIVE WHILE IN NEUTRAL	6	3-6
LOSS OF DRIVE FROM ONE OR TWO WHEELS (COMMANDED ACCELERATION ABNORMALLY LOW)	7	3-7
COMMANDED VEHICLE SPEED ABNORMALLY HIGH (SPEED NOT VARIABLE ON ONE OR MORE WHEELS)	8	3-8
LOSS OF DRIVE FROM ALL WHEELS	9	3-9
BRAKE WILL NOT RELEASE	10	3-10
LOSS OF VOICE COMM WITH MSFN	11	3-11

TABLE 3-1   MALFUNCTION PROCEDURES

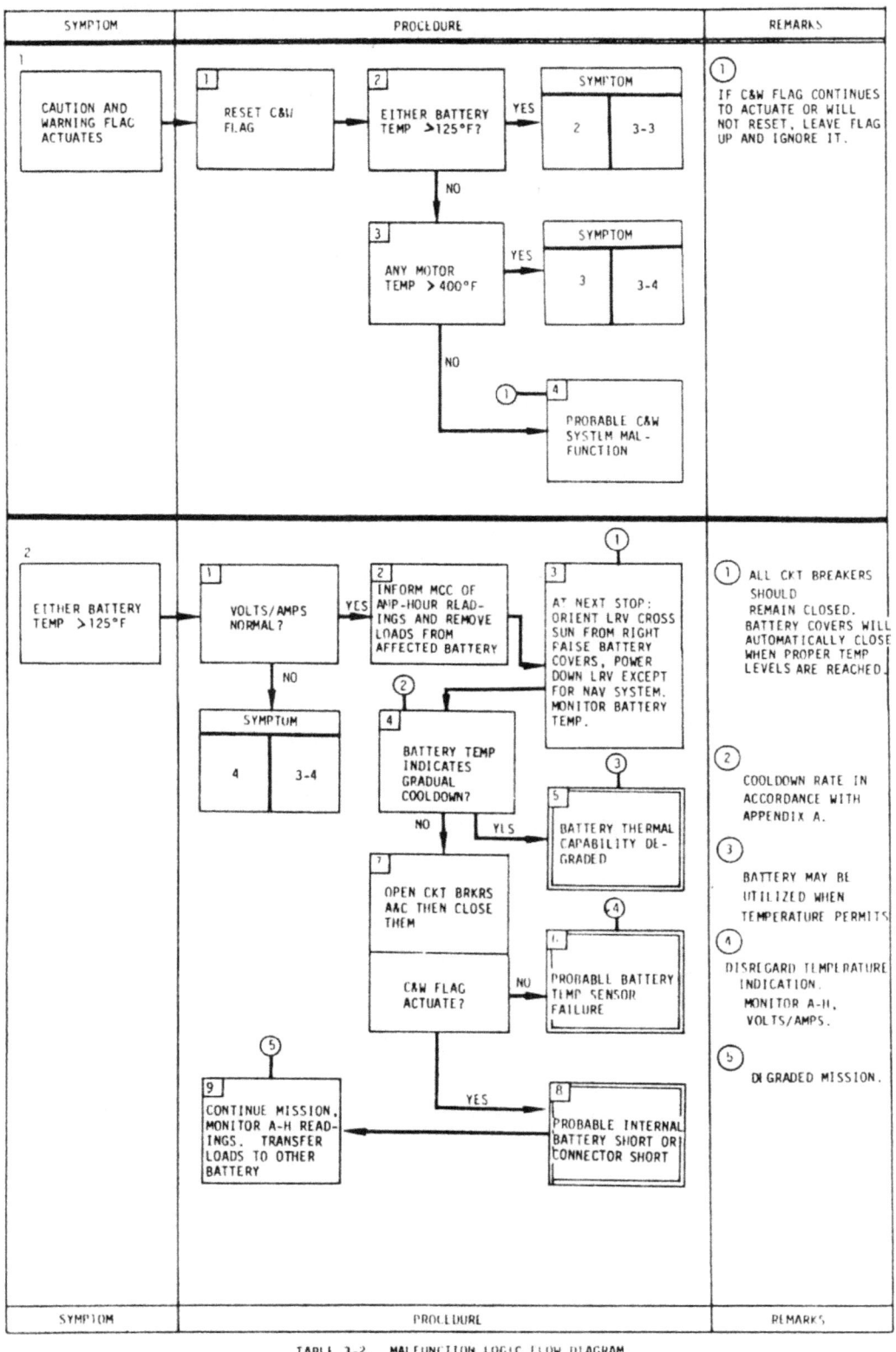

TABLE 3-2. MALFUNCTION LOGIC FLOW DIAGRAM

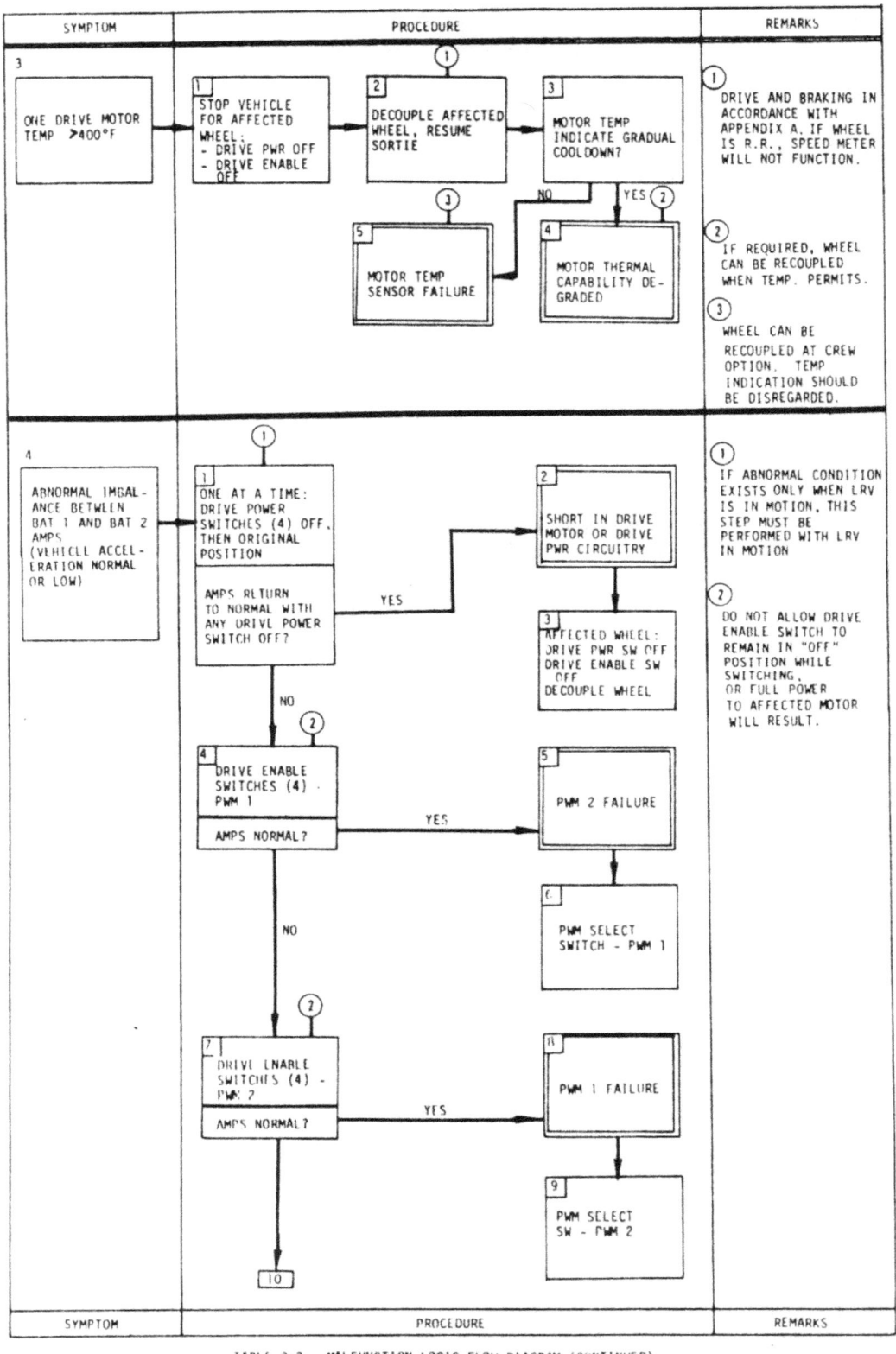

TABLE 3-2. MALFUNCTION LOGIC FLOW DIAGRAM (CONTINUED)

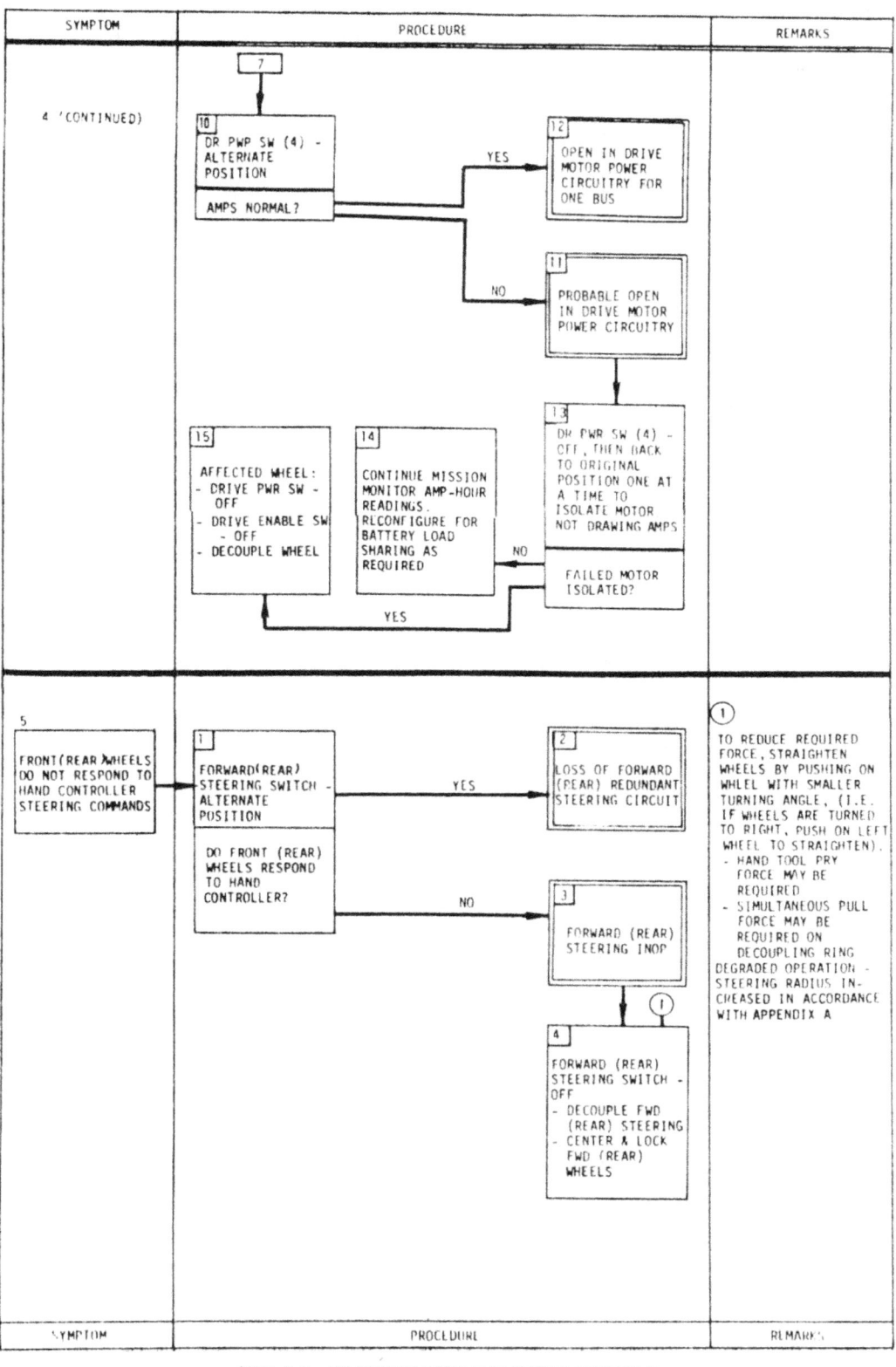

TABLE 3-2. MALFUNCTION LOGIC FLOW DIAGRAM (CONTINUED)

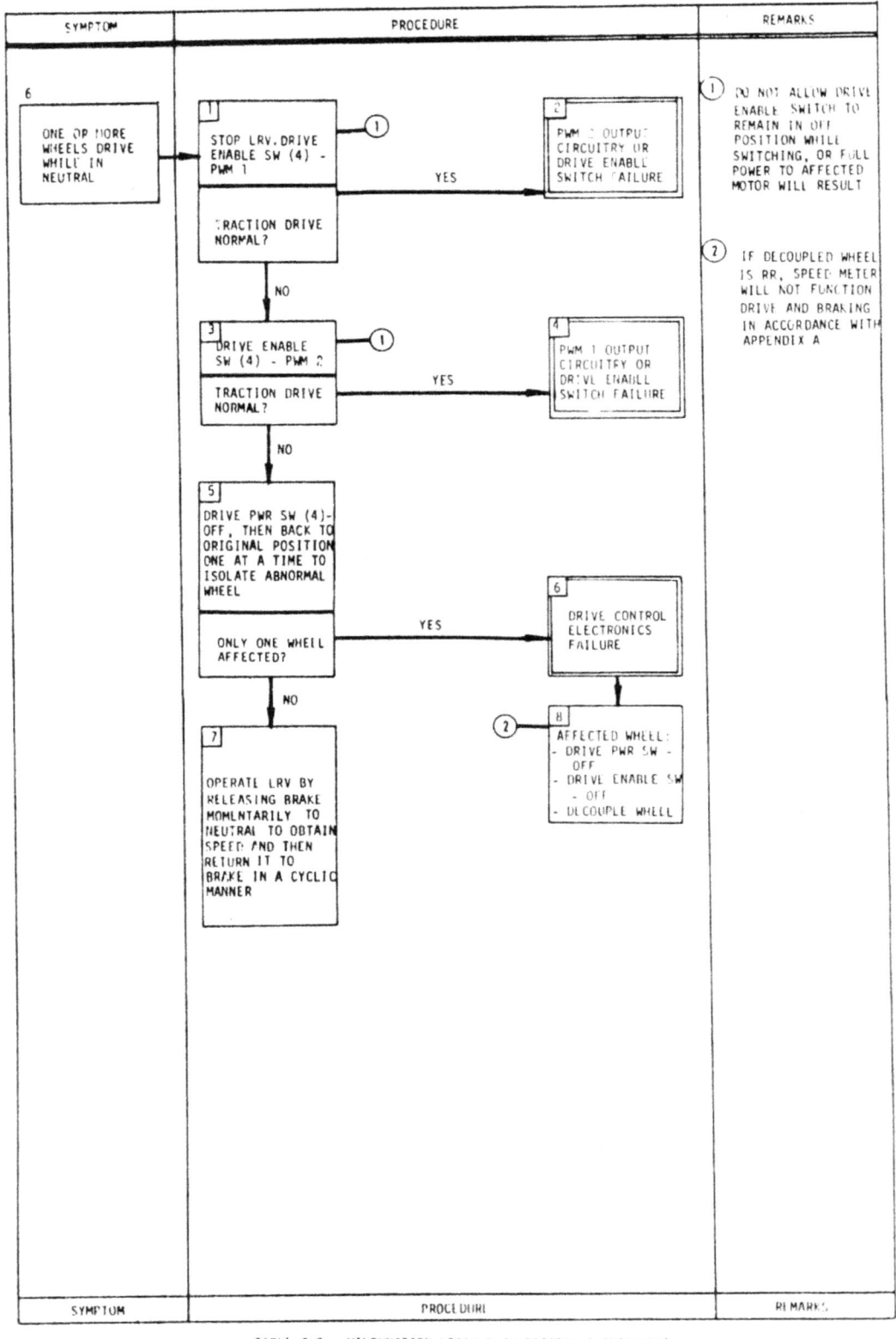

TABLE 3-2. MALFUNCTION LOGIC FLOW DIAGRAM (CONTINUED)

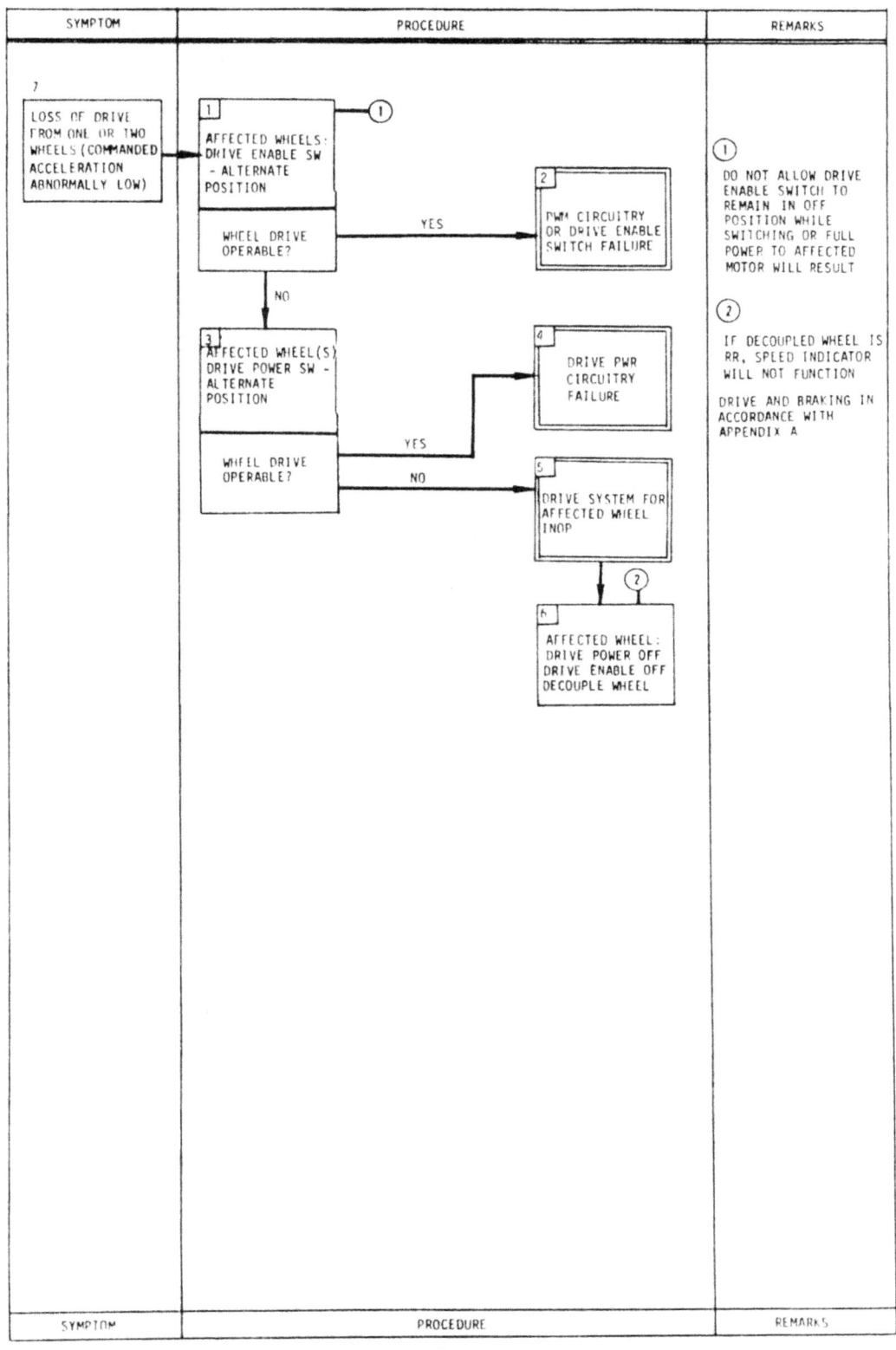

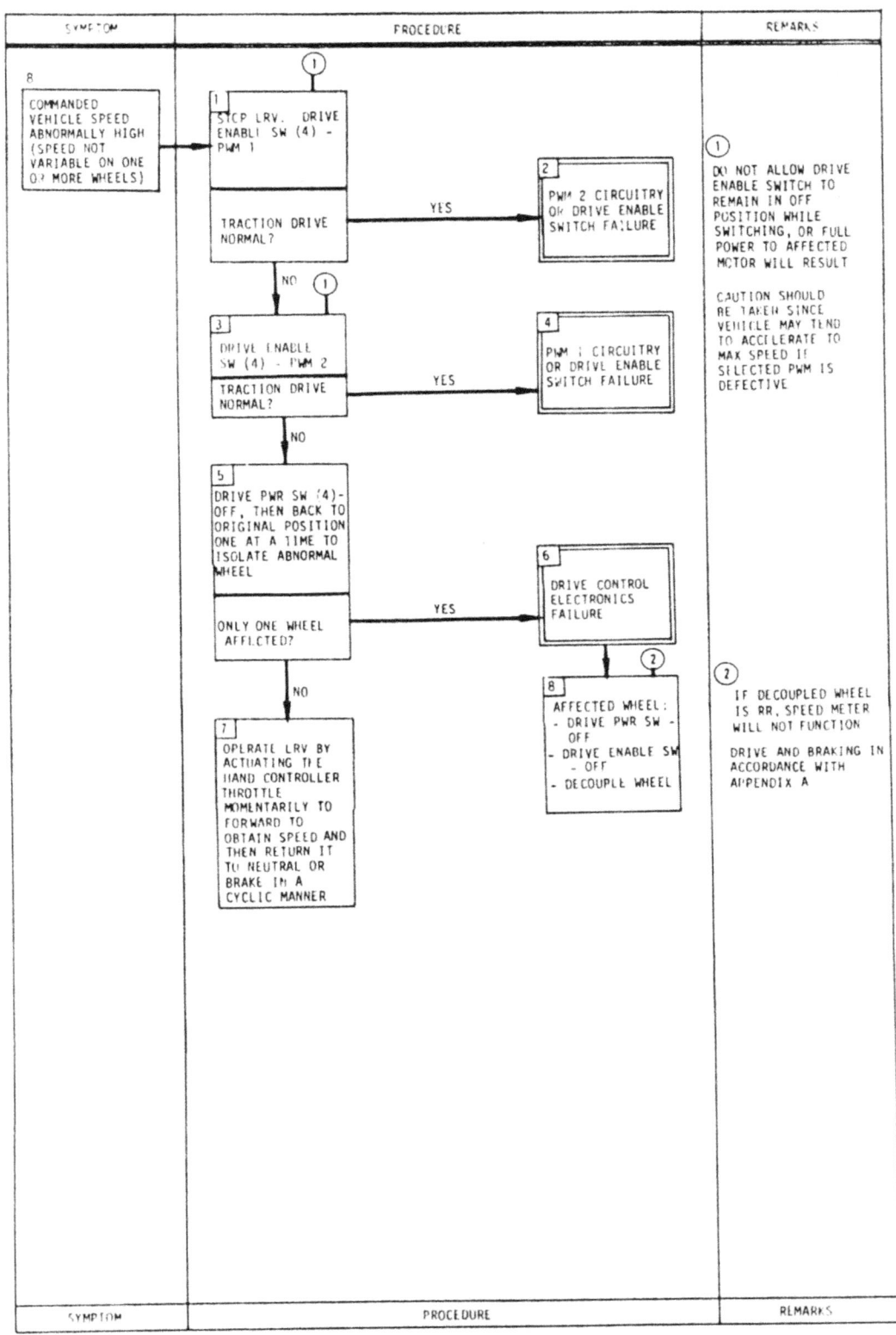

TABLE 3-2. MALFUNCTION LOGIC FLOW DIAGRAM (CONTINUED)

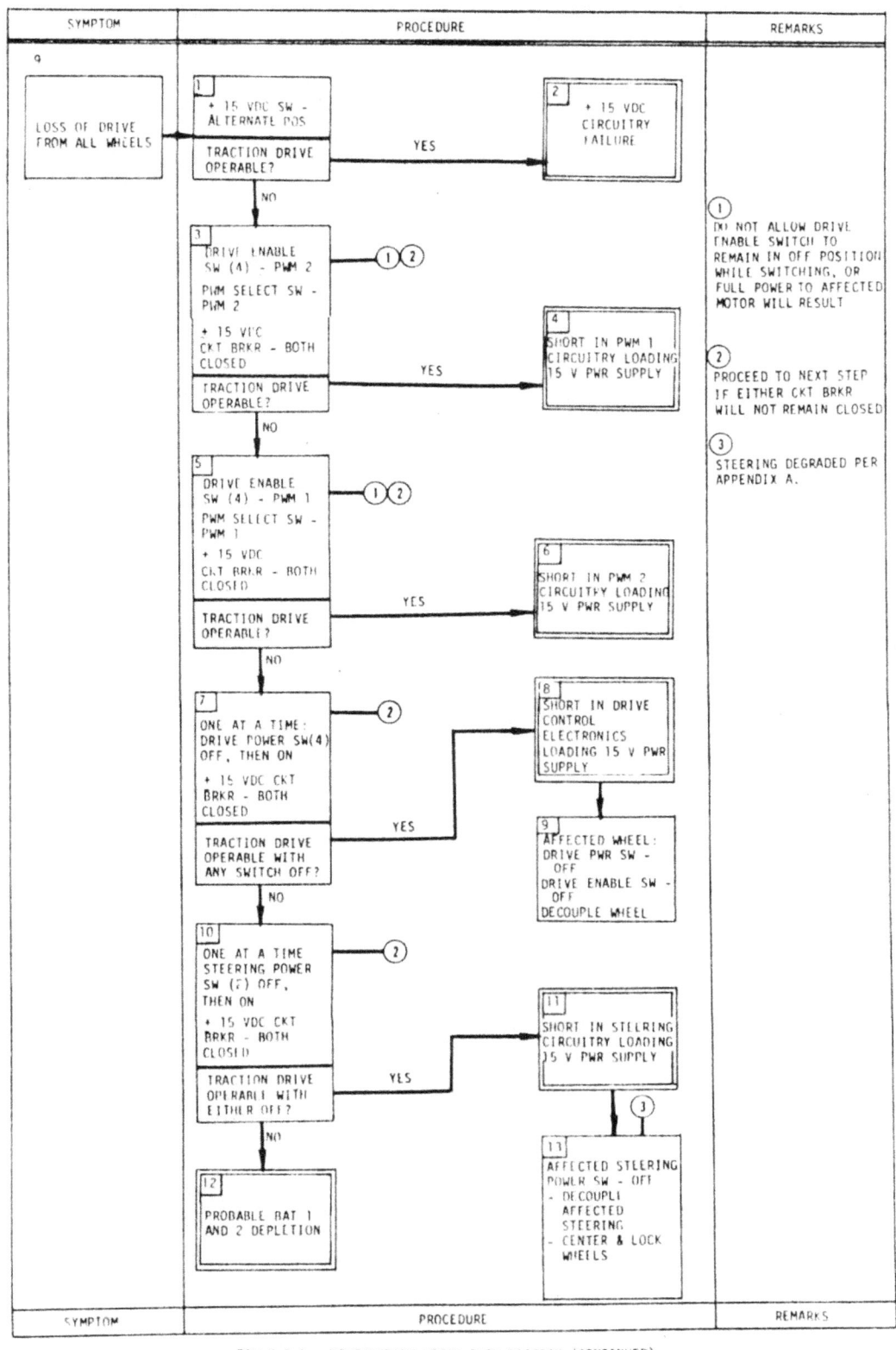

Table 3-2. Malfunction Logic Flow Diagram (Continued)

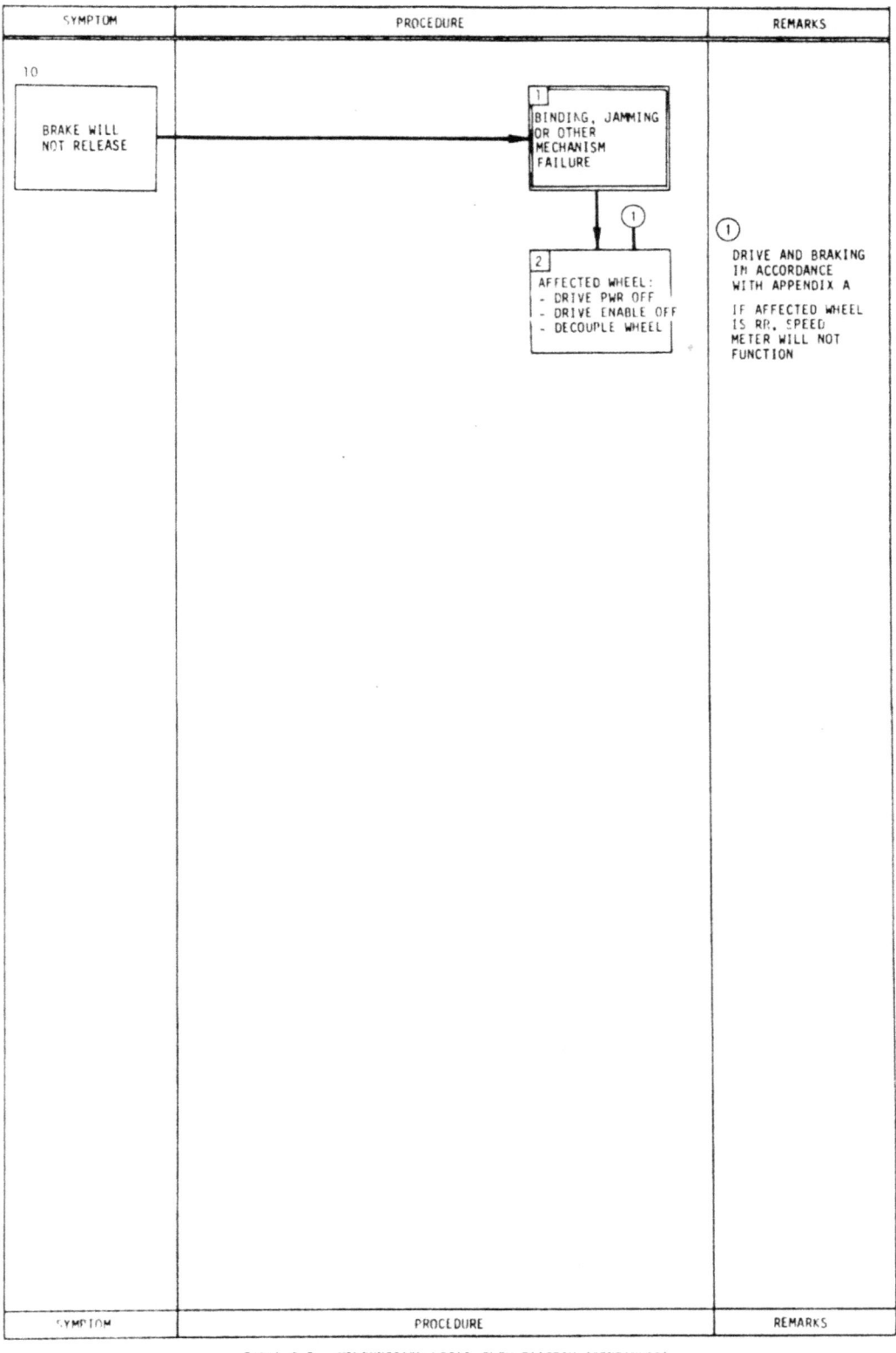

TABLE 3-2. MALFUNCTION LOGIC FLOW DIAGRAM (CONTINUED)

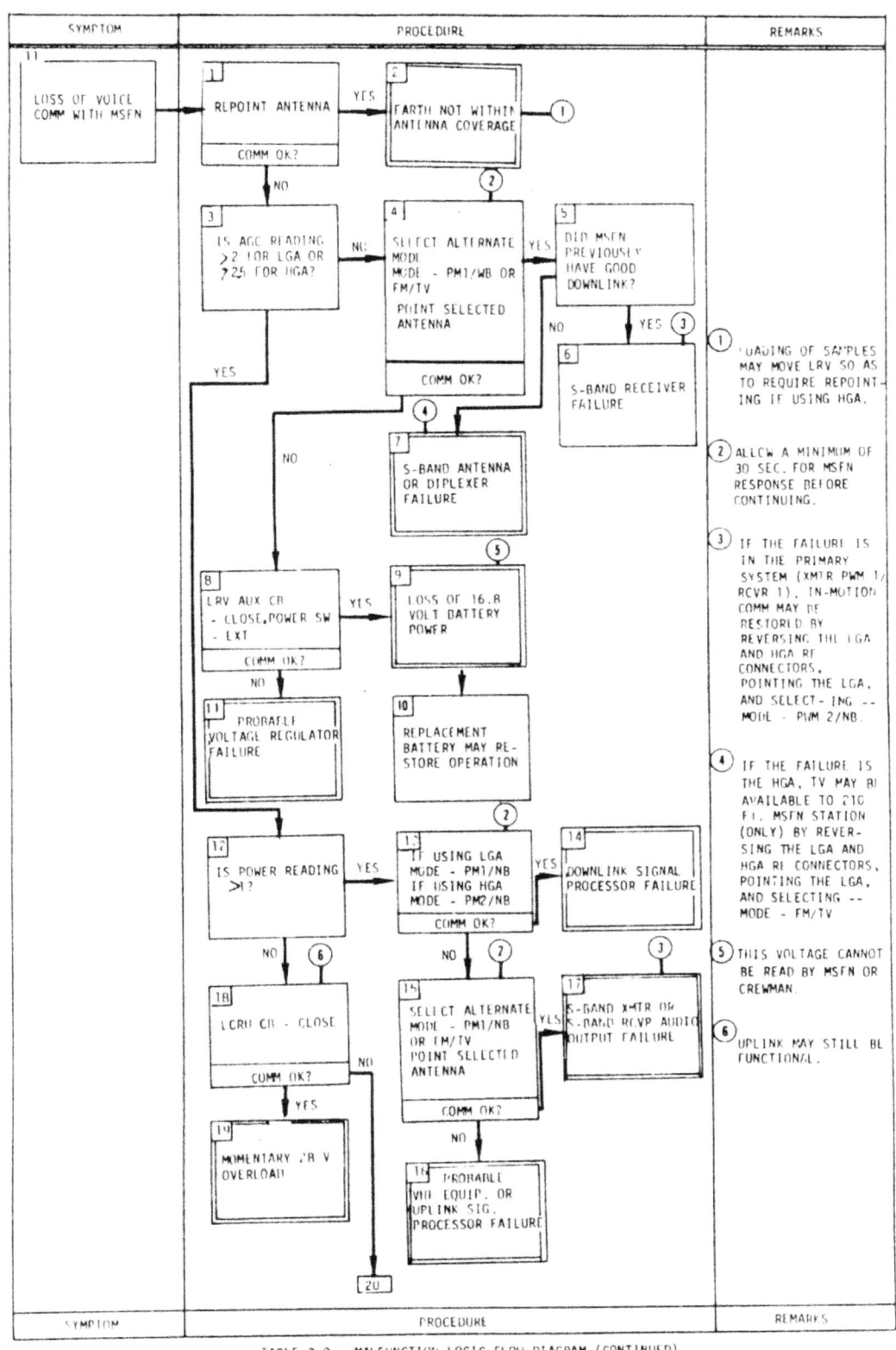

TABLE 3-2. MALFUNCTION LOGIC FLOW DIAGRAM (CONTINUED)

LUNAR ROVING VEHICLE
OPERATIONS HANDBOOK

SYMPTOM	PROCEDURE	REMARKS
11 (CONTINUED)	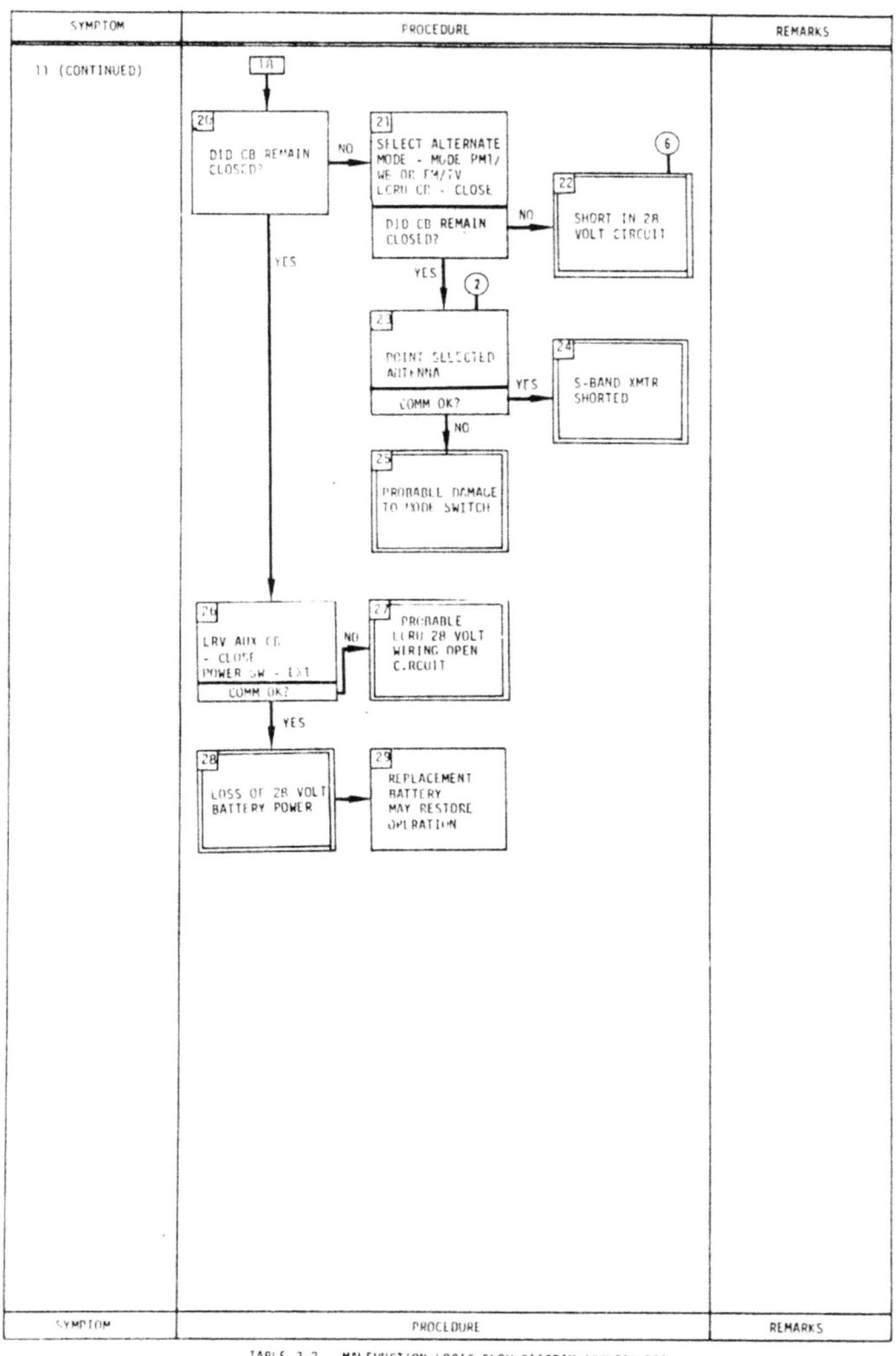	

TABLE 3-2. MALFUNCTION LOGIC FLOW DIAGRAM (CONTINUED)

Mission __J__ Basic Date __12/4/70__ Change Date __4/19/71__ Page __3-12__

LS006-002-2H
LUNAR ROVING VEHICLE
OPERATIONS HANDBOOK

MALFUNCTION PROCEDURES

SYMPTOM	NO.	PAGE
BATTERY TEMPERATURE MONITOR WILL NOT FUNCTION	1	3A-5
MOTOR TEMPERATURE MONITOR WILL NOT FUNCTION	2	3A-5
SPEED INDICATOR WILL NOT FUNCTION	3	3A-5
BEARING, DISTANCE OR RANGE INDICATORS INOPERATIVE	4	3A-6
SYSTEM RESET SWITCH WILL NOT FUNCTION	5	3A-6
GYRO TORQUING SWITCH WILL NOT FUNCTION	6	3A-7
SUN SHADOW DEVICE INOPERABLE	7	3A-7
NO OUTPUT FROM AUX PWR CONNECTOR	8	3A-7
BATTERY THERMAL DUST COVER WILL NOT CLOSE	9	3A-8
LOSS OF REMAINING STEERING WITH FRONT OR REAR STEERING INOPERATIVE	10	3A-8
BRAKE WILL NOT ACTIVATE ON ONE WHEEL	11	3A-8
DAMPER INOPERATIVE OR DEGRADED	12	3A-8
SLIDE SECTION OF DEPLOYABLE FENDER WILL NOT COMPLETELY DEPLOY	13	3A-8
TORSION BAR FAILURE	14	3A-8
WHEEL TRACTION POOR	15	3A-8

TABLE 3A-2 MALFUNCTION PROCEDURES

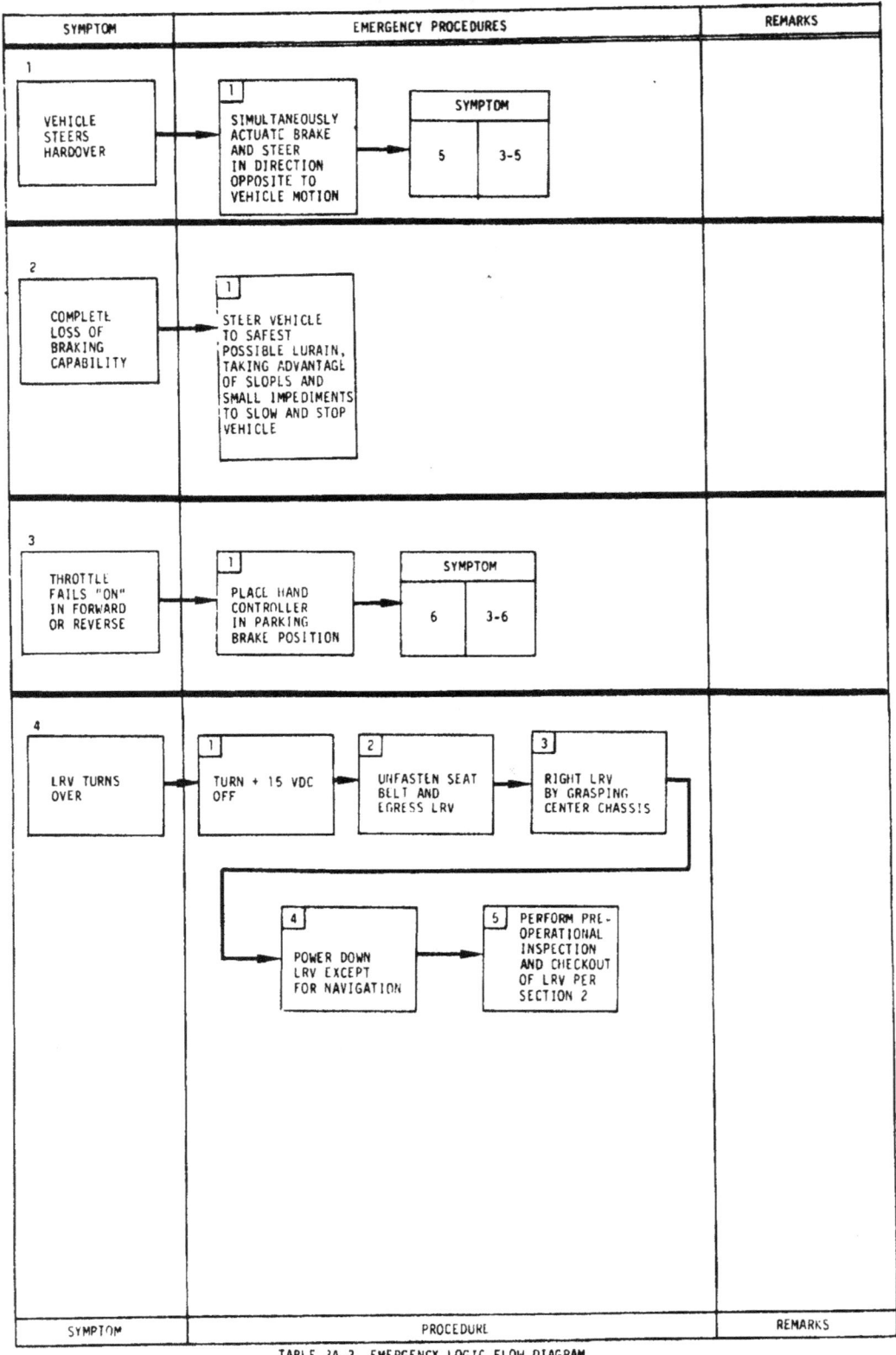

TABLE 3A-3 EMERGENCY LOGIC FLOW DIAGRAM

LS006-002-2H
LUNAR ROVING VEHICLE
OPERATIONS HANDBOOK

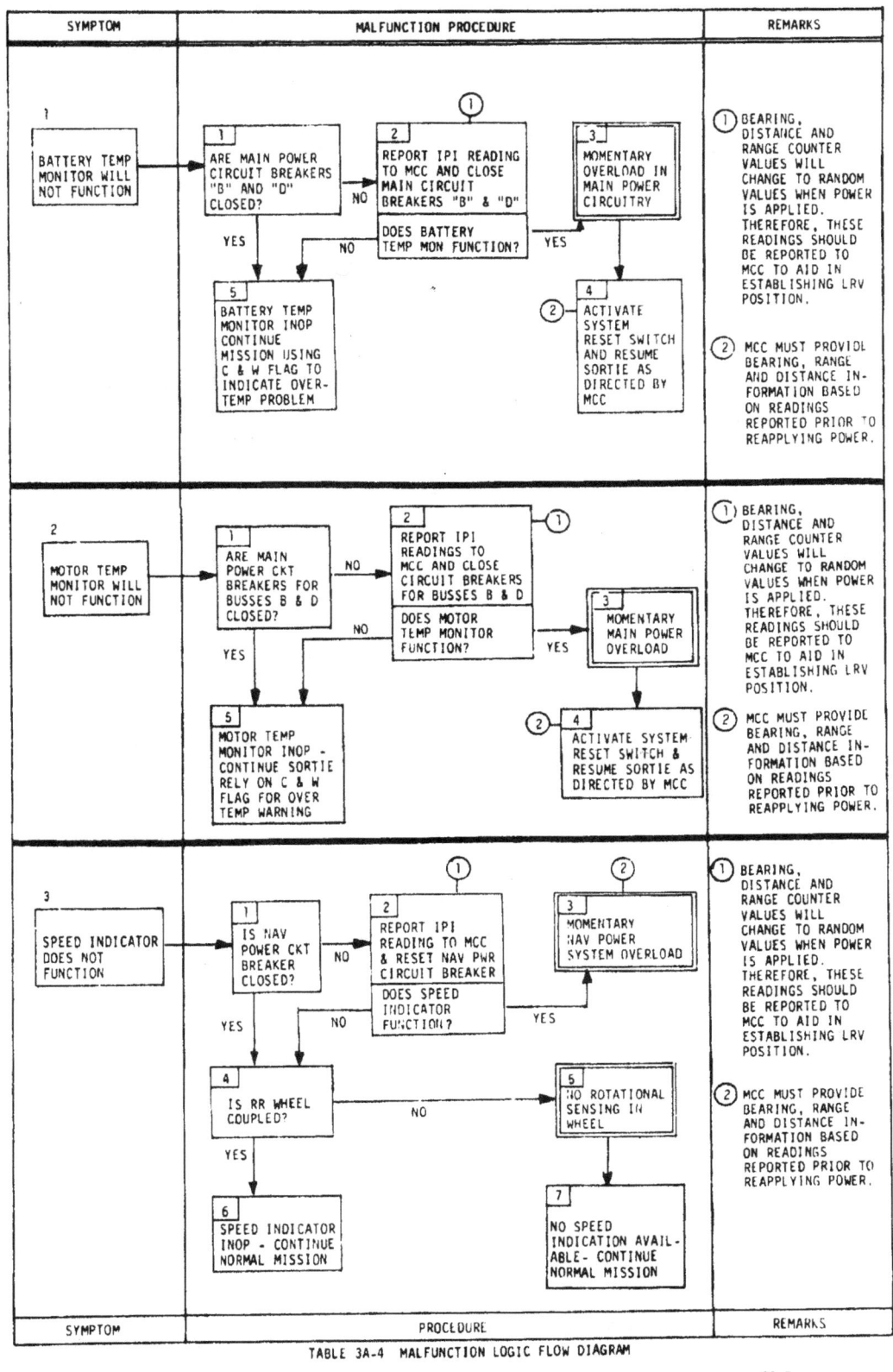

TABLE 3A-4 MALFUNCTION LOGIC FLOW DIAGRAM

Mission __J__ Basic Date __6/23/71__ Change Date _____ Page __3A-5__

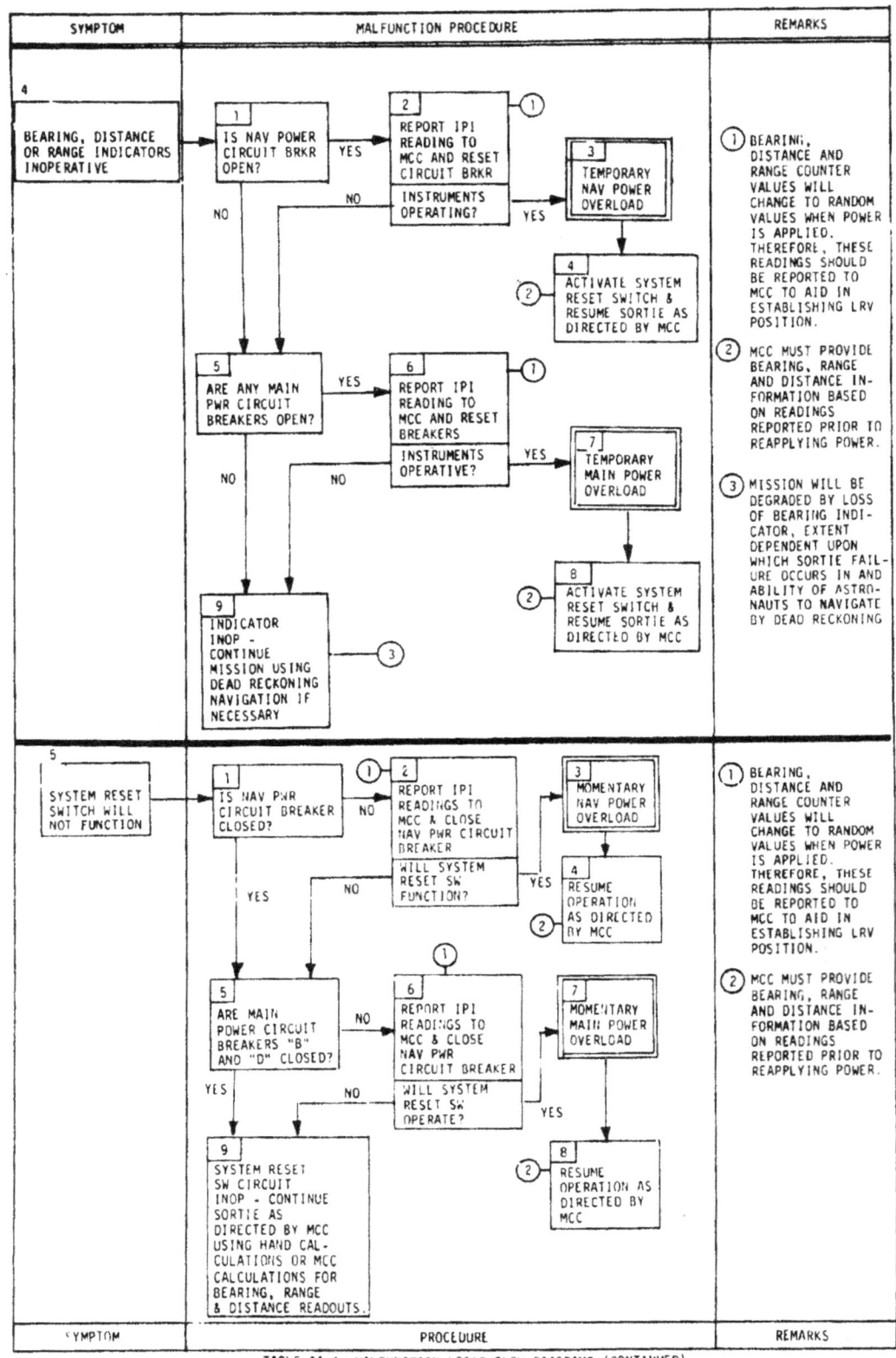

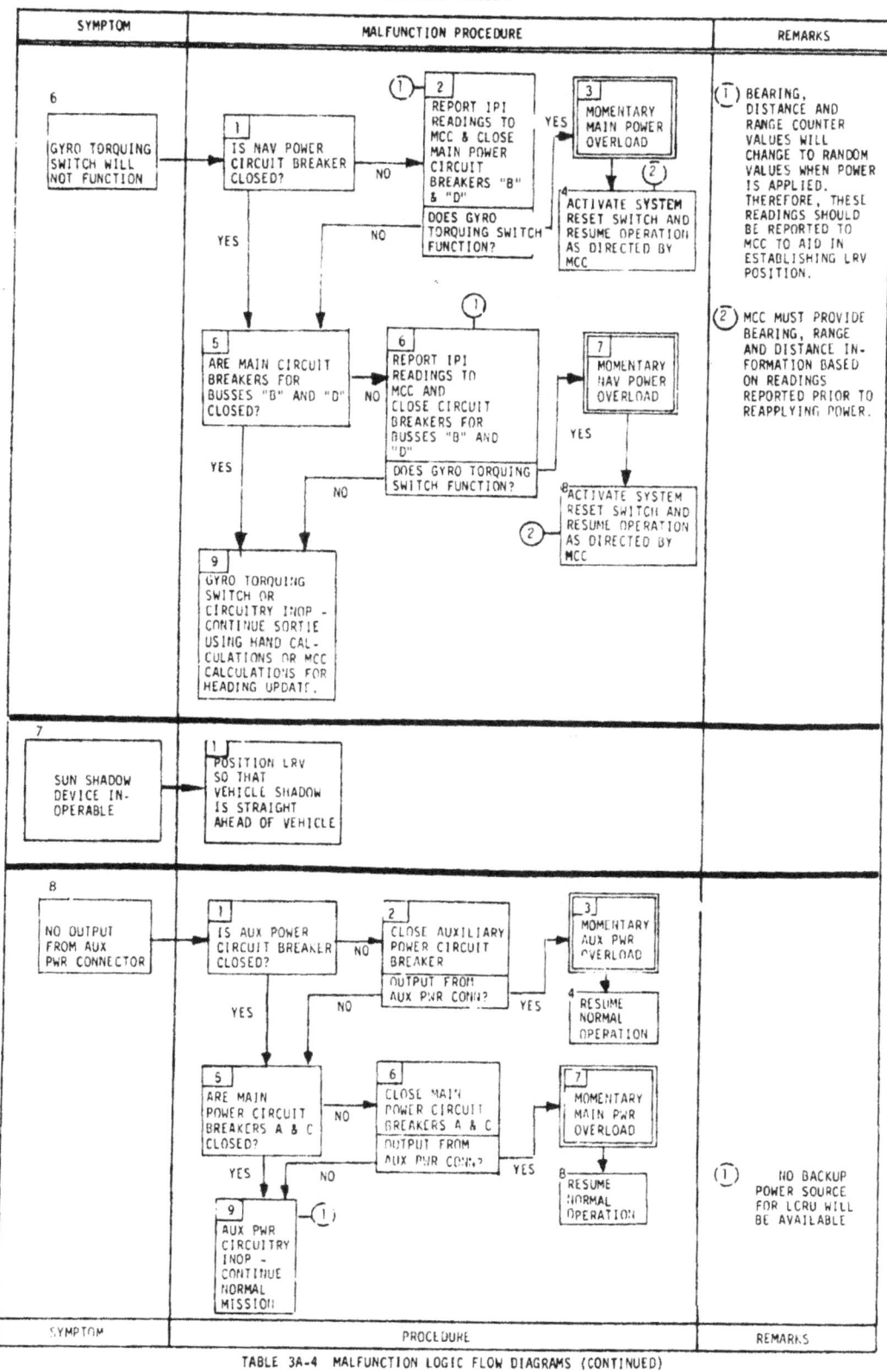

TABLE 3A-4 MALFUNCTION LOGIC FLOW DIAGRAMS (CONTINUED)

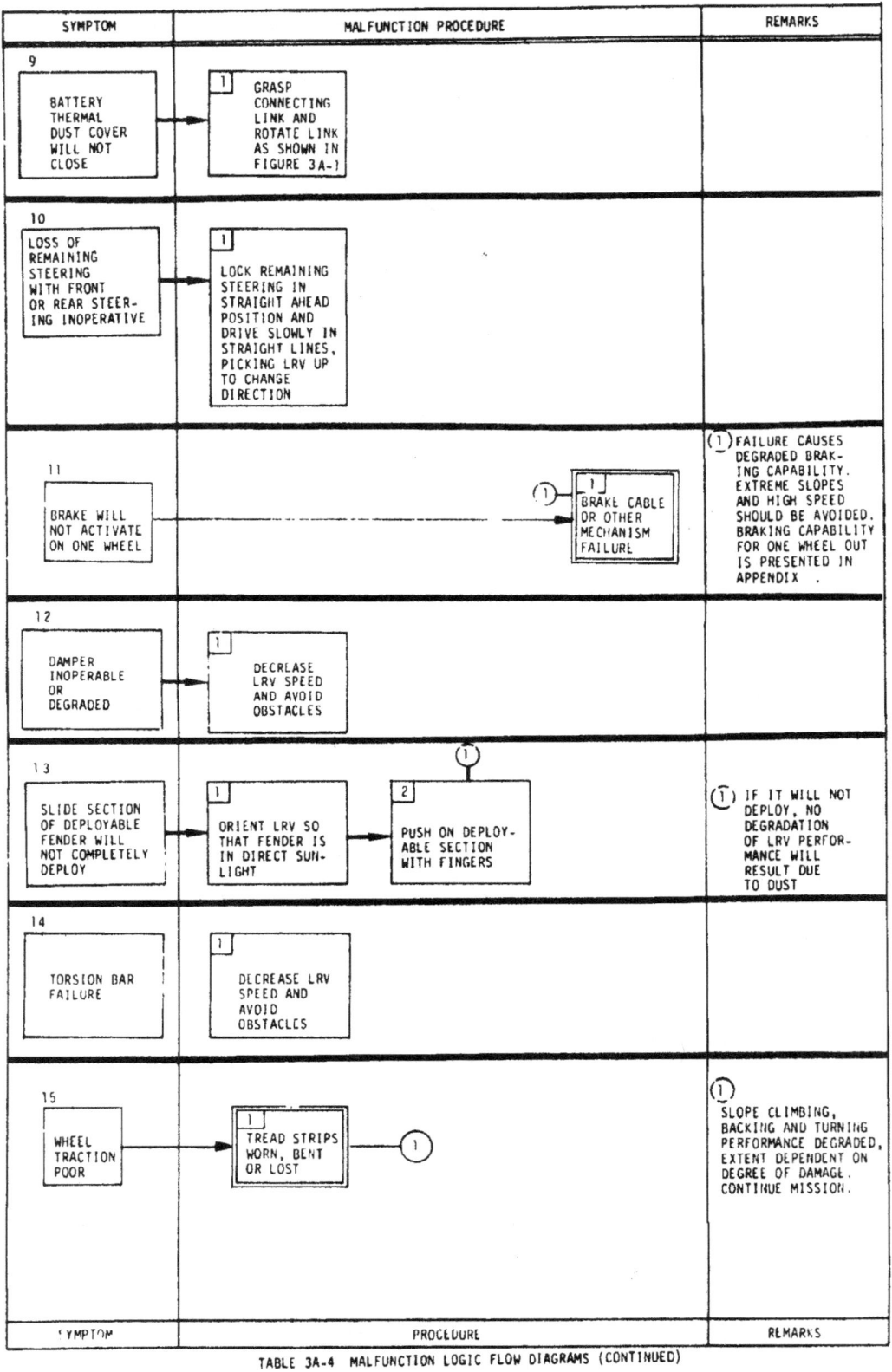

TABLE 3A-4 MALFUNCTION LOGIC FLOW DIAGRAMS (CONTINUED)

LS006-002-2H
LUNAR ROVING VEHICLE
OPERATIONS HANDBOOK

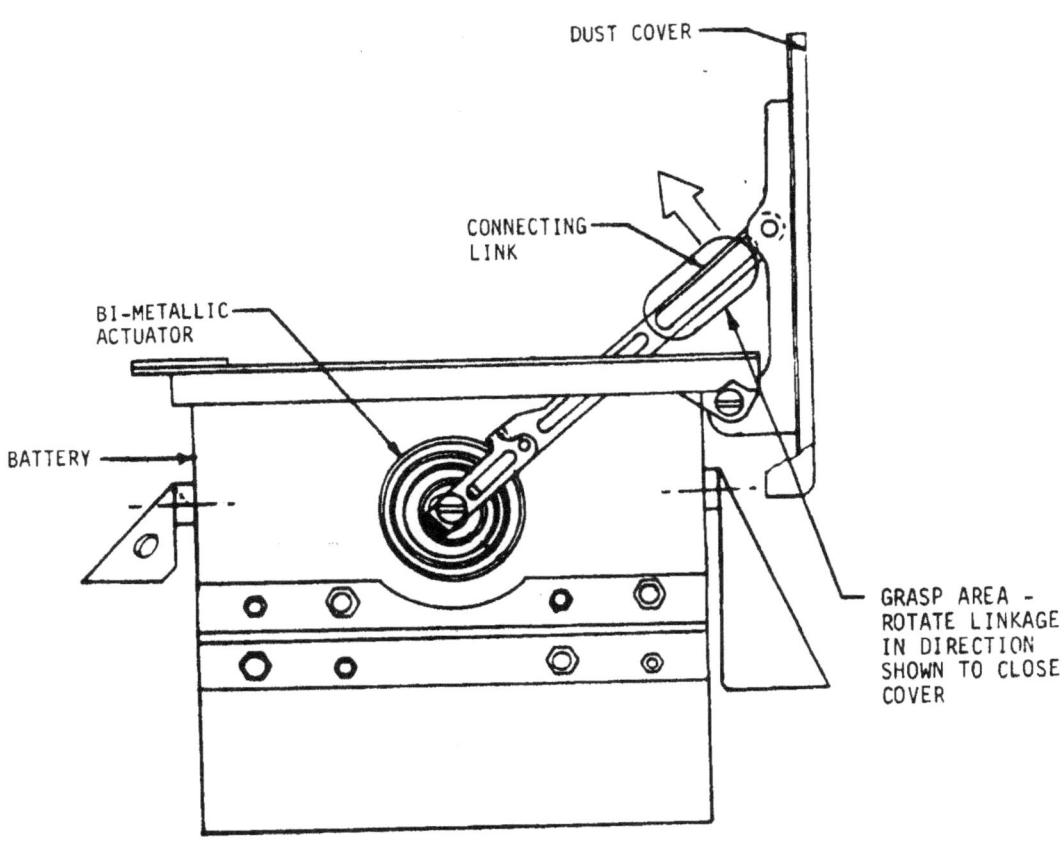

FIGURE 3A-1 CONTINGENCY DUST COVER CLOSURE

LS006-002-2H
LUNAR ROVING VEHICLE
OPERATIONS HANDBOOK

SECTION 4

AUXILIARY EQUIPMENT

4.0   INTRODUCTION

This section contains the LRV and 1G Trainer auxiliary equipment which includes provisions for transporting miscellaneous equipment for support of lunar activities including experiments, communications and photograph.

4.1   FORWARD CHASSIS PAYLOAD PROVISIONS

The forward chassis contains the equipment necessary to transport the LCRU, the high gain antenna, and the ground controlled television camera assembly (GCTA).

4.1.1   Lunar Communications Relay Unit (LCRU)

The LCRU is mounted in the two inboard receptacles on the forward chassis forward frame member as shown in figure 4-1. To conserve crew time on the lunar surface, the two LCRU support posts are installed in these receptacles at KSC before securing the LRV in the LM. In addition, the LRV/LCRU power cable is also connected to the LRV auxiliary connector before launch. The LCRU support posts and LRV/LCRU power cable are stowed on the LRV as shown in figure 4-2.

4.1.2   High Gain Antenna and GCTA

The high gain antenna is secured to the left outboard receptacle on the forward chassis forward frame member (figure 4-1). This receptacle is identical to the one on the right outboard side for the GCTA.

Provisions for securing the LCRU low gain antenna coax cable to the LRV are shown on figure 4-3.

4.2   CENTER CHASSIS PAYLOAD PROVISIONS

The center chassis has provisions to carry auxiliary equipment on the inboard handholds, under the crew seats and on the chassis floor.

4.2.1   Inboard Handhold Payload Receptacle

The inboard handholds are provided with receptacles for supporting the 16 mm Data Acquisition Camera and low gain antenna as shown in figure 4-4.

4.2.2   Under-Seat Stowage

One collapsible stowage bag is provided under each seat for transporting miscellaneous payload items. These bags are installed on the LRV before launch.

LS006-002-2H
LUNAR ROVING VEHICLE
OPERATIONS HANDBOOK

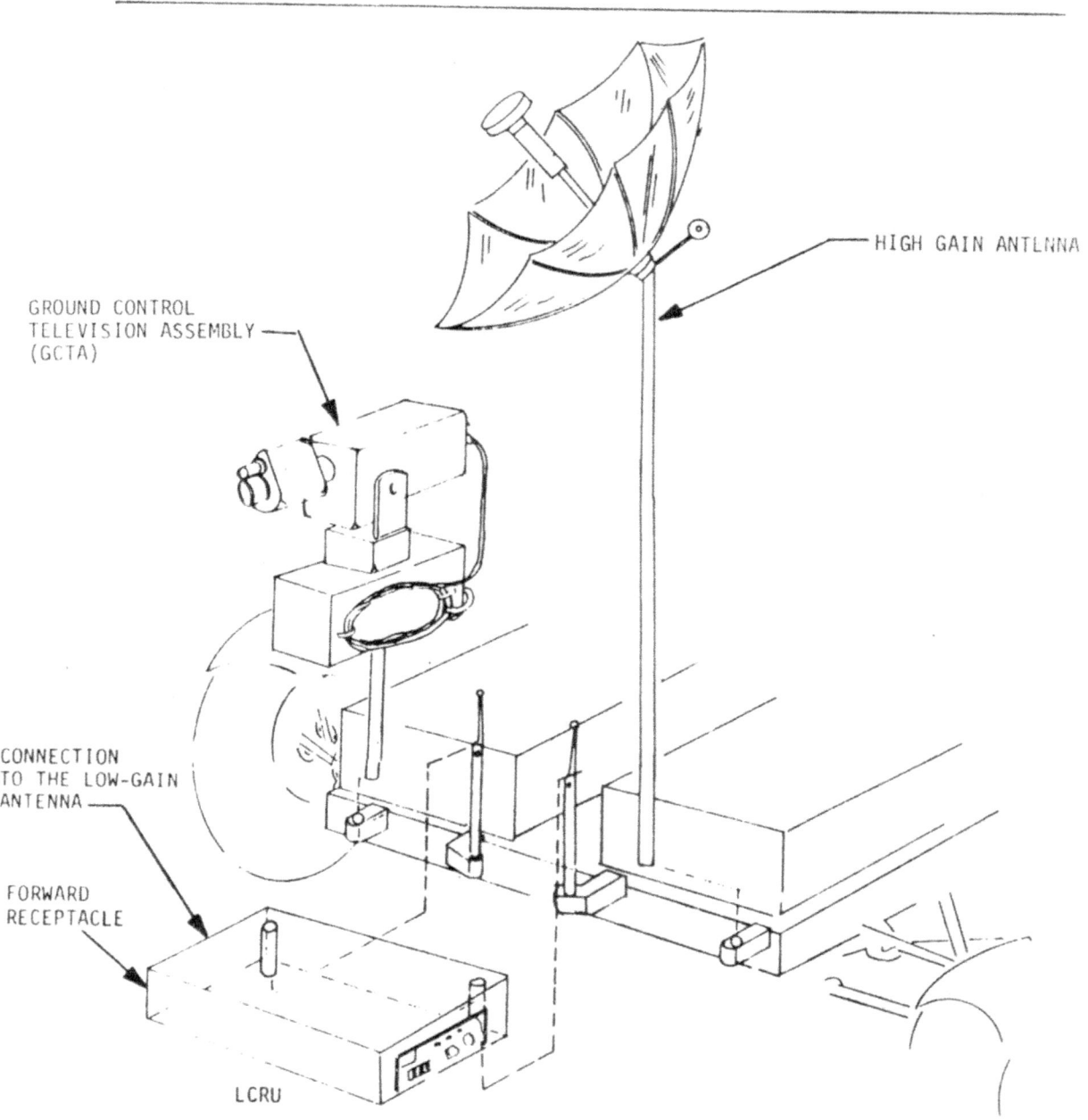

FIGURE 4-1. LCRU, HIGH GAIN ANTENNA, TV CAMERA INSTALLATION

LSU06-002-2H
LUNAR ROVING VEHICLE
OPERATIONS HANDBOOK

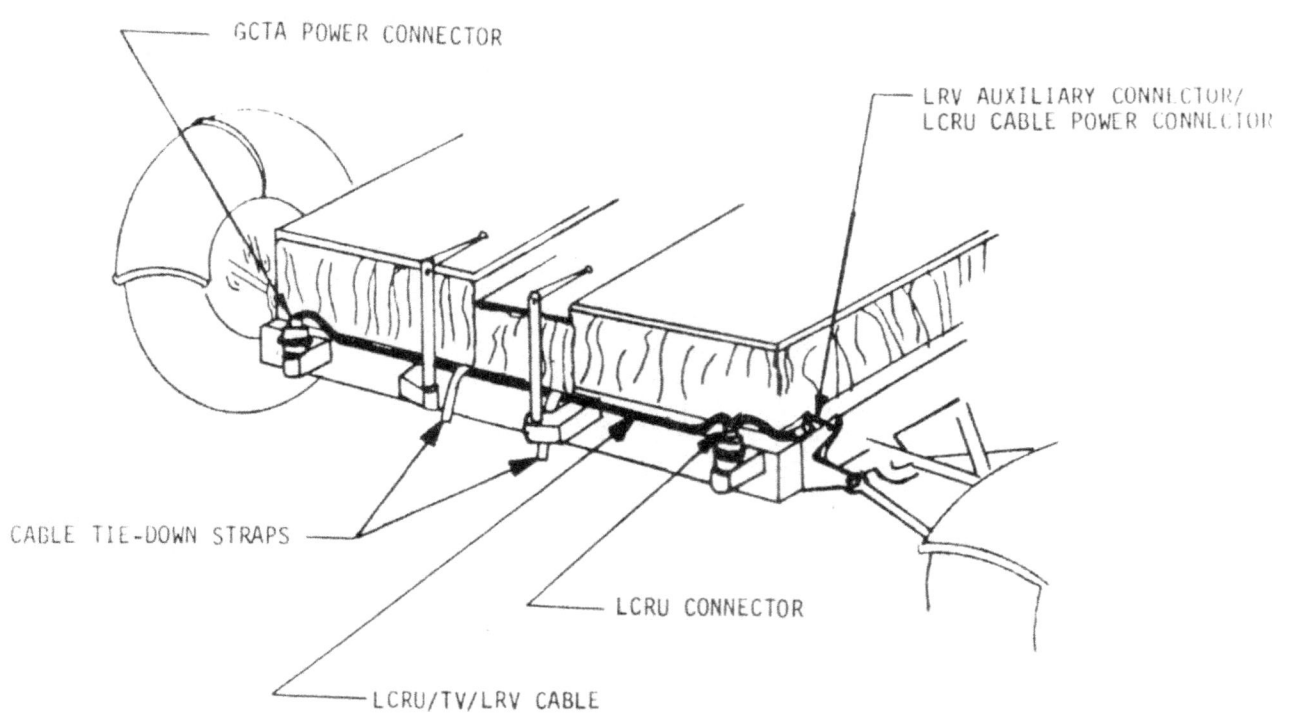

FIGURE 4-2. LCRU/TV/LRV CABLE STOWAGE

Mission __J__ Basic Date __12/4/70__ Change Date __4/19/71__ Page __4-3__

LS006-002-2H
LUNAR ROVING VEHICLE
OPERATIONS HANDBOOK

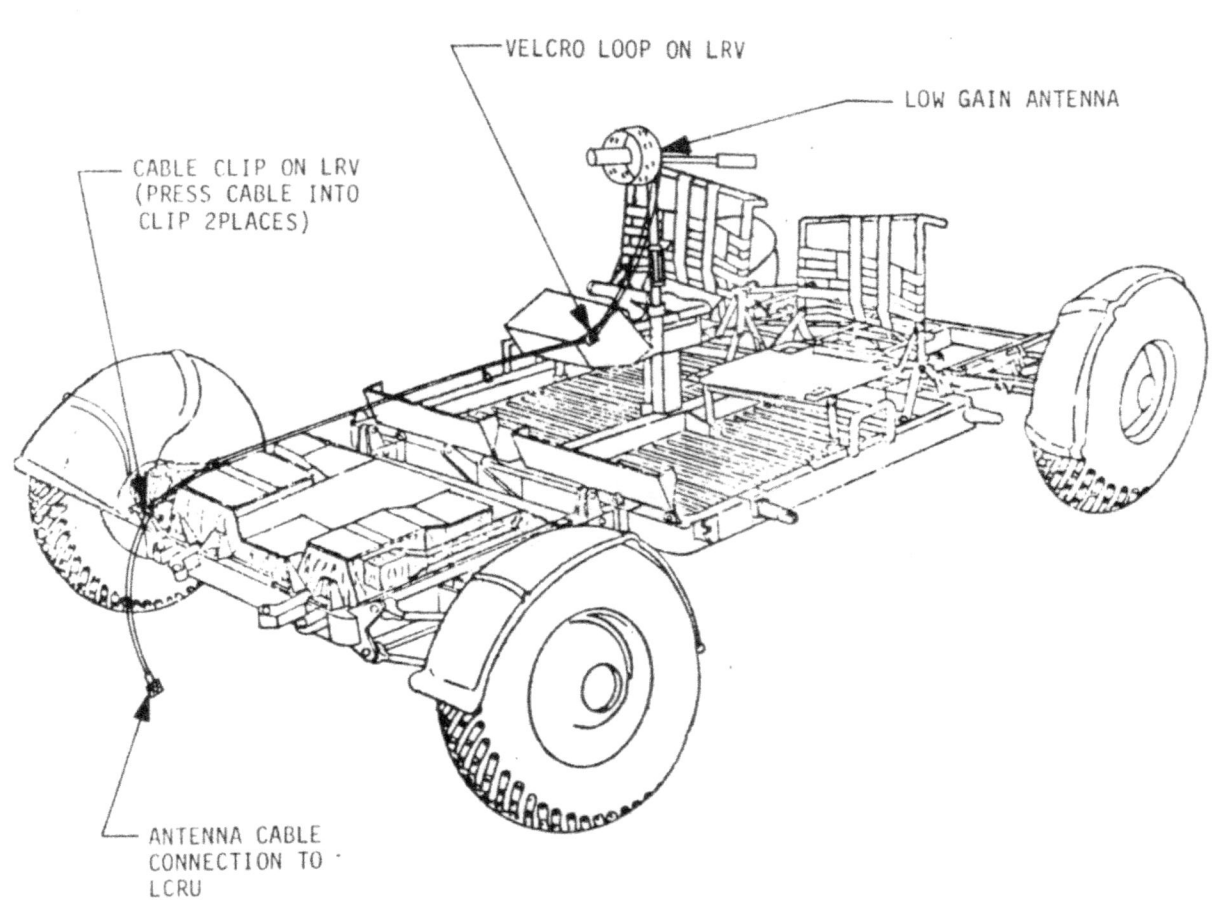

FIGURE 4-3 LCRU LOW GAIN ANTENNA CABLE
INSTALLATION ON LUNAR SURFACE

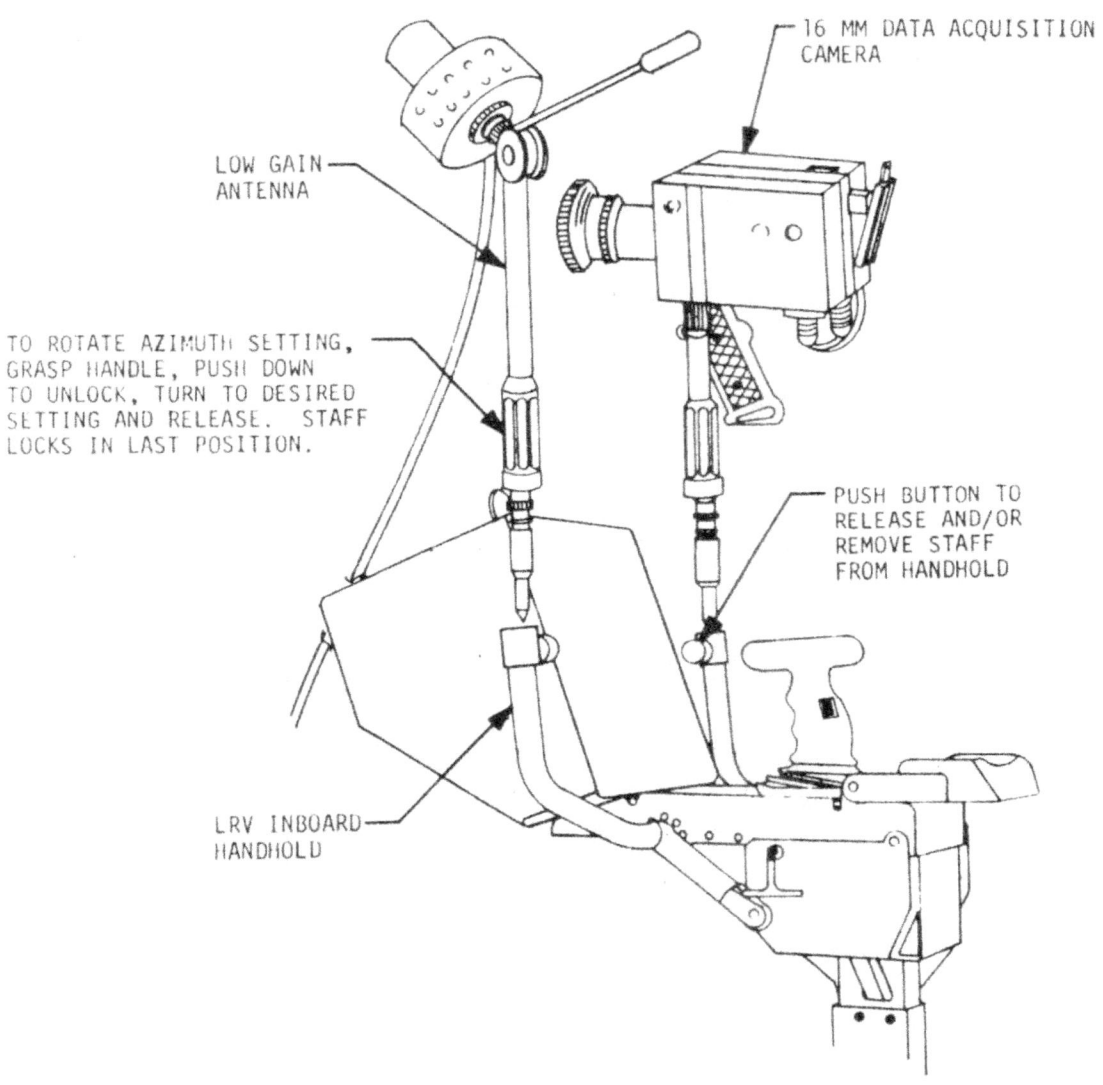

FIGURE 4-4. 16 MM DAC AND LOW GAIN ANTENNA INSTALLATION

LS006-002-2H
LUNAR ROVING VEHICLE
OPERATIONS HANDBOOK

4.2.2 (Continued)

The two bags are identical and are of the configuration shown in figure 4-5. The forward end of each bag is secured to the seat support frame. The bags are automatically erected to the useable position when the seat support frames are raised during LRV activation. The aft ends of the bags are held in place by springs attached to the rear member of the center chassis and by attachment to cross member on the seat back.
During operations, access to the stowage bags is gained by raising the seat off the seat support which exposes the entire bag and contents.

4.2.3 Floor Payload Stowage

When only one astronaut is operating the LRV, the area normally used by the second crewman may be used for payload stowage. This is accomplished by placing the seat in the operational stowage position shown in figure 4-6. The seat is secured in the stowed position by velcro straps.

NOTE: The under seat stowage bag must be removed to use the floor area as a stowage area.

4.2.4 Back-of-Seat Payload Stowage

The Buddy SLSS umbilical is carried in a bag attached to the back of the LRV right seat. Specific interface is shown in figure 4-7.

4.3 REAR CHASSIS PAYLOAD PROVISIONS

Payload stowage provisions for the rear chassis are shown in figures 4-8 and 4-9. The LH and RH adapters and pallet support posts are installed on the LRV before launch, arriving on the lunar surface in the configuration shown in figure 4-8. The payload pallet which interfaces with the adapters and support posts is stowed in LM Quadrant III, and arrives on the lunar surface with payload items already installed on the pallet. The crew removes the pallet from Quadrant III and installs the pallet, with attached payload, onto the support post and adapters on the LRV.

*** 1G Trainer Note ***

The rear chassis payload adapters for the 1G Trainer will allow identical astronaut functions to be performed, but the configuration of the adapters is not exactly identical to the LRV adapters.

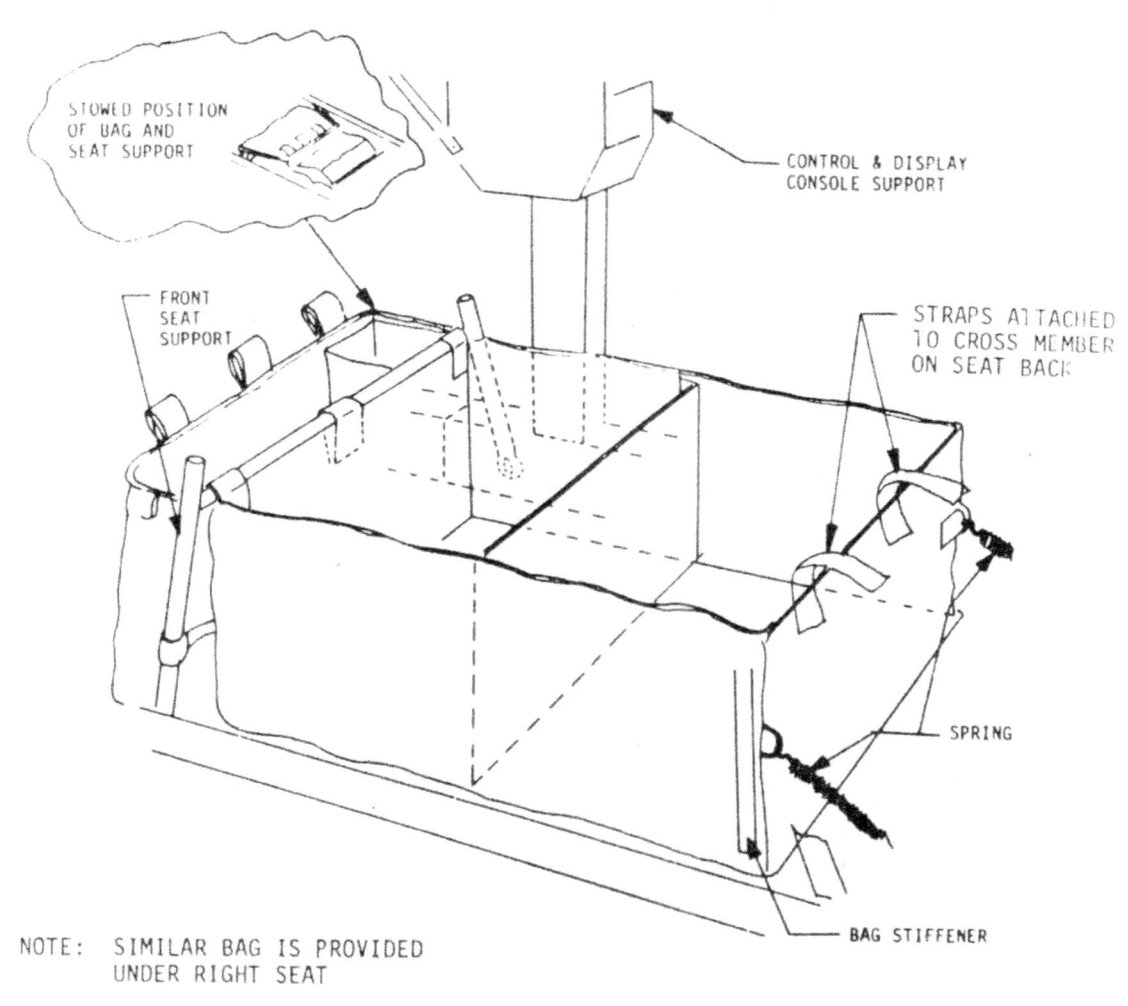

FIGURE 4-5 UNDER-SEAT STOWAGE BAG (LEFT SEAT)

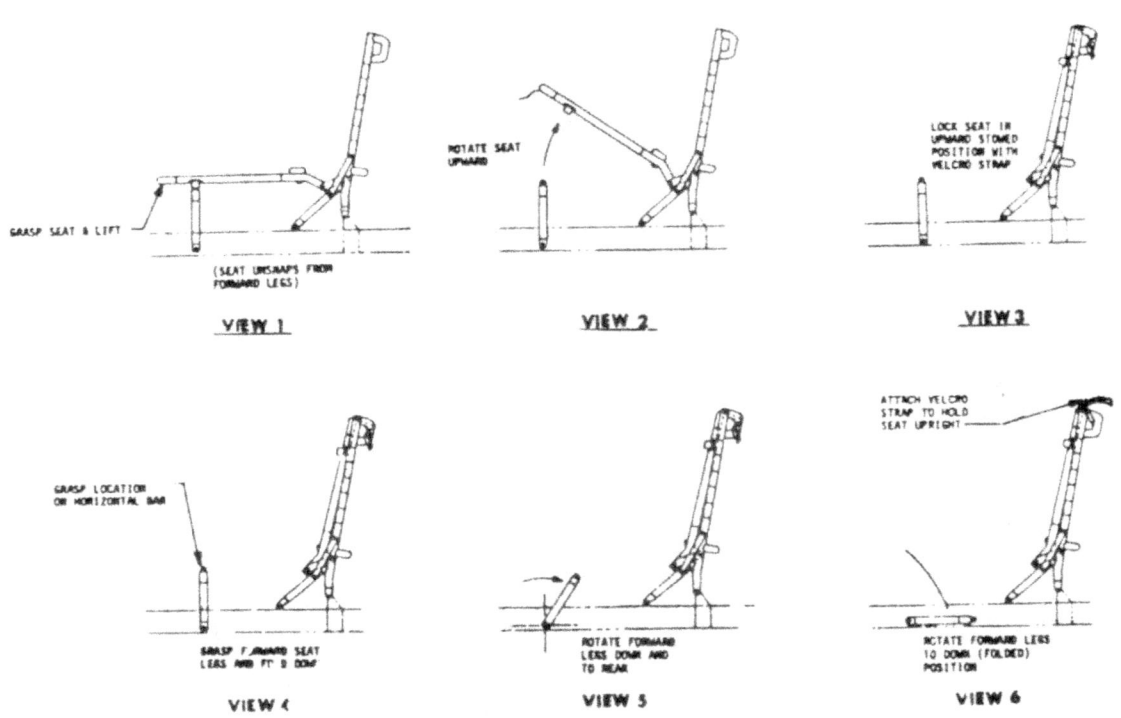

FIGURE 4-6 PASSENGER SEAT STOWAGE TO CREATE PAYLOAD AREA ON CENTER CHASSIS FLOOR

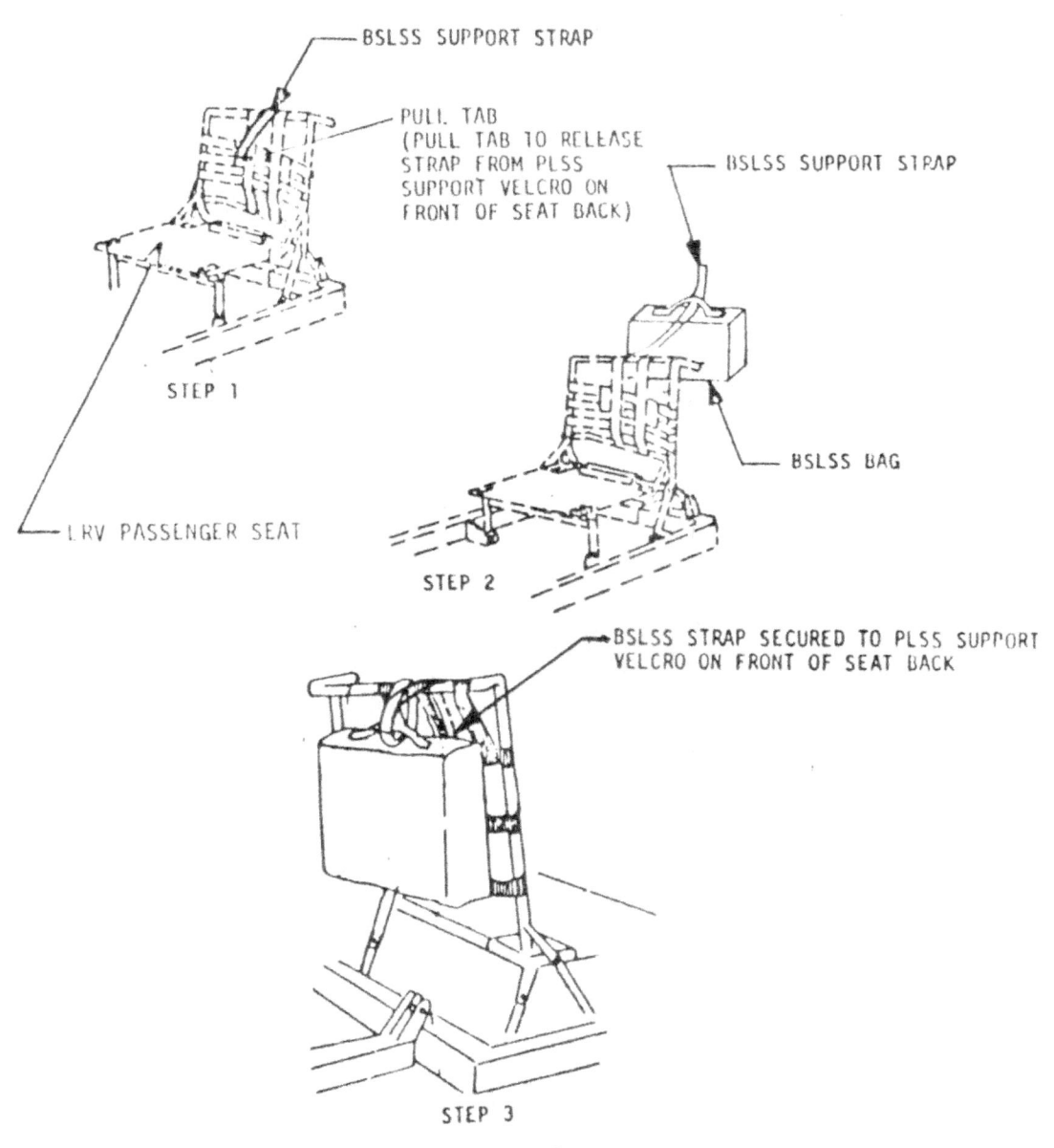

FIGURE 4-7. BUDDY SLSS INSTALLATION

LS006-002-2H
LUNAR ROVING VEHICLE
OPERATIONS HANDBOOK

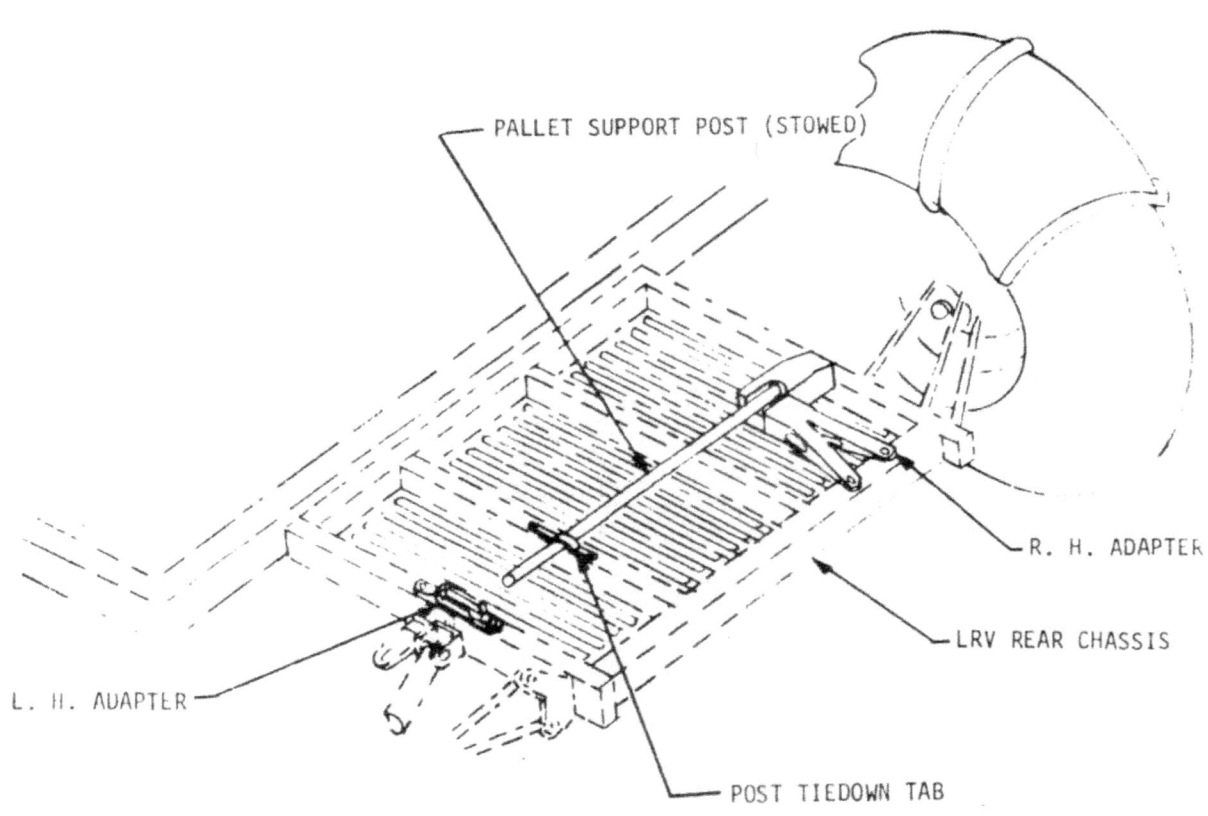

FIGURE 4-8 LRV REAR PAYLOAD PALLET ADAPTERS

LS006-002-2H
LUNAR ROVING VEHICLE
OPERATIONS HANDBOOK

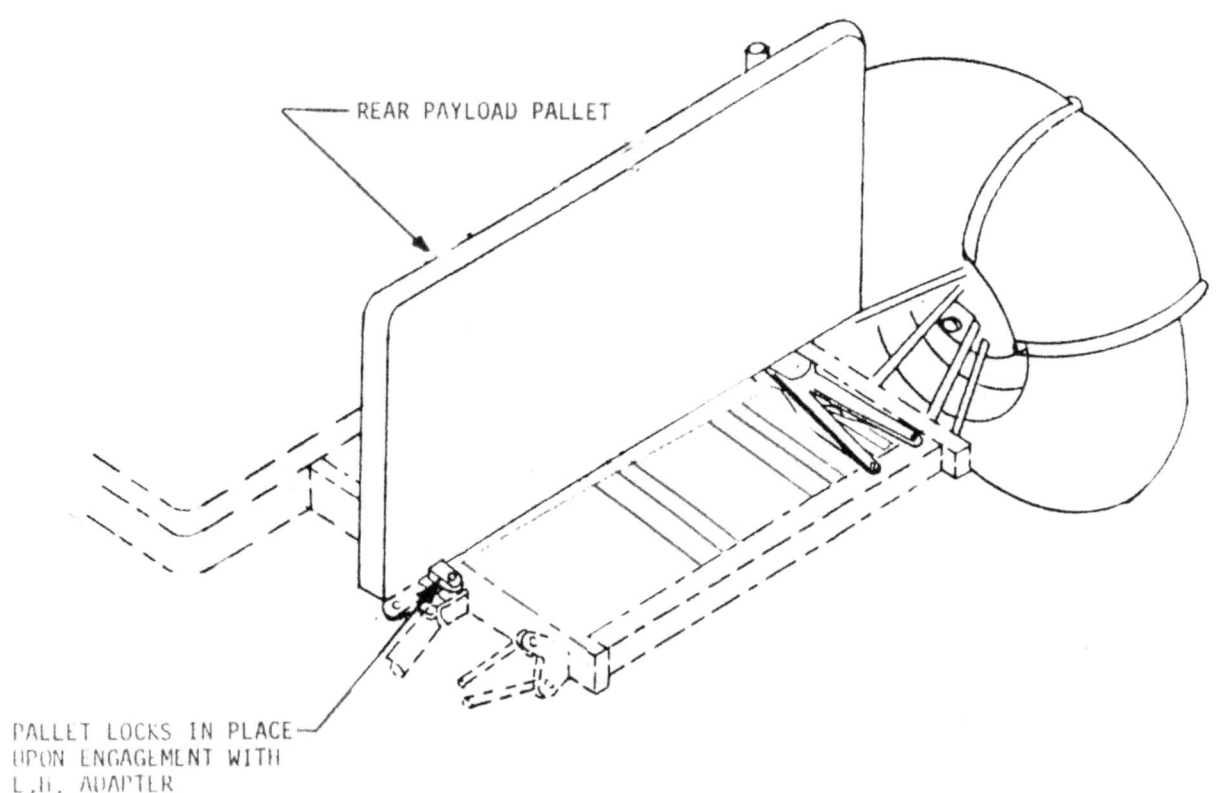

FIGURE 4-9 REAR PAYLOAD PALLET INSTALLED

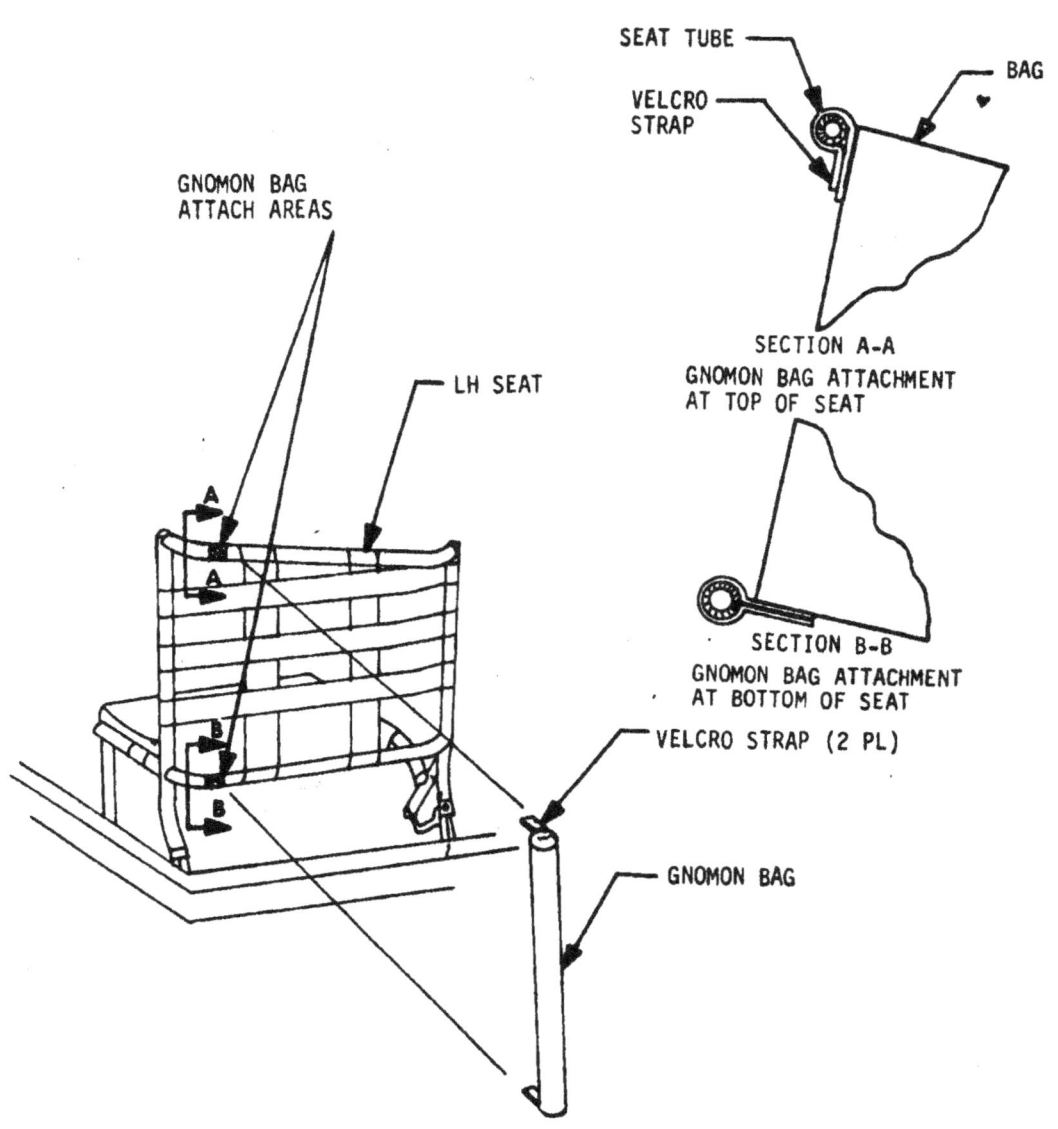

FIGURE 4-10 GNOMON BAG ATTACHMENT TO LRV

4.3   (Continued)

1G Trainer Note

The rear chassis payload adapters for the 1G Trainer will allow identical astronaut functions to be performed, but the configuration of the adapters is not exactly identical to the LRV adapters.

LS006-002-2H
LUNAR ROVING VEHICLE
OPERATIONS HANDBOOK

SECTION 5

OPERATING LIMITATIONS

5.0   INTRODUCTION

This section contains the LRV operating limitations.

5.1   PAYLOAD LIMITATIONS

The LRV is designed for lunar operation a total payload of 970 pounds earth weight distributed as defined by the LRV to Stowed Payload Interface Control Document, 13M07391. Loading the LRV beyond the 970 pound limit will cause the structural factor of safety to be lessened below the 1.5 design case.

The allowable center of gravity location for the total LRV, including payload, is shown in Figure 5-1. Loading the LRV such that the center of gravity falls outside the defined envelope will cause degradation of performance, including:

a. Possible steering discontinuity
b. Possible traction drive discontinuity
c. Possible periods of instability

1G TRAINER NOTE

The 1G Trainer is designed for a gross payload of 800 pounds. Performance degradation will occur if overloaded.

5.2   PARKING LIMITATIONS

To achieve proper thermal control of the LRV and stowed payload during between-EVA parking periods, the LRV must be oriented per figure 5-2. Parking the LRV outside these limits will result in display and control component overheating or LCRU overheating. There are no orientation constraints imposed on short-term parking during EVA's.

1G TRAINER NOTE

There are no parking limitations for the 1G Trainer.

5.3   SORTIE LIMITATIONS

The LRV is designed for EVA's of 6 hours duration. The thermal design is based on 3 hours of mobility operation in the 6-hour EVA, with the navigation system and controls and displays remaining on during the entire 6-hour EVA. Operation of the vehicle beyond these time durations will cause thermal limits to be exceeded.

The LRV is designed for continuous operation in shadows for not more than 2 hours due to temperature limits on the Control and Display components. See Appendix A for temperature rise and cooldown times.

LS006-002-2H
LUNAR ROVING VEHICLE
OPERATIONS HANDBOOK

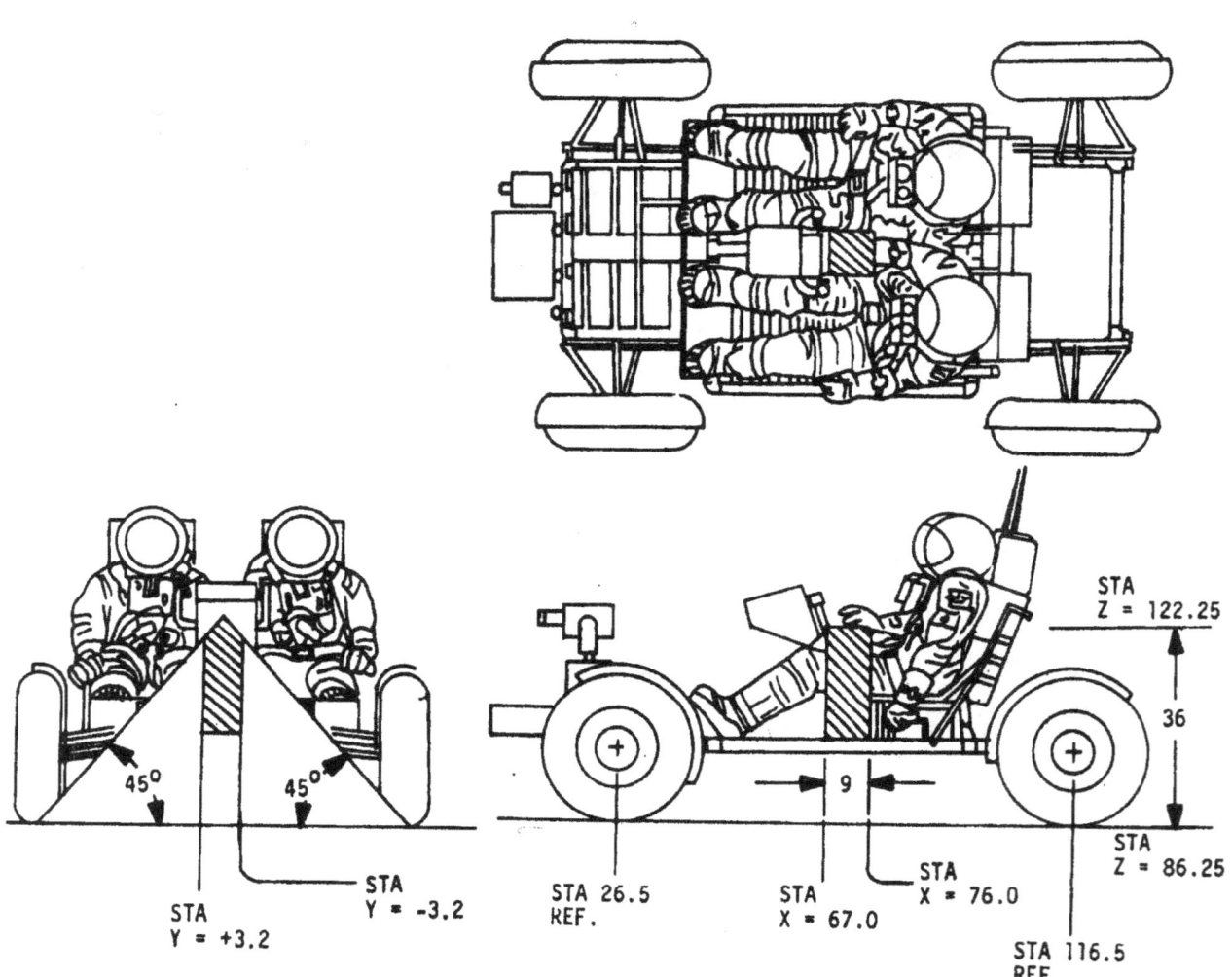

FIGURE 5-1. ALLOWABLE C.G. ENVELOPE FOR VEHICLE FULLY LOADED

LS006-002-2H
LUNAR ROVING VEHICLE
OPERATIONS HANDBOOK

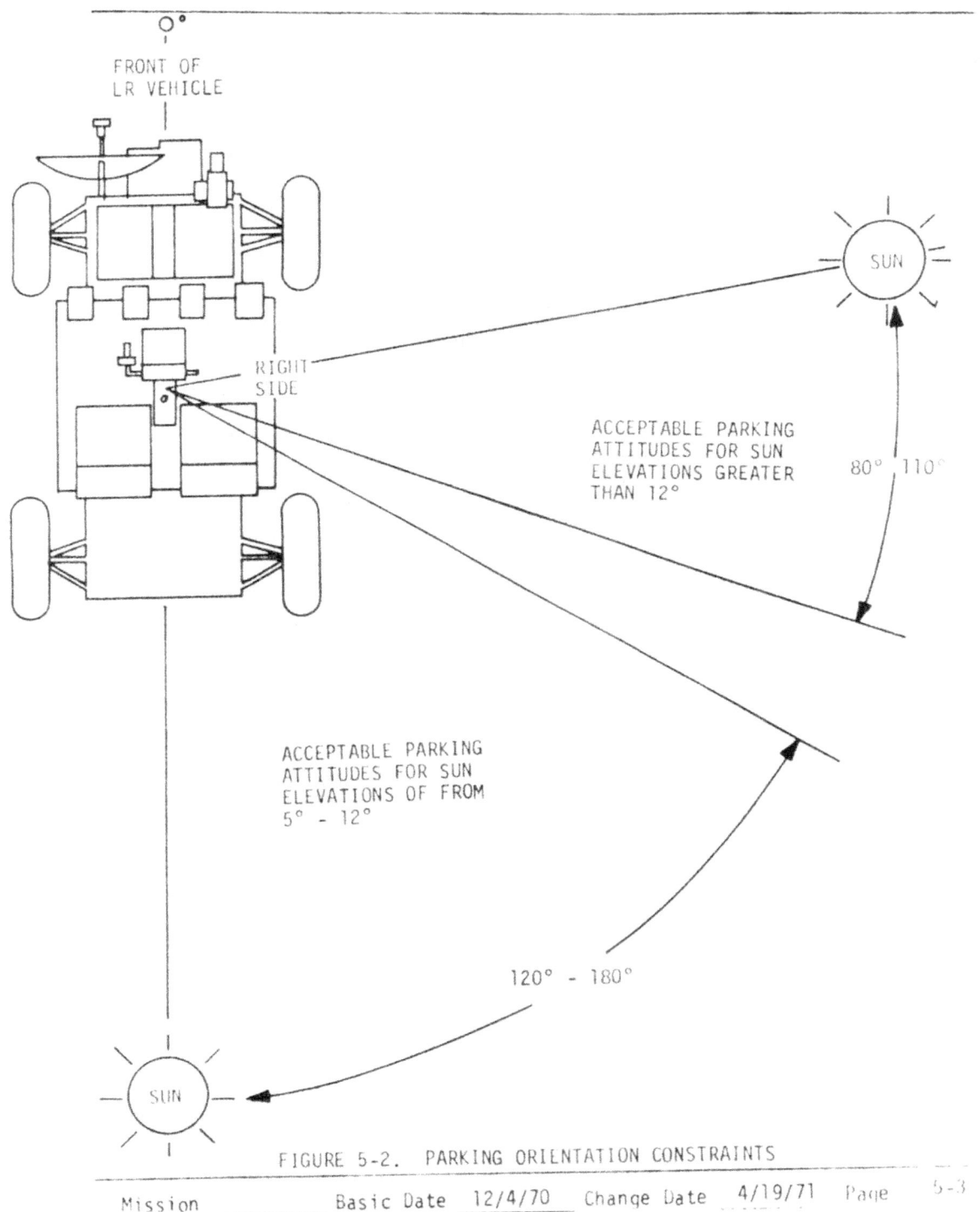

FIGURE 5-2. PARKING ORIENTATION CONSTRAINTS

LS006-002-2H
LUNAR ROVING VEHICLE
OPERATIONS HANDBOOK

5.4 NAVIGATION SYSTEM LIMITATIONS

The following limitations are placed on operating the LRV and 1G Trainer Navigation System.

a. The Navigation System is to be on for a minimum of three minutes before initialization to allow the gyro to reach operating speed.

b. The GYRO TORQUING Switch is not to be kept in the LEFT or RIGHT position for more than two minutes. After two minutes on, the switch must be kept OFF for a minimum of five minutes to prevent damage to the gyro torquing motor.

c. The navigation input voltage must not be allowed to be less than 30 VDC to prevent excessive computation and display errors and to prevent damage to navigation equipment if the under-voltage situation is prolonged. Therefore, it is imperative that the NAV POWER circuit breaker be open if the VOLTS indicator indicates less than 60. The VOLTS indicator should be checked periodically (at least each 15 minutes) to verify readings of not less than 60.

LS006-002-2H
LUNAR ROVING VEHICLE
OPERATIONS HANDBOOK

SECTION 6

OPERATING TIMELINES

6.0 INTRODUCTION

This section defines the approximate times for performing LRV functions on the lunar surface and 1G Trainer functions on earth.

1G TRAINER NOTE

Timelines for 1G Trainer operation are identical to those shown for LRV lunar surface operations except for traction drive decoupling and steering decoupling.

Included in this section are timelines for:

    Figure 6-1 - LRV Deployment
           6-2 - Post-Deployment Checkout
           6-3 - Pre-Sortie Checkout and Preparation
           6-4 - Post-Sortie Shutdown
           6-5 - Navigation Update
           6-6 - Traction Drive Decoupling
           6-7 - Steering Decoupling
           6-8 - Rear Steering Recoupling
           6-9 - 1G Trainer Battery Changeout
           6-10 - 1G Trainer Traction Drive Decoupling
           6-11 - 1G Trainer Steering Decoupling

LS006-002-2H
LUNAR ROVING VEHICLE
OPERATIONS HANDBOOK

- PULL VELCRO TAPE LOOSE FROM LEFT TRIPOD AND REMOVE INSULATION FROM AROUND LEFT LOWER SUPPORT ARM. (.1 MIN)
  - RELEASE CENTER BRAKED REEL DEPLOYMENT TAPE FROM NYLON BAG ATTACHED TO LEFT LOWER SUPPORT ARM; DRAPE TAPE OVER LM LANDING STRUT. (.2 MIN)
    - RELEASE DEPLOYMENT CABLE; DEPLOY FULL LENGTH OF CABLE AT 45° ANGLE FROM QUAD I TOWARD DESCENT LADDER. (.2 MIN)
      - PULL VELCRO TAPE LOOSE FROM RIGHT TRIPOD AND REMOVE INSULATION FROM AROUND RIGHT LOWER SUPPORT ARM. PULL INSULATION BLANKET LOOSE FROM VELCRO ON LOWER END OF LRV CENTER CHASSIS. (.2 MIN)
        - RELEASE DOUBLE BRAKED REEL DEPLOYMENT TAPE FROM NYLON BAG ATTACHED TO RIGHT LOWER SUPPORT ARM; DRAPE TAPE OVER CONVENIENT PROTRUSION. (.2 MIN)
          - VISUALLY INSPECT BOTH LOWER SUPPORT ARMS. (.1 MIN)
            - ASCEND TO LM PLATFORM TO DEPLOY LRV; OTHER CREWMAN GRASPS DEPLOYMENT CABLE AND MONITOR DEPLOYMENT. (.2 MIN)
              - PULL D-HANDLE TO RELEASE LRV FROM LM AND VISUALLY VERIFY THAT LRV MOVED OUTWARD FROM LM ABOUT 4°. (.2 MIN)
                - CREWMAN DESCENDS LM LADDER AND RETRIVES DOUBLE BRAKED REEL TAPE AT RIGHT SIDE OF VEHICLE. (.2 MIN)
                  - PULL TAPE FROM DOUBLE BRAKED REEL; VERIFY THAT LRV ROTATES OUTWARD FROM LM AND CAM FITTING ENGAGE LOWER SUPPORT ARMS. CONTINUE PULLING TAPE UNTIL AFT CHASSIS UNFOLDS ABOUT 45°; THEN VERIFY THAT AFT CHASSIS IS LOCKED INTO POSITION, THAT REAR WHEELS HAVE UNFOLDED AND THE TETHERED WHEEL STRUTS HAVE FALLEN FREE. ALSO VERIFY THAT THE FORWARD CHASSIS IS RELEASED FROM THE CONSOLE POST AND HAS RETURNED TO THE 35° POSITION. (.5 MIN)

FIGURE 6-1   LRV DEPLOYMENT TIMELINE

Mission __J__   Basic Date __12/4/70__   Change Date __4/19/71__   Page __6-2__

LS006-002-2H
LUNAR ROVING VEHICLE
OPERATIONS HANDBOOK

- CONTINUE PULLING TAPE FROM DOUBLE BRAKED REEL UNTIL AFT WHEELS CONTACT (2.0 MIN) LUNAR SURFACE. VERIFY THAT THE AFT WHEELS SLIDE ALONG THE LUNAR SURFACE, AS THE CREWMAN PULLS THE TAPE, UNTIL THE CENTER CHASSIS LIFTS OFF THE LOWER STRUT ARMS AND THE FORWARD CHASSIS UNFOLDS AND LOCKS INTO POSITION. VERIFY THAT THE FORWARD WHEELS HAVE UNFOLDED AND THE TETHERED WHEEL STRUTS HAVE FALLEN FREE. CONTINUE TO PULL THE TAPE UNTIL THE DOUBLE BRAKED REEL CABLES ARE SLACK.
- STOW DOUBLE BRAKED REEL TAPE OUT OF WORK AREA; RELEASE OUTER BRAKED (.2 MIN) REEL CABLE AT RR OF CENTER CHASSIS AND STOW OUT OF WORK AREA.
- RELEASE OUTER BRAKED REEL CABLE AT LR OF CENTER CHASSIS AND STOW (.1 MIN) OUT OF WORK AREA.
- RETRIEVE AND PULL TAPE OF CENTER BRAKED REEL UNTIL THE FORWARD (2.0 MIN) END OF THE LRV CONTACTS THE LUNAR SURFACE. STOW TAPE FROM CENTER BRAKED REEL OUT OF WORK AREA.
- OTHER CREWMAN TO RELEASE DEPLOYMENT CABLE AND STOW OUT OF (.2 MIN) WORK AREA.
- RETURN TO RF HINGE LATCH OF LRV; VERIFY RIGHT FRONT (.2 MIN) HINGE LATCHED.
- DEPLOY RF FENDER EXTENSION. (.1 MIN)
- REMOVE PINS 9 AND 10 AND TOSS CLEAR OF WORK AREA. (.2 MIN)
- GRASP APEX OF TRIPOD WITH RIGHT HAND AND PULL PIN 11. (.1 MIN)
- DISCARD TRIPOD MAIN MEMBERS CLEAR OF DEPLOYMENT AREA. (.1 MIN)
- GRASP RIGHT TRIPOD CENTER MEMBER IN RIGHT HAND (.2 MIN) PULL PIN 12 AND DISCARD.
- INSERT RIGHT TRIPOD CENTER MEMBER IN FRAME FOR (.1 MIN) TOEHOLD.

FIGURE 6-1 LRV DEPLOYMENT TIMELINE (CONTINUED)

Mission __J__ Basic Date __12/4/70__ Change Date __4/19/71__ Page __6-3__

LS006-002-2H
LUNAR ROVING VEHICLE
OPERATIONS HANDBOOK

- RELEASE RIGHT FOOTREST RESTRAINT AND ERECT RIGHT FOOTREST AND VERIFY LATCHED IN POSITION. (.1 MIN)

- PULL AND TURN C/D CONSOLE LATCH P13, 90° CW. WHEN CONSOLE DEPLOYS, RAISE INBOARD HANDHOLDS AND LOCK IN OPERATIONAL POSITION, ROTATE P13 90° CW AND VERIFY CONSOLE LOCKED IN POSITION (.1 MIN)

- RELEASE RIGHT SEAT BELT FROM STOWAGE POSITION AND STOW IN TEMPORARY LOCATION. (.1 MIN)

- GRASP FRONT OF RIGHT SEAT FRAME AND LIFT TO STABLE OVERCENTER POSITION AND ERECT RIGHT SEAT FRONT LEGS. VERIFY SEAT STOWAGE BAG ERECTS. (.1 MIN)

- PULL SEAT PAN FRAME UP AND POSITION ENDS OF FRAME UNDER BACK REST SUPPORT MEMBER LOWER SEAT FRAME TO ENGAGE FRONT LEGS AND VERIFY LATCHED. (.1 MIN)

- VERIFY RIGHT REAR HINGE LATCHED. (.1 MIN)

- VISUALLY VERIFY REAR STEERING DECOUPLING RING SEAL HAS NOT BEEN BROKEN. (.1 MIN)

- DEPLOY RIGHT REAR FENDER EXTENSION. (.1 MIN)

- VERIFY LEFT REAR HINGE LATCHED AND DEPLOY LEFT REAR FENDER EXTENSION. (.2 MIN)

- RELEASE INBOARD HANDHOLD VELCRO TIEDOWN STRAP. (.1 MIN)

- RELEASE LEFT SEAT BELT FROM STOWAGE POSITION AND STOW IN TEMPORARY LOCATION. (.1 MIN)

- GRASP FRONT OF LEFT SEAT FRAME AND LIFT TO STABLE OVERCENTER POSITION AND ERECT LEFT SEAT FRONT LEGS. VERIFY SEAT STOWAGE BAG ERECTS. (.1 MIN)

FIGURE 6-1   LRV DEPLOYMENT TIMELINE (CONTINUED)

Mission   J   Basic Date   12/4/70   Change Date   4/19/71   Page   6-4

LS006-002-2H
LUNAR ROVING VEHICLE
OPERATIONS HANDBOOK

- PULL SEAT PAN FRAME UP AND POSITION ENDS OF FRAME UNDER BACK REST SUPPORT MEMBER. LOWER SEAT FRAME TO ENGAGE FRONT LEGS AND VERIFY LATCHED. (.1 MIN)

- FOLD INBOARD ARMREST DOWN. (.1 MIN)

- SUPPORT CONSOLE WITH LEFT HAND; GRASP C/D CONSOLE PANEL "T" HANDLE P7, TURN 90° CW TO LOCKED POSITION. (.1 MIN)

- ROTATE INBOARD HANDHOLD TO LOCKED POSITION WHILE ROTATING CONSOLE DOWNWARD WITH LEFT HAND. (.1 MIN)

- ROTATE T-HANDLE P7 90° CW AND FOLD FLUSH WITH CONSOLE BOX AND SECURE INTO POSITION WITH VELCRO STRAP. (.1 MIN)

- REMOVE ATTITUDE INDICATOR LOCK PIN AND DISCARD. (.1 MIN)

- REMOVE C&W FLAG LOCK PIN AND DISCARD. (.1 MIN)

- PULL PINS P3 AND 4 AND TOSS CLEAR OF WORK AREA. (.1 MIN)

- GRASP TRIPOD APEX WITH LEFT HAND AND PULL PIN P5; DISCARD PIN P5 AND TRIPOD MAIN MEMBERS. (.1 MIN)

- GRASP TRIPOD CENTER MEMBER IN LEFT HAND; PULL PIN P6 AND DISCARD. (.2 MIN)

- USE HOOKED END OF SHORT TRIPOD MEMBER TO PULL CABLE P2 AND VERIFY THAT TELESCOPING RODS SADDLE FALLS AWAY. (.1 MIN)

- INSERT TRIPOD CENTER MEMBER IN LRV FRAME FOR TOEHOLD. (.1 MIN)

- RELEASE LEFT FOOTREST RESTRAINT DEVICE AND ERECT LEFT FOOTREST AND VERIFY LATCHED POSITION. (.1 MIN)

FIGURE 6-1   LRV DEPLOYMENT TIMELINE (CONTINUED)

Mission __J__   Basic Date __12/4/70__ Change Date __4/19/71__ Page __6-5__

LS006-002-2H
LUNAR ROVING VEHICLE
OPERATIONS HANDBOOK

- MOVE TO LEFT FRONT HINGE AREA AND VERIFY LEFT FRONT HINGE LATCHED. (.1 MIN)

- DEPLOY LEFT FRONT FENDER EXTENSION. (.1 MIN)

- INSPECT AND VERIFY BATTERY NO. 1 AND SPU DUST COVERS ARE CLOSED AND SECURED. (.1 MIN)

- VERIFY THAT THE FORWARD STEERING DECOUPLING RING SEAL HAS NOT BEEN BROKEN. (.1 MIN)

- MOVE TO RIGHT SIDE OF VEHICLE AND VERIFY THAT BATTERY NO. 2 DUST COVER IS CLOSED AND SECURED. (.2 MIN)

- AT RIGHT SIDE OF VEHICLE ROTATE "T" HANDLE P13 90° CW AND FOLD "T" HANDLE FLUSH WITH CONSOLE BOX AND SECURE. (.2 MIN)

LRV DEPLOYMENT COMPLETE.

TOTAL TIME 11.0 MINUTES

FIGURE 6-1.   LRV DEPLOYMENT TIMELINE (CONTINUED)

Mission   J   Basic Date   12/4/70   Change Date   4/19/71   Page   6-6

LS006-002-2H
LUNAR ROVING VEHICLE
OPERATIONS HANDBOOK

- INGRESS LRV AND FASTEN SEAT BELT, AND VERIFY HAND CONTROLLER IN PARKING BRAKE POSITION. (.2 MIN)

- VERIFY SWITCHES AND CIRCUIT BREAKERS IN PRE-LAUNCH POSITIONS. (.2 MIN)

- CLOSE BUS A, BUS B, BUS C, BUS D CIRCUIT BREAKERS. (.1 MIN)

- SET BATTERY SWITCH TO VOLTS X 1/2 AND REPORT VOLTS INDICATED. (.1 MIN)

- SET BATTERY SWITCH TO AMPS, CLOSE THE + 15 DC PRIM CIRCUIT BREAKER, CLOSE STEERING FORWARD AND REAR CIRCUIT BREAKERS, CLOSE DRIVE POWER LF, RF, LR, RR CIRCUIT BREAKERS, SET FRONT DRIVE ENABLE SWITCHES TO PWM 2, SET REAR DRIVE ENABLE TO PWM 1, AND SET ± 15 DC SWITCHES TO SEC. (.3 MIN)

- SET STEERING FORWARD SWITCH TO BUS C, SET STEERING REAR SWITCH TO BUS B, SET FRONT DRIVE POWER SWITCHES TO BUS C, SET REAR DRIVE POWER SWITCHES TO BUS B. (.1 MIN)

- SET HAND CONTROLLER REVERSE INHIBIT SWITCH IN UP POSITION AND CALL OTHER CREWMAN TO DIRECT AND MONITOR BACKING OPERATIONS AND TO VERIFY THAT DRIVE MOTORS AND STEERING MOTORS ARE WORKING. (.1 MIN)

- RECEIVE CLEAR SIGNAL FROM OTHER CREWMAN, RELEASE PARKING BRAKE AND BACK LRV CLEAR OF LM, STOP LRV AND SET PARKING BRAKE. RESET REVERSE INHIBIT SWITCH TO DOWN POSITION. (.4 MIN)

- RELEASE PARKING BRAKE AND DRIVE FORWARD TO MESA PARKING AREA FOR EQUIPMENT LOADING. VERIFY THAT THE VEHICLE RESPONDS TO HAND CONTROLLER COMMANDS FOR LEFT AND RIGHT STEERING, SPEED CONTROL, BRAKING AND THAT ALL FOUR WHEELS ARE ROTATING (NOT SLIDING). (1.0 MIN)

- STOP LRV AND SET HAND CONTROLLER IN THE PARKING BRAKE POSITION; NEUTRAL THROTTLE. (.1 MIN)

FIGURE 6-2. LRV POST DEPLOYMENT CHECKOUT TIMELINE

Mission J   Basic Date 12/4/70   Change Date 4/19/71   Page 6-7

LS006-002-2H
LUNAR ROVING VEHICLE
OPERATIONS HANDBOOK

▶ SHUT DOWN LRV POWER AND VERIFY THAT THE HAND CONTROLLER IS IN THE PARKING BRAKE POSITION, THAT THE FRONT AND REAR DRIVE POWER SWITCHES ARE OFF, THAT (.2 MIN) THE FORWARD AND REAR STEERING SWITCHES ARE OFF, THAT THE ± 15 VDC SWITCH IS OFF AND THAT THE NAV POWER CIRCUIT BREAKER IS OPEN.

▶ RELEASE AND STOW SEAT BELT; EGRESS VEHICLE. (.3 MIN)

(TOTAL TIME 3.1 MIN.)

FIGURE 6-2. LRV POST DEPLOYMENT CHECKOUT TIMELINE
(CONTINUED)

Mission __J__ Basic Date __12/4/70__ Change Date __4/19/71__ Page __6-8__

LS006-002-2H
LUNAR ROVING VEHICLE
OPERATIONS HANDBOOK

- PERFORM VISUAL INSPECTION TO VERIFY THAT THE BATTERY AND SPU DUST COVERS ARE CLOSED. (.1 MIN)
  - LOAD LRV WITH EQUIPMENT SELECTED FOR SORTIE. (TIME TO BE SUPPLIED BY NASA/MSC)
    - LRV DRIVER INGRESS LRV LEFT SEAT AND FASTEN SEAT BELT; VERIFY BRAKE SET. (.3 MIN)
      - OTHER CREWMAN INGRESS LRV RIGHT SEAT AND FASTEN SEAT BELT. (.3 MIN)
        - ACTIVATE LRV ELECTRICAL SYSTEM. (.4 MIN)
          - REPORT AMP-HR INDICATION FOR EACH BATTERY. (.1 MIN)
            - REPORT AMPS INDICATION FOR EACH BATTERY. (.1 MIN)
              - SET BATTERY SWITCH TO VOLTS x 1/2, REPORT VOLTS INDICATION FOR EACH BATTERY AND RETURN BATTERY SWITCH TO AMPS SETTING. (.1 MIN)
                - RELEASE PARKING BRAKE AND DRIVE TO LEVEL AREA NEAR THE LM. (.5 MIN)
                  - DEPLOY SUN SHADOW DEVICE AND VEHICLE ATTITUDE INDICATOR TO READ ROLL. PARK WITHIN 3° OF DOWN SUN (PER SSD), AND LEVEL WITH $\pm 6°$ ROLL, THEN SET BRAKE. (.2 MIN)
                    - REPORT THE SUN AZIMUTH ANGLE, AND PITCH AND ROLL ANGLES. FOLD SSD. (.2 MIN)
                      - PULL SYSTEM RESET SWITCH FROM DETENT AND MOVE TO RESET POSITION. RETURN TO OFF. (.1 MIN)

FIGURE 6-3. PRE-SORTIE CHECKOUT AND PREPARATION TIMELINE

Mission   J   Basic Date 12/4/70 Change Date 4/19/71 Page 6-9

LS006-002-2H
LUNAR ROVING VEHICLE
OPERATIONS HANDBOOK

- FOLD VEHICLE ATTITUDE INDICATOR TO DRIVE POSITION. (.1 MIN)

- VERIFY THAT BEARING, DISTANCE, AND RANGE INDICATORS ARE ZERO; THEN RETURN SYSTEM RESET SWITCH TO OFF POSITION. (.1 MIN)

- RECEIVE CORRECTED HEADING FROM MCC, PULL GYRO TORQUING SWITCH FROM DETENT AND OPERATE TO CORRECT HEADING INDICATION; THEN TURN GYRO TORQUING SWITCH TO OFF. (.8 MIN)

- REPORT BATTERY AND DRIVE MOTOR TEMPERATURES. (.1 MIN)

- REPORT BATTERY CURRENT WITH VEHICLE IN MOTION. (.1 MIN)

- ADVISE MCC THAT PRE-SORTIE PREPARATION AND CHECKOUT IS COMPLETE; REQUEST CLEARANCE FOR SORTIE. (.1 MIN)

(TOTAL TIME 3.7 MIN. PLUS
EQUIPMENT LOADING TIME)

FIGURE 6-3. PRE-SORTIE CHECKOUT AND PREPARATION TIMELINE
(CONTINUED)

Mission ___J___ Basic Date __12/4/70__ Change Date __4/19/71__ Page __6-10__

LS006-002-2H
LUNAR ROVING VEHICLE
OPERATIONS HANDBOOK

- STOP LRV IN SELECTED PARKING AREA AND PLACE HAND CONTROLLER IN PARKING BRAKE POSITION, THROTTLE IN NEUTRAL. (.2 MIN)
- REPORT BEARING, DISTANCE AND RANGE READINGS TO MCC. (.2 MIN)
- REPORT AMP-HR AND VOLTS INDICATIONS FOR EACH BATTERY; SET BATTERY SWITCH TO AMPS SETTING. (.2 MIN)
- REPORT BATTERY AND DRIVE MOTOR TEMPERATURES. (.2 MIN)
- SHUTDOWN LRV POWER. (.3 MIN)
- RELEASE SEAT BELT, STOW SEAT BELT, EGRESS LRV. (.4 MIN)
- ALIGN HIGH GAIN ANTENNA. (.2 MIN)
- SET LCRU MODE SWITCH TO TV RMT. (.1 MIN)
- OPEN LRV BATTERY AND SPU DUST COVERS. (.3 MIN)

TOTAL TIME 2.1 MIN

FIGURE 6-4. POST-SORTIE SHUTDOWN TIMELINE

Mission J Basic Date 12/4/70 Change Date 4/19/71 Page 6-11

LS006-002-2H
LUNAR ROVING VEHICLE
OPERATIONS HANDBOOK

- SELECT A LEVEL AREA AND PARK THE LRV HEADING DOWN SUN, AND SET PARKING BRAKE (PITCH LESS THAN $\pm 6°$, DOWN SUN WITHIN $\pm 3°$). (.2 MIN)
- DEPLOY ROLL ANGLE DEVICE AND NOTE READING (ROLL LESS THAN 6°). (.1 MIN)
- RELAY LRV HEADING AND ROLL ANGLE TO MCC. (.2 MIN)
- RECEIVE CORRECTED HEADING FROM MCC. (1 MIN)
- IF INDICATED HEADING DIFFERS FROM CALCULATED BY GREATER THAN 2°, PULL GYRO TORQUING TOGGLE FROM DETENT AND SLEW HEADING AS REQUIRED. RETURN GYRO TOGGLE TO OFF POSITION. (.2 MIN)
- NOTIFY MSFN "LRV READY TO RESUME SORTIE". (.1 MIN)

TOTAL TIME 1.8 MIN

FIGURE 6-5. NAVIGATION UPDATE TIMELINE

LS006-002-2H
LUNAR ROVING VEHICLE
OPERATIONS HANDBOOK

- STOP LRV, SET BRAKE. (.1 MIN)
- RELEASE AND STOW SEAT BELT.
- EGRESS VEHICLE. (.1 MIN)
- REMOVE TOEHOLD. (.1 MIN)
- MOVE TO DEFECTIVE WHEEL DRIVE UNIT. (.2 MIN)
- ENGAGE CONTINGENCY TOOL THROUGH EITHER WHEEL DECOUPLING HOOK. (.1 MIN)
- PULL OUT TO LIMIT OF HOOK TRAVEL AND ROTATE 90° COUNTERCLOCK- WISE. (.1 MIN)
- ENGAGE CONTINGENCY TOOL IN OTHER DECOUPLING HOOK, PULL OUT AND ROTATE. (.1 MIN)
- RE-INSTALL LEFT SIDE TOEHOLD. (.2 MIN)

1G TRAINER NOTE

TIMES ARE IDENTICAL FOR 1G TRAINER SIMULATED DECOUPLING. (SEE FIGURE 6-9 FOR ACTUAL TIME REQUIRED FOR TECHNICIAN TO EFFECT DECOUPLING)

TOTAL TIME (1.0 MIN)

FIGURE 6-6. LRV TRACTION DRIVE DECOUPLING TIMELINE (CONTINGENCY OPERATION)

LS006-002-2H
LUNAR ROVING VEHICLE
OPERATIONS HANDBOOK

STOP LRV AND SET BRAKE. TURN DRIVE POWER AND STEERING OFF	(.2 MIN)
RELEASE AND STOW SEAT BELT.	(.1 MIN)
EGRESS VEHICLE.	(.2 MIN)
MOVE TO LEFT FRONT OR RIGHT REAR DECOUPLING RING LOCATION AS APPROPRIATE.	(.2 MIN)
PULL STEERING DECOUPLING RING AS APPROPRIATE.	(.1 MIN)
MANUALLY STRAIGHTEN WHEELS OF DEFECTIVE DRIVE SYSTEM. MOVEMENT OF WHEEL MAY BE NECESSARY FOR DECOUPLING	(.3 MIN)

1G TRAINER NOTE

TIMES ARE IDENTICAL FOR 1G TRAINER
SIMULATED DECOUPLING. (SEE FIGURE
6-10 FOR ACTUAL TIME REQUIRED FOR
TECHNICIAN TO EFFECT DECOUPLING)

TOTAL TIME 1.1 MIN

FIGURE 6-7. LRV STEERING DECOUPLING TIMELINE
(CONTINGENCY OPERATION)

LS006-002-2H
LUNAR ROVING VEHICLE
OPERATIONS HANDBOOK

- STOP LRV AND SET BRAKE. TURN DRIVE POWER AND STEERING OFF (.2 MIN)
  - RELEASE AND STOW SEAT BELT (.1 MIN)
    - EGRESS VEHICLE (.2 MIN)
      - MOVE TO AREA FRONT OF RIGHT REAR FENDER (.2 MIN)
        - RELEASE RECOUPLING TOOL TIEDOWN VELCRO STRAP AND REMOVE RECOUPLING TOOL FROM STOWAGE BLOCK (.1 MIN)
          - PULL TAB TO OPEN REAR STEERING SECTOR DUST COVER AND PULL DUST COVER BACK (.1 MIN)
            - PUSH BUTTON ON TOP OF SECTOR GEAR TO RE-ENGAGE GEAR (.1 MIN)
              - INSERT RECOUPLING TOOL IN TOP OF RAISED AREA ON SECTOR GEAR, ROTATE TOOL TO ENGAGE LOCK PIN AND LIFT TOOL UNTIL PIN LOCKS IN RAISED POSITION. REMOVE TOOL. (.2 MIN)
                - REPLACE STEERING SECTOR DUST COVER (.1 MIN)
                  - REPLACE RECOUPLING TOOL IN STOWAGE BLOCK AND SECURE WITH VELCRO STRAP. (.1 MIN)

1G Trainer Note

RECOUPLING OF THE REAR STEERING FOR THE 1G
TRAINER WILL BE PERFORMED BY A TECHNICIAN.

TOTAL TIME 1.4 MINUTES

FIGURE 6-8. LRV REAR STEERING RECOUPLING TIMELINE (CONTINGENCY OPERATION)

Mission J   Basic Date 12/4/70   Change Date 4/19/71   Page 6-15

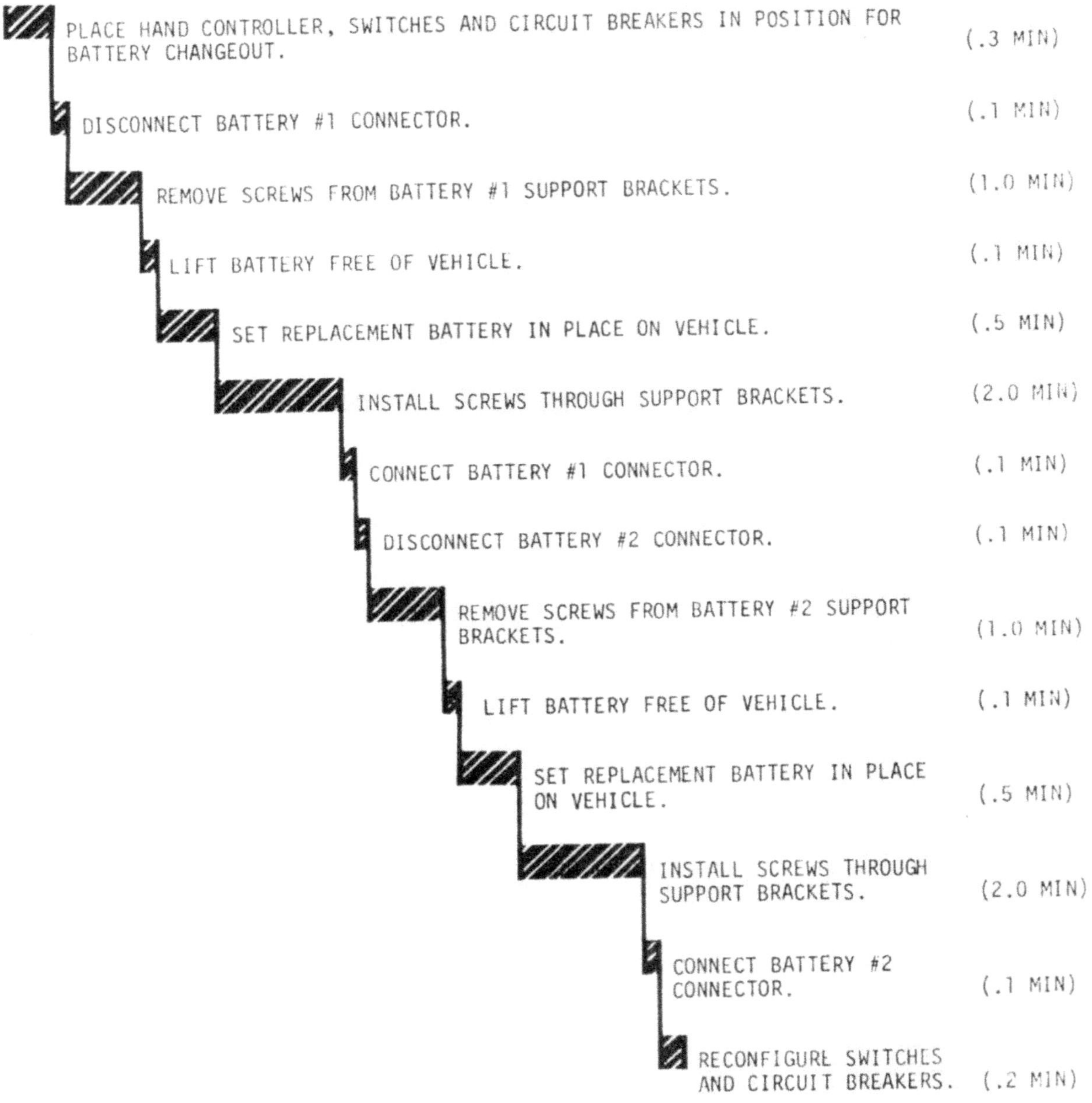

LS006-002-2H
LUNAR ROVING VEHICLE
OPERATIONS HANDBOOK

PLACE HAND CONTROLLER, SWITCHES AND CIRCUIT BREAKERS IN POSITION FOR BATTERY CHANGEOUT.	(.3 MIN)
DISCONNECT BATTERY #1 CONNECTOR.	(.1 MIN)
REMOVE SCREWS FROM BATTERY #1 SUPPORT BRACKETS.	(1.0 MIN)
LIFT BATTERY FREE OF VEHICLE.	(.1 MIN)
SET REPLACEMENT BATTERY IN PLACE ON VEHICLE.	(.5 MIN)
INSTALL SCREWS THROUGH SUPPORT BRACKETS.	(2.0 MIN)
CONNECT BATTERY #1 CONNECTOR.	(.1 MIN)
DISCONNECT BATTERY #2 CONNECTOR.	(.1 MIN)
REMOVE SCREWS FROM BATTERY #2 SUPPORT BRACKETS.	(1.0 MIN)
LIFT BATTERY FREE OF VEHICLE.	(.1 MIN)
SET REPLACEMENT BATTERY IN PLACE ON VEHICLE.	(.5 MIN)
INSTALL SCREWS THROUGH SUPPORT BRACKETS.	(2.0 MIN)
CONNECT BATTERY #2 CONNECTOR.	(.1 MIN)
RECONFIGURE SWITCHES AND CIRCUIT BREAKERS.	(.2 MIN)

TOTAL TIME 8.1 MIN

FIGURE 6-9. 1G TRAINER BATTERY CHANGEOUT

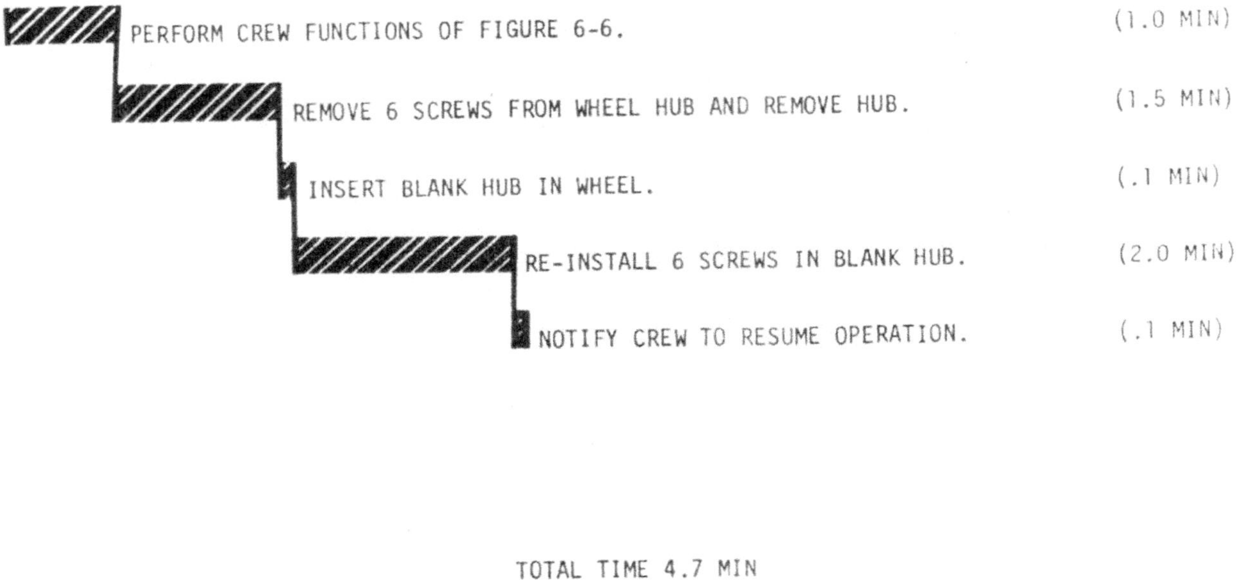

LS006-002-2H
LUNAR ROVING VEHICLE
OPERATIONS HANDBOOK

PERFORM CREW FUNCTIONS OF FIGURE 6-6.	(1.0 MIN)
REMOVE 6 SCREWS FROM WHEEL HUB AND REMOVE HUB.	(1.5 MIN)
INSERT BLANK HUB IN WHEEL.	(.1 MIN)
RE-INSTALL 6 SCREWS IN BLANK HUB.	(2.0 MIN)
NOTIFY CREW TO RESUME OPERATION.	(.1 MIN)

TOTAL TIME 4.7 MIN

FIGURE 6-10. 1G TRAINER TRACTION DRIVE DECOUPLING TIMELINE

Mission ___J___ Basic Date _12/4/70_ Change Date _4/19/71_ Page _6-17_

LS006-002-2H
LUNAR ROVING VEHICLE
OPERATIONS HANDBOOK

PERFORM CREW FUNCTIONS OF FIGURE 6-7. (.8 MIN)

INSTALL CLAMP ON LEFT STEERING ARM, BUTTED AGAINST CHASSIS FRAME. (1.0 MIN)

INSTALL CLAMP ON RIGHT STEERING ARM, BUTTED AGAINST CHASSIS FRAME. (1.0 MIN)

NOTIFY CREW TO RESUME OPERATION. (.1 MIN)

TOTAL TIME 2.9 MIN

FIGURE 6-11. 1G TRAINER STEERING DECOUPLING TIMELINE

Mission ___J___ Basic Date __12/4/70__ Change Date ___4/19/71___ Page __6-18__

LS006-002-2H
LUNAR ROVING VEHICLE
OPERATIONS HANDBOOK

SECTION 7
OPERATING PROFILES

7.1 LRV OPERATING PROFILE

7.1.1 Normal Operating Profile

The LRV is designed for nominal operation in accordance with the profile shown in figure 7-1 for a lunar surface stay time of 78 hours.

7.1.1.1 Sortie Profile

o The normal sortie profile consists of a maximum of 3 hours of driving time and 3 hours of station time, to total 6 hours per LRV sortie.

o The normal sortie will be accomplished with all four traction drives and both steering assemblies active.

o During the sortie, stops will be made at periodic intervals to update LRV navigation system. At these stops, the crew will report indicator readouts per 2.6.

NOTE
Navigation updates will be performed only if indicator HEADING differs from MCC Calculated heading by more than 2°.

o The LRV navigation system and console displays will remain energized throughout the six hour sortie duration.

o At stops exceeding five minutes, the power to the traction drive units, steering motors, and drive controller will be turned off (accomplished by placing STEERING FRONT and REAR switches OFF; LF, RF, LR, RR DRIVE POWER switches OFF, and ± 15VDC switch OFF).

o Circuit breakers and switch settings will be set to utilize both batteries at approximately the same rate, (e.g., steering for the front wheels powered from Battery No. 1 and steering for the rear wheels from Battery No. 2).

o Power from the auxiliary connector will be supplied only as required by the LCRU for special cases.

o Driving speed during the sortie will be varied at the crew's discretion. The speed profile for a sortie leg would consist of beginning at zero, accelerating to a desired driving speed, repeating cycles of decelerating to a slower speed, re-accelerating to the desired driving speed, and decelerating to stops. Speed will vary between zero and 14 km/hr.

o Operation of the LCRU high gain antenna and the television camera will be conducted only when the LRV is stopped. The low gain antenna will be manually oriented during LRV traverses.

LS006-002-2H
LUNAR ROVING VEHICLE
OPERATIONS HANDBOOK

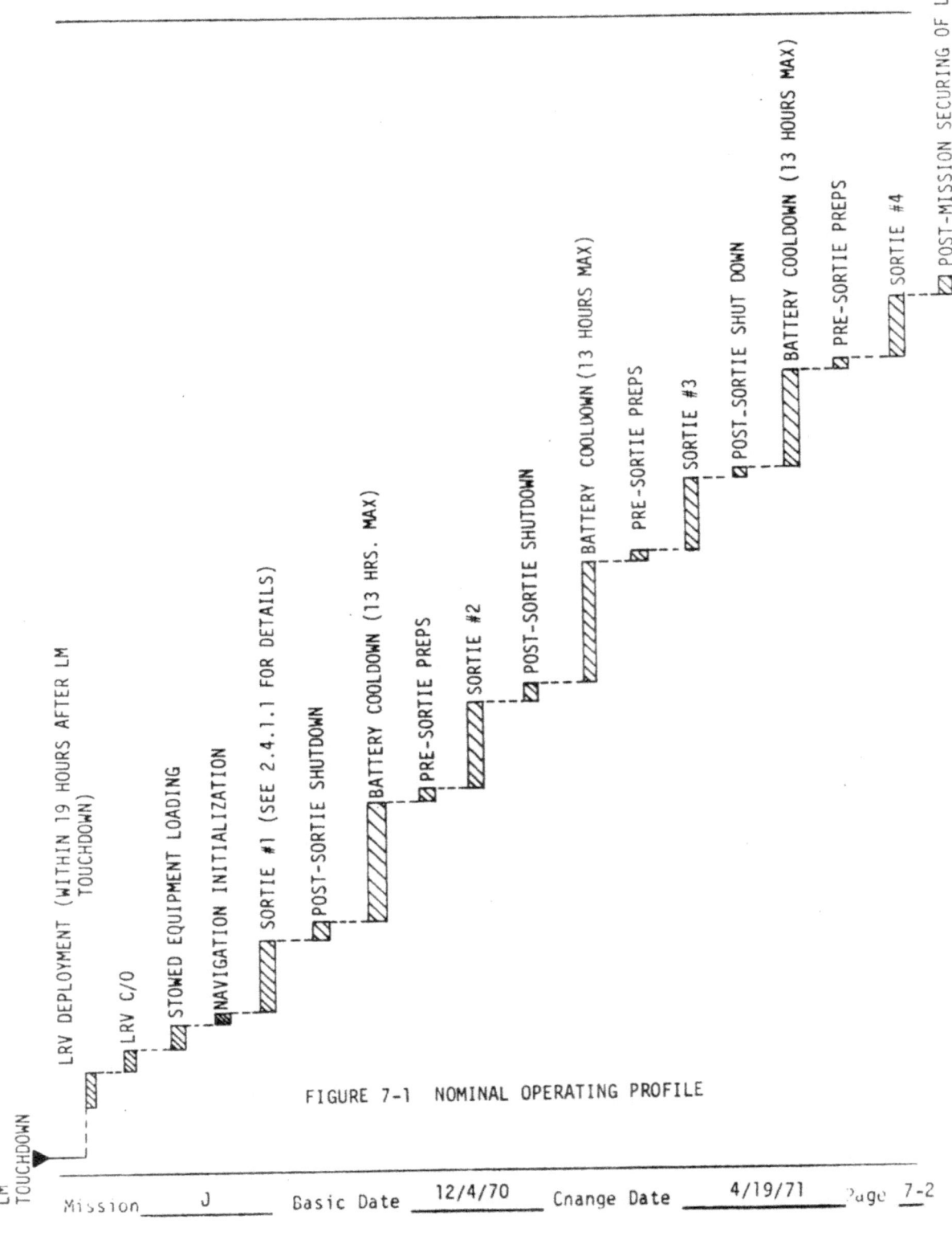

FIGURE 7-1 NOMINAL OPERATING PROFILE

LS006-002-2H
LUNAR ROVING VEHICLE
OPERATIONS HANDBOOK

7.1.1.1 (Continued)

o Science experiment equipment will be transported by the LRV only. Operation of the science equipment will be done only when the LRV is stopped.

7.1.2 Contingency Operating Profiles

Operating profiles for contingencies depend upon the specific contingency experienced. The following paragraphs contain operating profiles for a selected number of cases.

7.1.2.1 LRV Operating Profile with Failure of Traction Drive Units

The operating profile for a contingency caused by failure of a traction drive assembly (motor or harmonic drive) would be the same as that for normal operation with the following exceptions:

a. The specific traction drive would be uncoupled to allow "free wheeling" of that traction drive and drive power and control to that traction drive would be switched off. (Braking for the uncoupled wheel would be lost).

b. Speed/slope capability would be limited to that shown in Appendix A.

NOTE

1. If the right rear traction drive is decoupled, the speed indication on the console speedometer will not function.

2. If two traction drives are decoupled, the navigation odometer will not function.

7.1.2.2 LRV Operating Profile with Failure of One Steering Motor

The operating profile for a contingency caused by a failure of one steering motor would be the same as for the normal profile with the following exceptions:

a. Power to the disabled steering motor would be switched off.

b. The steering decoupling mechanism for the disabled steering motor would be activated by the crew. (This action is reversible for the rear steering motor only).

c. The minimum turning radius of the vehicle will be 6.2 meters as opposed to 3.1 meters with both steering motors operable.

7.1.2.3 LRV Operating Profile with Failure of One Battery

In the event of failure of one battery, all switches would be set to select the appropriate busses being supplied power from the remaining operable battery. For example, if Battery No. 1 fails, all control panel selections showing use of Bus A would be switched to select Bus C, since Bus C is supplied power from Battery No. 2 Each battery is capable of carrying the entire power load of the LRV.

LS006-002-2H
LUNAR ROVING VEHICLE
OPERATIONS HANDBOOK

7.1.2.3   (Continued)

The temperature rise in the remaining operable battery would be greater, however, causing the battery temperature to reach its upper limit in a shorter time as shown in Appendix A.

Range would be less, the specific amount depending on the point in the mission at which the failure occurs. Consult Appendix A for range capabilities.

7.2   1G TRAINER OPERATING PROFILE

7.2.1   Normal 1G Trainer Operating Profile

The 1G Trainer is designed for nominal operation in accordance with the profile shown in figure 7-2.

7.2.1.1

o Sortie begins with fully charged batteries installed in the vehicle and enough charged batteries or chase vehicle power available to support the duration and/or length of the sortie. (See Appendix B for battery capability).

o Science stops should be scheduled at points in the sortie corresponding timewise, to battery change-out requirements.

o Battery change-out will be accomplished without removing power to the navigation system when changeout must be done in mid-sortie.

o Sortie time is not limited by trainer capability except for battery life limitations, and sun angle if sun is used for gyro update.

NOTE

Pre-calibrated check points can be used in place of sun for heading reference. This option allows use of the 1G Trainer on cloudy days, etc. and does not require ephemeris data.

o The normal sortie will be accomplished with all four traction drives and both steering assemblies active.

o During the sortie, stops will be made at periodic intervals to update the navigation system. At these stops the crew will also report temperatures of traction drives and batteries and state of charge of each battery.

NOTE

Navigation updates will be performed only if indicated heading differs from calculated heading by more than 2°.

o The navigation system will remain energized throughout the sortie.

o Driving speed during the sortie will be varied at the crew's discretion. The speed profile for the sortie leg would begin at zero, accelerate and decelerate to avoid obstacles, maintain constant speed over very smooth surfaces,

Mission ___J___   Basic Date __12/4/70__   Change Date __4/19/71__   Page _7-4_

LS006-002-2H
LUNAR ROVING VEHICLE
OPERATIONS HANDBOOK

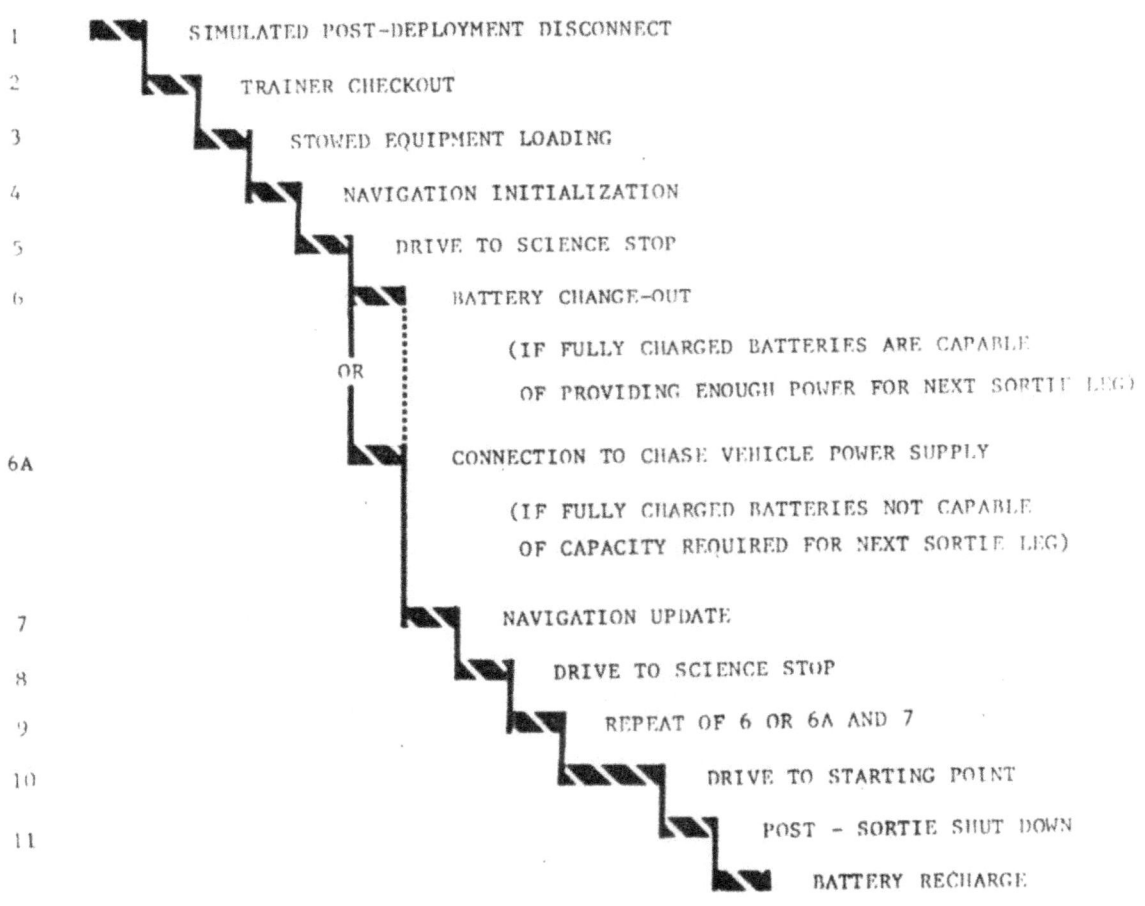

FIGURE 7-2. NOMINAL OPERATING PROFILE FOR 1G TRAINER

LS006-002-2H
LUNAR ROVING VEHICLE
OPERATIONS HANDBOOK

7.2.1.1 (Continued)

decelerate to observe geologic features, reaccelerate for driving and decelerating to a stop at science stations. Speed will vary between zero and 16 km/hr.

7.2.2 Contingency Operating Profiles

Operating profiles for contingencies depend upon the specific contingency experienced. The following paragraphs contain operating profiles for a selected number of cases.

7.2.2.1 Operating Profile with Failure of Traction Drive Unit

The operating profile for a contingency caused by failure of a traction drive assembly (motor or gear box) would be the same as that for normal operations with the following exceptions:

a. The specific traction drive would be uncoupled to allow free wheeling of that traction drive and drive power and control to that traction drive would be switched off.

NOTE

1. Decoupling of 1G Trainer traction drives requires a mechanic to physically do the decoupling. Simulated decoupling devices are provided for simulation of astronaut interface, but will not actually provide the decoupling. See Section 8.

2. If the right rear traction drive is decoupled, the speed indicator on the console will not function.

3. If two traction drives are decoupled, the navigation odometer will not function.

7.2.2.2 Operating Profile with Failure of One Steering Motor/Gear Reducer

The operating profile for a contingency caused by a failure of one steering motor would be the same as for the normal profile with the following exceptions:

a. Power to the disabled steering motor/gear reducer would be switched off.

b. The steering arms would be clamped by a technician.

NOTE

The simulated steering decoupling rings provided for crew interface will not effect steering decoupling.

c. Steering radius would be twice as great as with both motors operable. The driver should exercise greater caution when avoiding obstacles.

Mission J   Basic Date 12/4/70   Change Date 4/19/71   Page 7-6

LS006-002-2H
LUNAR ROVING VEHICLE
OPERATIONS HANDBOOK

SECTION 8

1G TRAINER NON-CREW
PROCEDURES

8.0 INTRODUCTION

This section contains procedures to be performed by personnel other than crew members in support of 1G Trainer Operations. Electrical block diagrams for the 1G Trainer are provided in Figures 8-1 through 8-8.

8.1 GENERAL PROCEDURES

8.1.1 Visual Inspection

Prior to, and at the conclusion of, each sortie or mission, visually inspect the vehicle for the following:

a. Finish or surface damage
b. Structural integrity of parent materials, welds, and other mechanical joints
c. Loose fasteners
d. Electrical cable abrasion, fraying, temperature damage, shorting, loose connector
e. Hydraulic line damage
f. Evidence of mechanical interference
g. Dust or debris in suspension and steering joints and bushings
h. Brake, battery, or shock absorber leakage
i. General configuration

CAUTION

Any discrepant item noted during this inspection must be corrected prior to vehicle operation.

8.1.2 General Repair

8.1.2.1 Finish or Surface Damage

All unpainted aluminum surfaces may be touched up if required by use of the processes specified in MIL-C-5541 Type I, Grade A or B, Class I. Information covering painted surfaces is carried on the appropriate piece-part drawing. Sharp nicks or any local surface deformation which could be a stress riser should be blended to the surrounding surface and refinished.

8.1.2.2 Structural Integrity

Loss of structural integrity or permanent deformation in the chassis, suspension, or any load bearing member should be considered grounds for immediate

LS006-002-2H
LUNAR ROVING VEHICLE
OPERATIONS HANDBOOK

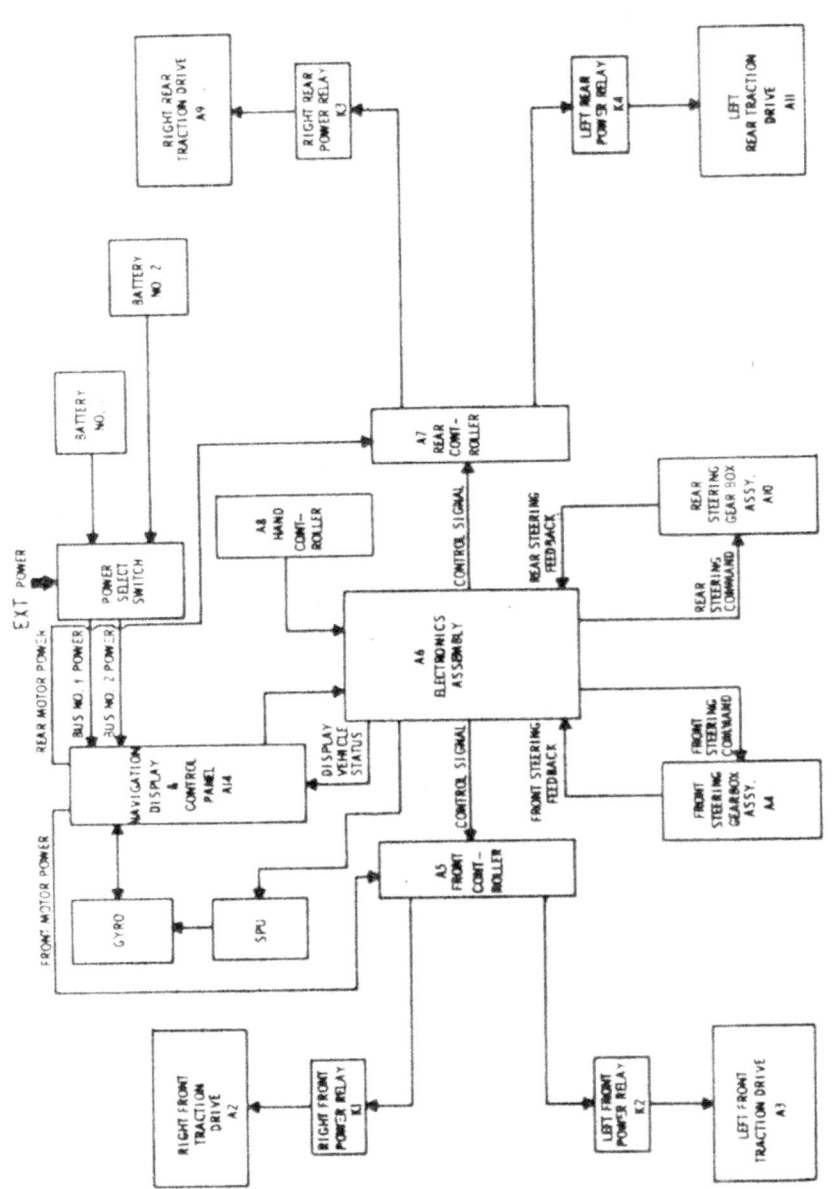

FIGURE 8-1

1G TRAINER BASIC VEHICLE BLOCK DIAGRAM

LS006-002-2H
LUNAR ROVING VEHICLE
OPERATIONS HANDBOOK

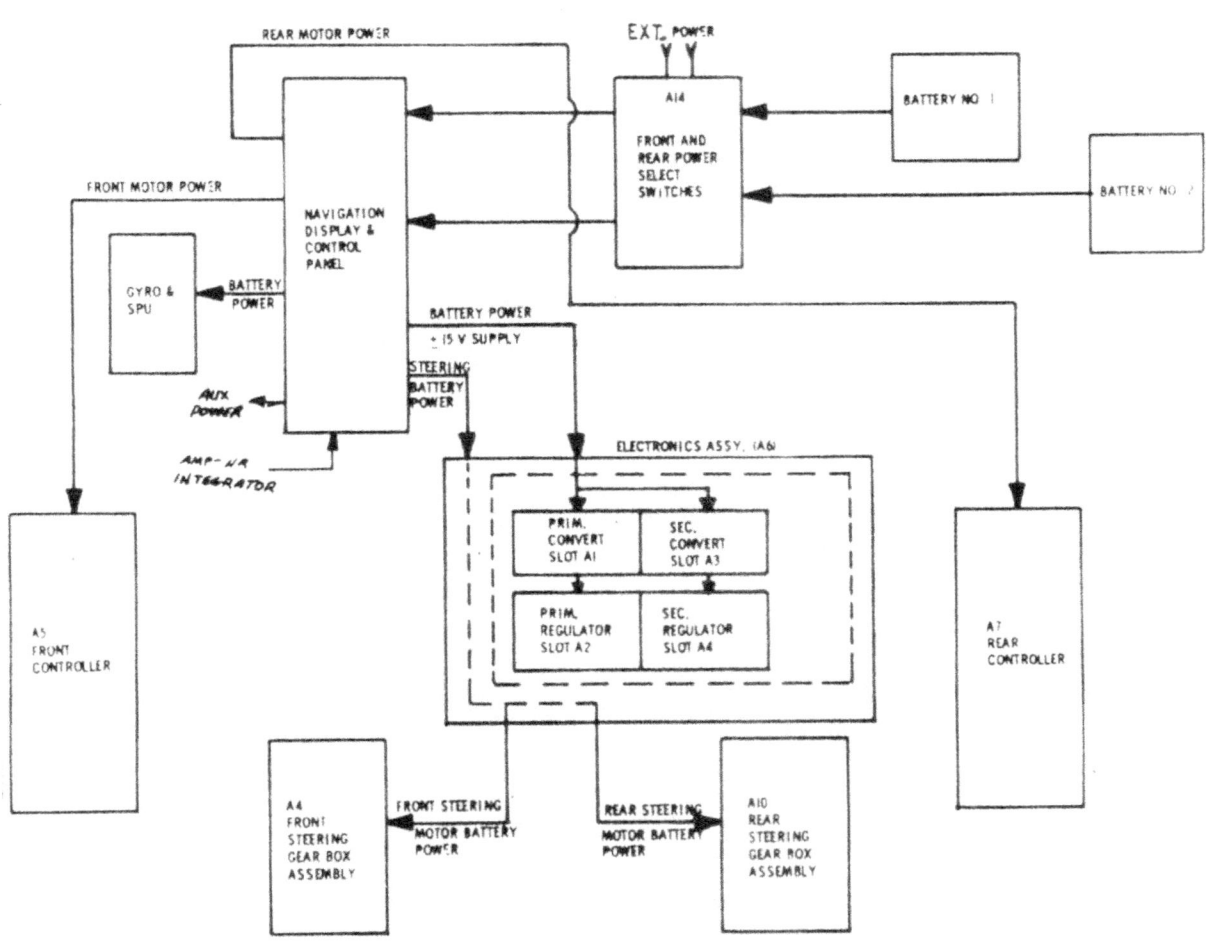

FIGURE 8-2   1G TRAINER VEHICLE POWER DISTRIBUTION BLOCK DIAGRAM

LS006-002-2H
LUNAR ROVING VEHICLE
OPERATIONS HANDBOOK

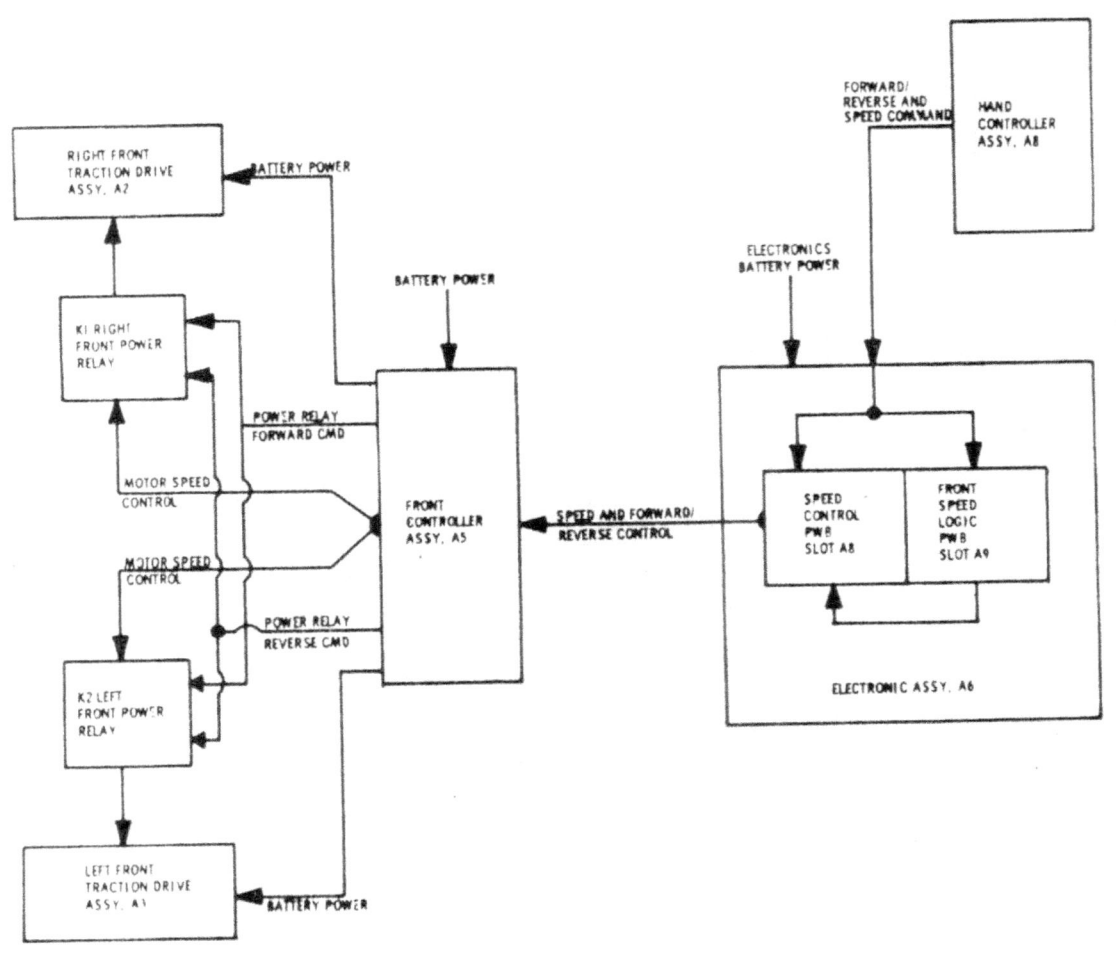

FIGURE 8-3

1G TRAINER VEHICLE FRONT TRACTION DRIVE
ELECTRICAL SIGNAL ROUTING BLOCK DIAGRAM

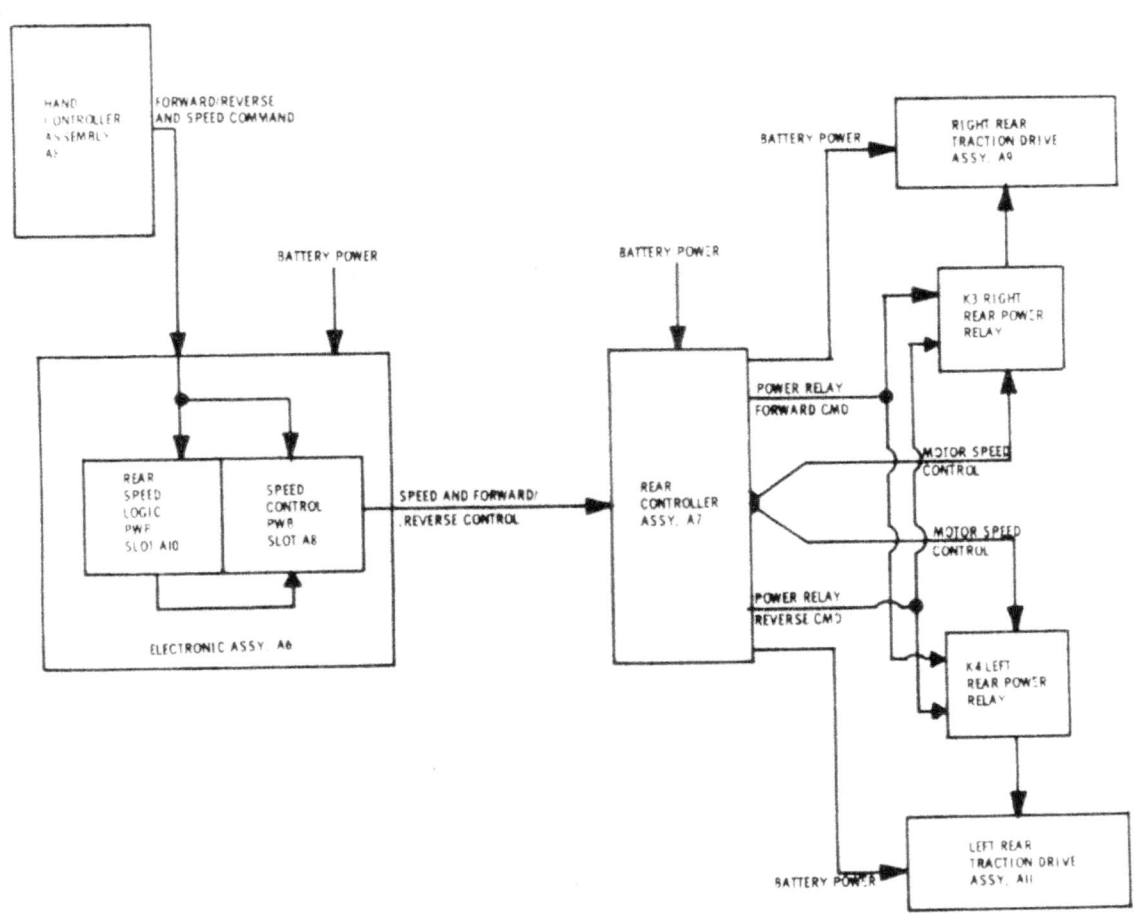

FIGURE 8-4 1G TRAINER VEHICLE REAR TRACTION DRIVE ELECTRICAL SIGNAL ROUTING BLOCK DIAGRAM

LS006-002-2H
LUNAR ROVING VEHICLE
OPERATIONS HANDBOOK

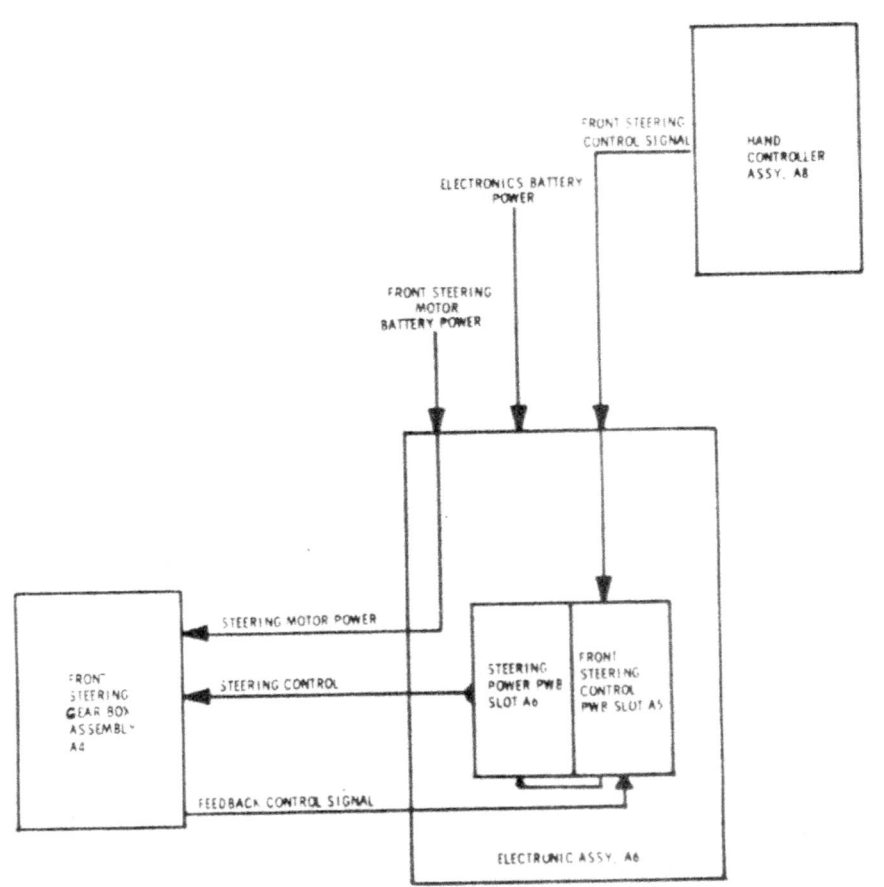

FIGURE 8-5 1G TRAINER VEHICLE FRONT STEERING
ELECTRICAL SIGNAL ROUTING BLOCK DIAGRAM

LS006-002-2H
LUNAR ROVING VEHICLE
OPERATIONS HANDBOOK

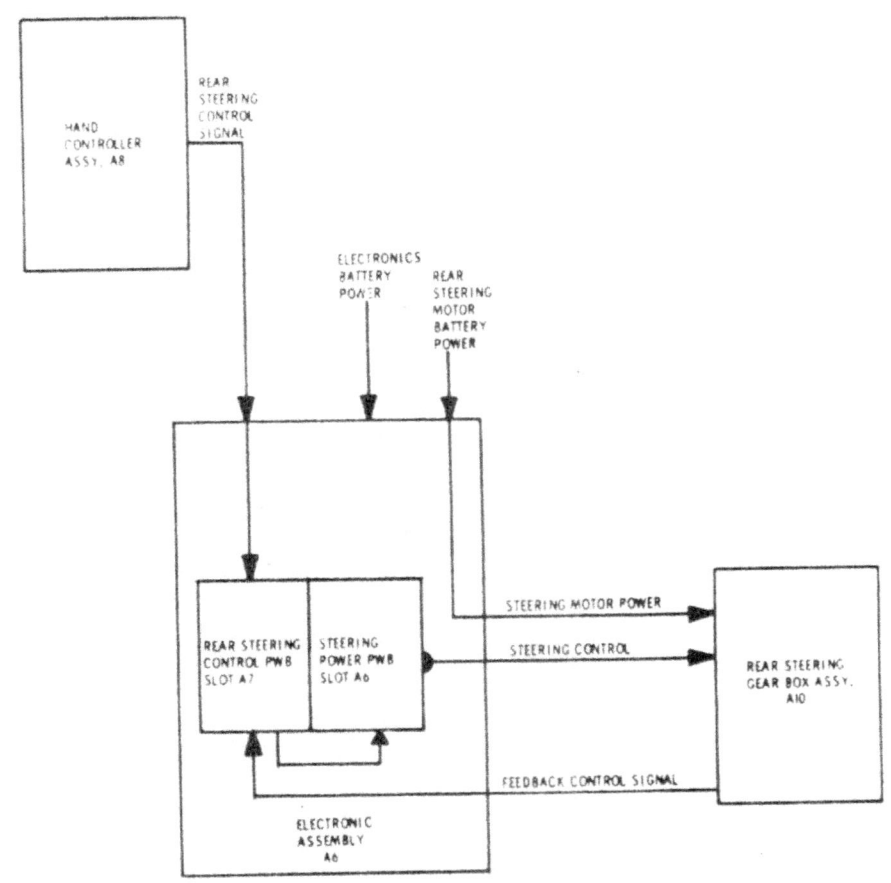

FIGURE 8-6 1G TRAINER VEHICLE REAR STEERING
ELECTRICAL SIGNAL ROUTING BLOCK DIAGRAM

LS006-002-2H
LUNAR ROVING VEHICLE
OPERATIONS HANDBOOK

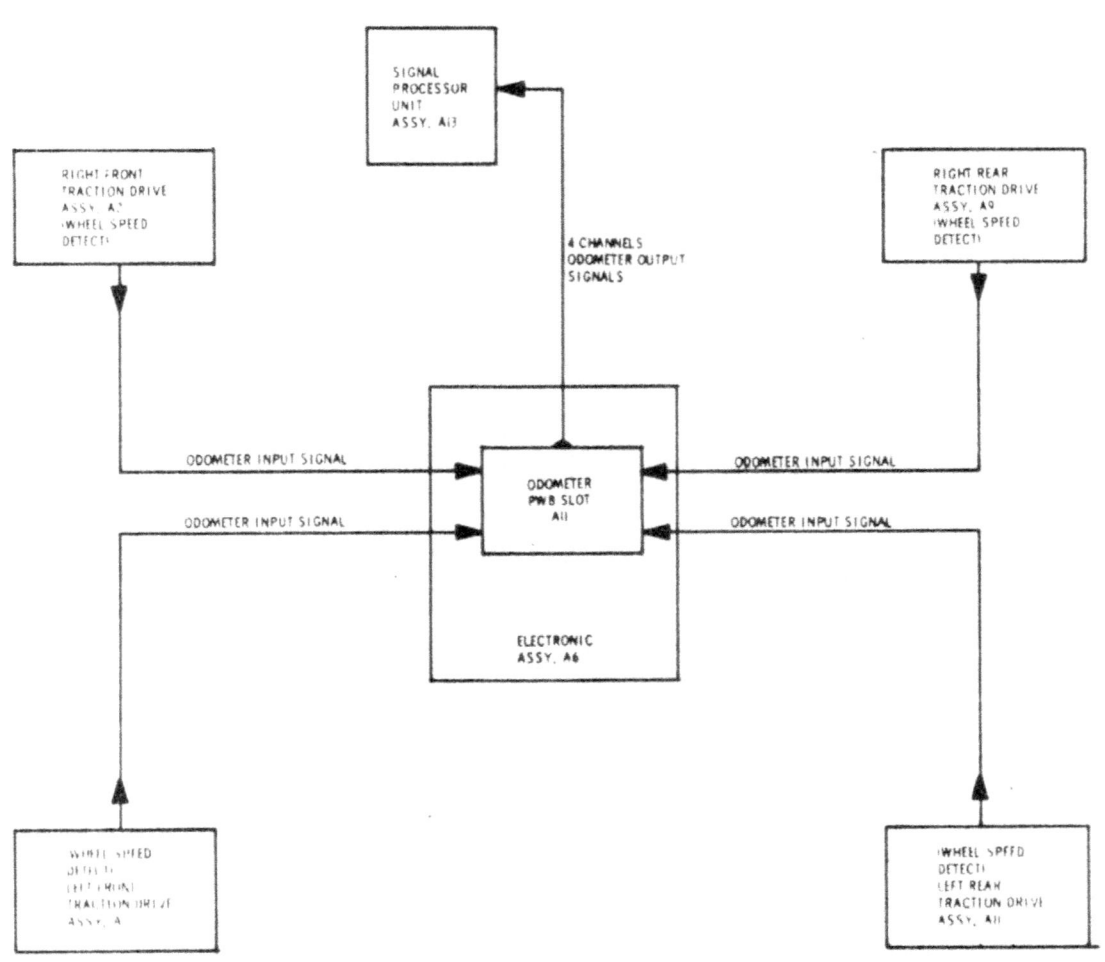

FIGURE 8-7   1G TRAINER VEHICLE ODOMETER
              ELECTRICAL SIGNAL ROUTING BLOCK DIAGRAM

LS006-002-2H
LUNAR ROVING VEHICLE
OPERATIONS HANDBOOK

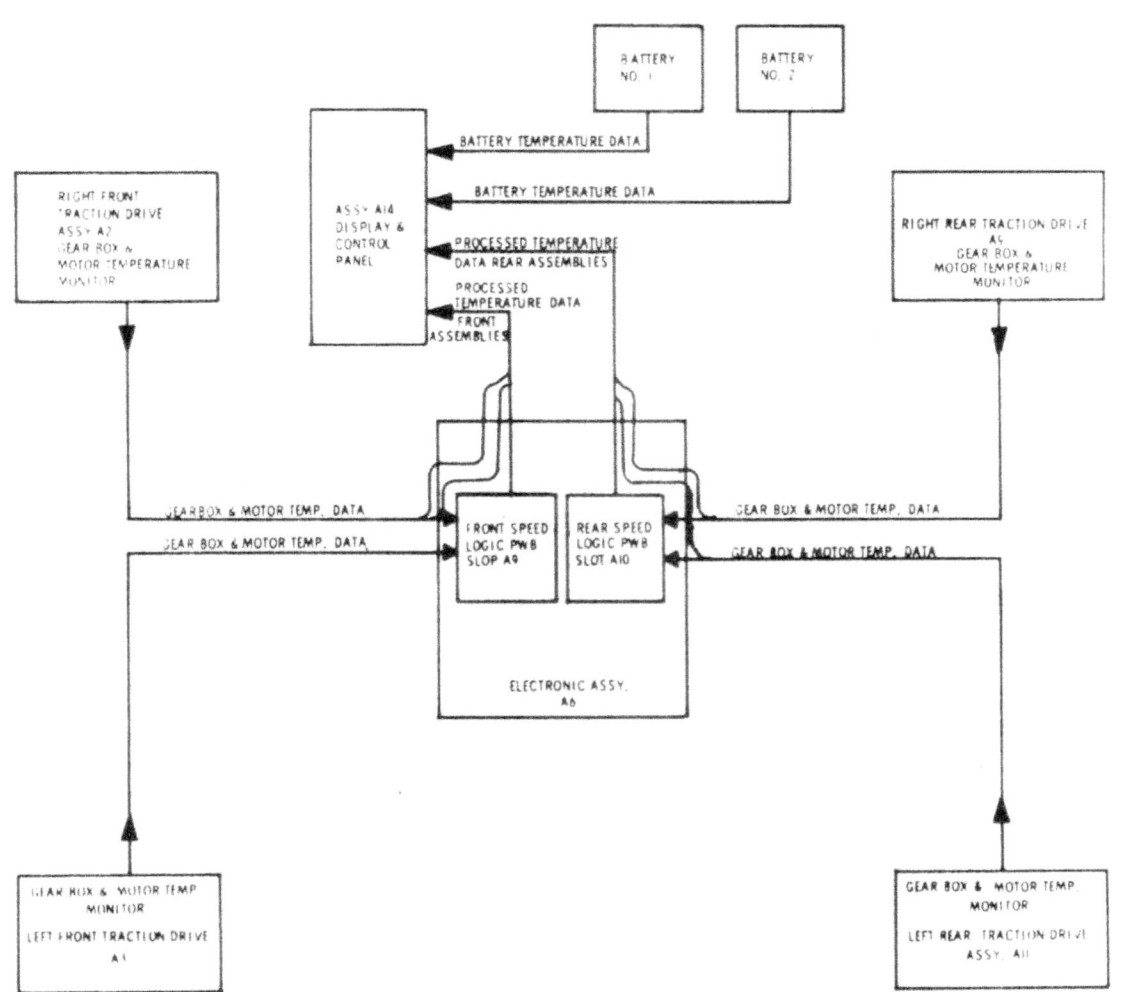

FIGURE 8-8 1G TRAINER VEHICLE TEMPERATURE DIAGNOSTICS
ELECTRICAL SIGNAL ROUTING BLOCK DIAGRAM

Mission ___J___ Basic Date __12/4/70__ Change Date __4/19/71__ Page 8-9

8.1.2.2 (Continued)

discontinuance of vehicle operation. Engineering support should be solicited for specific repair instructions.

8.1.2.3 Loose Fasteners

Retighten fasteners. Replace any damaged threaded insert. Verify that the appropriate locking medium was used.

8.1.2.4 Cable Damage

Replace the wire if conductor damage has occurred. If damage is limited to insulation, repair with tape, shrink tubing or a like material with insulating and moisture resistant properties similar to the original material.

8.1.2.5 Hydraulic Line Damage

Repair using conventional techniques.

8.1.2.6 Mechanical Interferences

Steering/suspension interference may be caused by improper gain setting (Section 8.2.3.1). Inspect interference to determine cause and treat any permanent structural deformation as a loss of structural integrity. Replace any damaged component or subassembly that has caused or resulted from the interferences.

8.1.2.7 Dust or Debris

(See cleaning.)

8.1.2.8 Leakage

Replace leaking component on subassembly - see cleaning.

8.1.2.9 Configuration

The vehicle configuration should be as described in the top assembly drawing.

8.1.3 Cleaning

At the conclusion of each vehicle mission, remove any accumulation of sand, dust, or other foreign material. Clean hydraulic fluid or like contamination with a Freon Degreaser, or equivalent. Battery electrolite (KOH) should be treated in accordance with specific instructions (Section 8.2.8.5). Do not apply adhesive backed tape to the electrical cables because later removal may also remove the silver coating.

LS006-002-2H
LUNAR ROVING VEHICLE
OPERATIONS HANDBOOK

8.1.4 Storage

a. <u>Long Term Storage</u>. The vehicle should be stored inside a controlled access area wherein the ambient temperature is 70 + 20°F and the relative humidity does not exceed 90 percent. Batteries should be removed and stored separately. Periodic visual inspection should be made of the vehicle as defined in Section 2.1.1 and the ambient conditions should also be monitored.

b. <u>Short Term Storage</u>. For overnight or other short term storage the vehicle should be covered; however, batteries need not be removed. Ambient conditions must be within a range of -20°F to +120°F, with relative humidity less than 100 percent.

8.1.5 Safety Considerations

Vehicle operation should not be attempted by untrained personnel because of its unique control and handling characteristics. Driving skills must be developed only after a period of verbal direction and checkout, followed by actual operation under the surveillance of an instructor. Under no circumstances shall vehicle operation be other than specified in this Operation Manual.

CAUTION

Vehicle design specifications required that a minimum of three wheels support the vehicle at all times. Operation of the vehicle over terrain wherein two wheels support the entire vehicle may cause serious structural damage.

LS006-002-2H
LUNAR ROVING VEHICLE
OPERATIONS HANDBOOK

8.2   SPECIFIC PROCEDURES

8.2.1   Chassis

Visually inspect in accordance with Section 8.1.1. No additional adjustment or maintenance is required.

8.2.2   Hand Controller

No periodic adjustment or maintenance is required on the unit. Hand controller characteristics are shown in Figures 1-12, 1-13 and 1-14.

8.2.2.1 Lubrication and Cleaning of the Hand Controller

This operation should be done at a maximum of 300 hour intervals. Cleaning and relubricating has to be done with the hand controller removed. The slide attached to the boot should be removed from the hand controller assembly during this operation to prevent damage to the boot.

With the hand controller removed (see Section 8.3.1), clean the unit thoroughly. Clean all gear meshes and rubbing surfaces with "Freon Degreaser" cleaning fluid (dispensed from a pressurized can) or equivalent. Relubricate areas with Dow Corning "MOLYKOTE" or equivalent.

8.2.3   Suspension

Suspension clearance and alignment procedures will be required only if removal or replacement procedures have been performed on suspension and related traction drive or steering linkages. No suspension maintenance is required other than as part of the pre and post sortie visual inspection of the vehicle (Section 8.1.1).

8.2.3.1 Clearance and Alignment Adjustments

a. Locate vehicle on level surface.

b. Load vehicle with 520 pounds distributed as follows:

        245 lb. driver seat
        245 lb. passenger seat
        130 lb. equally distributed over rear

NOTE: If the vehicle includes any of the following: LCRU, high gain antenna, camera, rear-deck payload packages, the stated 130 pound rear-deck payload shall be decreased accordingly.

8.2.3.1 (Continued)

c. Use torsion bar bracket screws to position adjustable torsion bar retainer until 35 cm minimum ground clearance exists when measured at the crew compartment corners.

d. If inadequate adjustment exists remove the torsion bar (see Section 8.3.2) and rotate one additional spline.

e. Apply power to both steering systems and position at electrical zero.

f. Lay a straight edge across the center of the drive cover of both left side traction drive assemblies.

g. If a clearance of more than 1/16 inch exists between the cover and straight edge, unlock one (or both if required) steering tie rod and adjust until straight edge is flat on both hubs.

h. Repeat the procedure using right side traction drive assemblies.

i. Lock tie rod nuts.

j. Locate vehicle wheels on teflon pads or other low friction material.

k. Attach a protractor to the drive cover of both front traction drive assemblies.

l. Apply front steering power and electrically zero the steering.

m. Zero the protractor, allowing adequate clearance for complete steering movement.

n. Record steering displacement at both wheels while applying full steering command in both left and right turn direction.

CAUTION

Interference may occur between the wheel, fender and steering linkage.

o. If interference occurs, adjust potentiometer R16 of printed wiring board RTV 20217 in the CW direction.

p. Trim potentiometer R16 until inside wheel displacement is 48° minimum.

q. Repeat procedure for rear steering.

8.2.3.1 (Continued)

Acceptance test data on steering operation and hand controller position versus wheel angles is presented in Table 8-1 for reference only.

8.2.4 Traction Drive

No Traction Drive adjustments are required; however, each unit can be mechanically decoupled to simulate LRV degraded operation or to permit towing without back-driving the motors (see Section 8.2.5.3, a through c). The preventive maintenance described in sections 8.2.4.1 through 8.2.4.3 is necessary.

8.2.4.1 Pre and Post Mission

Visually inspect per Section 8.1.1.

8.2.4.2 Every 50 Hours of Operation

Clean air filter as follows:

a. Remove traction drive blower.
b. Remove filter material.
c. Clean or wash in water.
d. Dry and re-install.

8.2.4.3 Every 200 Hours of Operation

Relubricate gearbox as follows:

a. Remove traction drive assembly from the vehicle.
b. Remove hub assembly with drive cover and outer bearing by removing the eight brake disc screws.
c. Secure brake disc to king pin by tieing two places, using hub mounting screw holes.
d. Disconnect gearbox thermostat retainer band.
e. Remove any cable clamps, ties or restraints as needed and work gearbox thermostat cable through gearbox flange as far as possible.
f. Remove gearbox mounting screws.

CAUTION

Do not remove the brake disc and seal or damage the thermostat cable.

g. Remove the three planetary stages and degrease.
h. Degrease the gear housing.

LS006-002-2H
LUNAR ROVING VEHICLE
OPERATIONS HANDBOOK

Hand Controller Angle	Steering Proportionality			
	Degrees Rotation of Wheel from Neutral Position			
	LF Wheel	RF Wheel	LR Wheel	RR Wheel
Right Soft Stop (S.S)	15	26	28	16-1/2
Right H.S.	19-1/2	49-1/4	48	21-1/4
Left S.S.	29	17-1/2	16	26
Left H.S.	48-1/2	21-1/2	21	51-1/2

NOTES:
1. DATA IS FOR REFERENCE ONLY
2. DATA IS FROM 1G TRAINER ACCEPTANCE TEST

TABLE 8-1 1G TRAINER STEERING OPERATION DATA

Mission __J__ Basic Date __12/4/70__ Change Date __4/19/71__ Page __8-15__

LS006-002-2H
LUNAR ROVING VEHICLE
OPERATIONS HANDBOOK

8.2.4.3 (Continued)

i. Repack all planetary gears and bearings with RTV 21119-001 lubricant.
j. Reassemble gearbox and verify that the output stage bearing is seated against the retaining ring.
k. Re-assemble.

8.2.5 Wheels

Wire wheels or pneumatic tires may be interchanged as sets.

8.2.5.1 Wire Wheels

Visually inspect in accordance with Section 8.1.1. Discontinue wheel use when wire breakage approaches 200 wires. Loose tread strips may be wired in place, since disassembly is not practical.

8.2.5.2 Pneumatic Tires

Visually inspect in accordance with Section 8.1.1. Repair, using conventional commercial tire techniques. For most operations inflate the tire to 30 psig - approximately 13.9 inch rolling radius (as measured from the hub center to the operating surface). Do not exceed 40 psig. Air pressure may be reduced for soft soil operation.

8.2.5.3 Wheel Decoupling

This procedure is for decoupling any of the four wheels. Nonfunctional decoupling clips are installed on each wheel for LRV simulation. The actual decoupling is accomplished by replacing drive hubs with blanks.

a. Hand Controller - parking brake position with throttle control in neutral.
b. DRIVE POWER Switches (4) - OFF.
c. STEERING Switches (2) - OFF.
d. + 15 VDC Switch - OFF.
e. Remove the six Phillips-head screws from the hub of the wheel to be decoupled (figure 8-9).
f. Remove the drive hub.
g. Obtain a blank "decoupling" hub.
h. Insert the blank hub into the wheel.
i. Install the six screws into the blank hub to secure the hub to the wheel.
j. Notify crew to resume operation.

8.2.6 Brakes

8.2.6.1 Brake Adjustments - Maximum Stopping Effort

a. Install long master cylinder actuating levers on both cylinders and connect external springs.

LS006-002-2H
LUNAR ROVING VEHICLE
OPERATIONS HANDBOOK

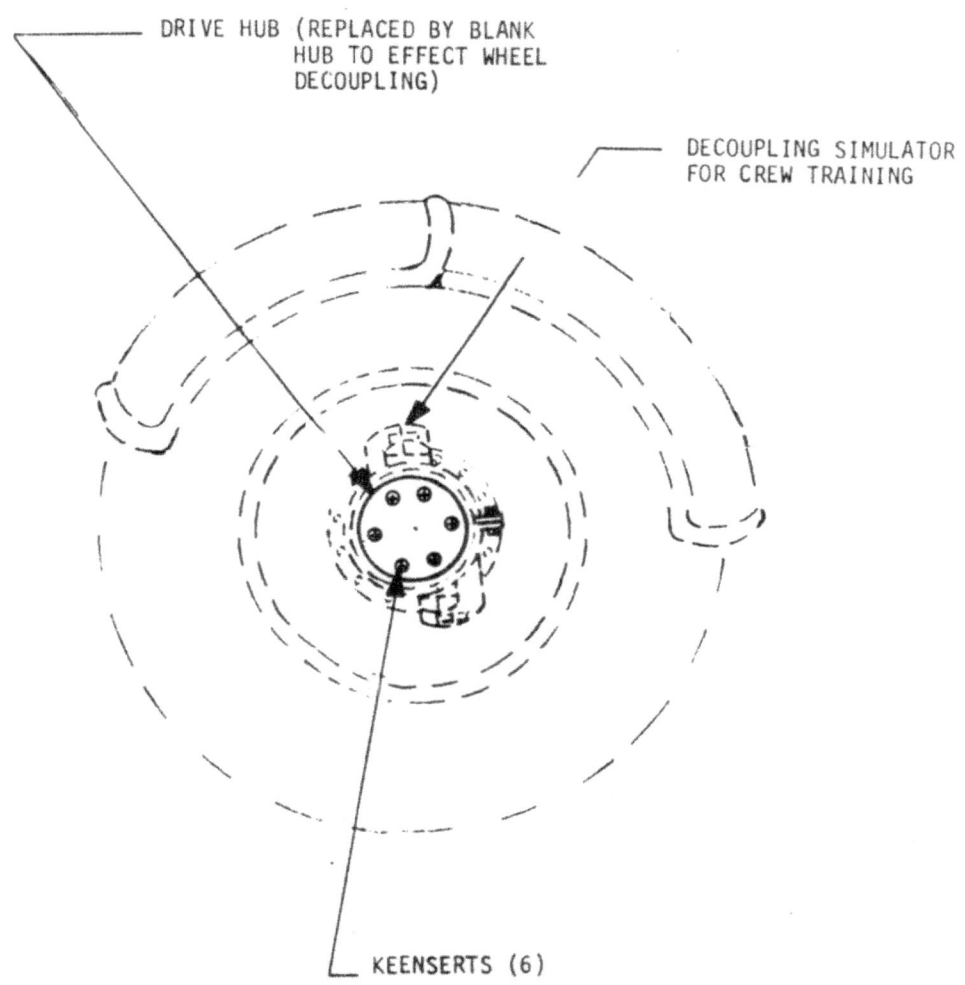

TYPICAL FOR ALL FOUR WHEELS

FIGURE 8-9   1G TRAINER WHEEL DECOUPLING

LS006-002-2H
LUNAR ROVING VEHICLE
OPERATIONS HANDBOOK

8.2.6.1 (Continued)

NOTE: If lever operation results in a spongy effect or in brakes with leak down, bleed the system of entrapped air before proceeding.

b. Connect cable clevis to the top lever holes.

c. With the hand controller forward (brakes off), concurrently adjust cable lengths at the hand controller yoke for both master cylinders.

   (1) Maintain horizontal position of the yoke (equalized load).
   (2) Adjust length until the brake pulley position is as shown in Figure 8-10. Maintain corresponding relative rotational position between pulleys.
   (3) Maintain clearance between the actuation lever and the master cylinder piston with clevis adjustment.

d. Remove master cylinder covers and verify fluid level.

e. Concurrently adjust the actuation lever position (with the clevis) until no clearance exists between the lever and piston. Continue to adjust until application and release of the brakes with the hand controller results in the absence of an oil spout in the fluid reservoir.

f. Readjust the clevis(es) the minimum amount necessary for consistent occurrence of the oil spout.

g. Re-verify fluid level and install the reservoir cover.

h. Braking effort may be verified as follows:

   (1) Adapt a 200 ft. lb. capacity torque wrench to the center of a blank traction drive cover, RTV 20613. (A 5/8 diameter high strength socket head bolt may be used).
   (2) Install adapted cover in place of any existing drive cover.
   (3) Remove vehicle weight from the traction drive.
   (4) Set the hand controller to the park position.
   (5) Attach torque wrench with 7/8 inch socket to the adapter cover.
   (6) Torque with wrench until the wheel rotates - indicated value should be approximately 195 ft. lbs.

NOTE: If substantially less torque is measured the nonlinear pulley may be incorrectly positioned and the procedure should be repeated starting at Step 8.2.6.1.c.

Mission   J   Basic Date   12/4/70   Change Date   4/19/71   Page   8-18

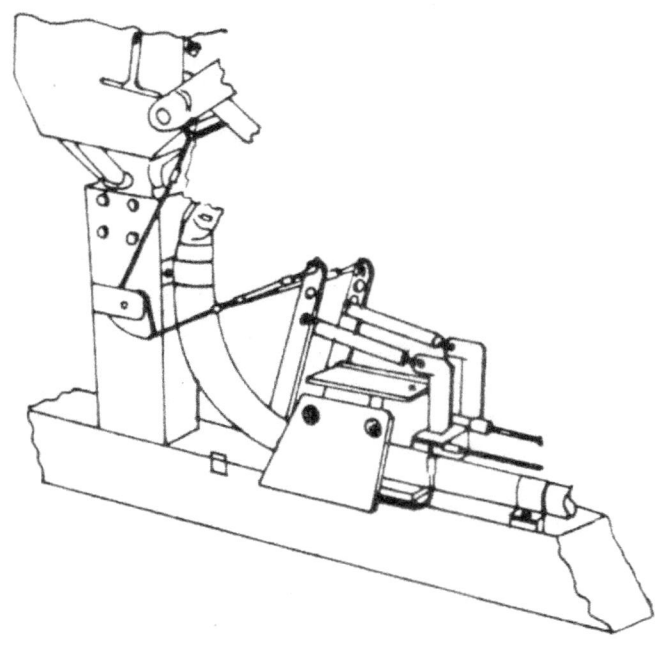

FIGURE 8-10 1G TRAINER BRAKE LINKAGE

LS006-002-2H
LUNAR ROVING VEHICLE
OPERATIONS HANDBOOK

8.2.6.2   Brake Adjustments - Degraded Operation

a. Install long master cylinder actuating levers on both cylinders and connect external springs. Verify reservoir fluid level.

   NOTE: If lever operation results in a spongy effect or in brakes which leak down, bleed the system of entrapped air before proceeding.

b. Connect cable clevis to the lower lever holes.

c. With the hand controller forward (brakes off), concurrently adjust cable lengths at the hand controller yoke for both master cylinders.

   (1) Maintain horizontal position of the yoke (equalized load).
   (2) Adjust length until the brake pulley position is as shown in Figure 8-10. Continue adjustment, shortening cable length to maximum amount possible. Maintain corresponding relative rotational position between pulleys.
   (3) Maintain clearance between the actuation lever and the master cylinder piston with clevis adjustment.

d. Concurrently adjust the actuation level position (with the clevis) until no clearance exists between the lever and piston.

e. Braking effort may be verified as follows:

   (1) Adapt a 200 ft. lb. capacity torque wrench to the center of a blank traction drive cover, RTV 20613. (A 5/8 inch diameter high strength socket head bolt may be used).
   (2) Install adapted cover in place of any existing drive cover.
   (3) Remove vehicle weight from the traction drive.
   (4) Set the hand controller to the park position.
   (5) Attach torque wrench with 7/8 inch socket to the adapter cover.
   (6) Torque with wrench until the wheel rotates - indicated value should be approximately 60 ft. lbs.

   NOTE: If substantially more torque is measured, install short master cylinder actuating levers and repeat the procedure.

8.2.7   Steering Unit

No adjustments are required to the steering gearboxes after installation.

8.2.7.1   Clean and Relubricate Steering Gearboxes

This operation should be done at not less than 50 hour or more than 300 hour intervals. Because of the inaccessibility of the front steering unit, clean and lubrication procedure should be accomplished, if the front cover panel

LS006-002-2H
LUNAR ROVING VEHICLE
OPERATIONS HANDBOOK

8.2.7.1 (Continued)

has to be removed for other service operations (not to exceed 50 hours as mentioned previously).

Cleaning the steering gearboxes should be done by first blowing out any dirt or dust with compressed air. Gears are to be cleaned with "Freon Degreaser" fluid or equivalent. Relubricate the gear teeth with Dow Corning "MOLYKOTE" or equivalent.

8.2.7.2 Steering Decoupling

a. Verify the crew has manually positioned the forward or rear wheels (whichever is to be simulated as the decoupled wheels) to the straight ahead position.

b. Hand Controller - parking brake position with throttle control in neutral.
c. DRIVE POWER Switches (4) - OFF.
d. STEERING Switches - OFF.
e. Obtain two steering arm clamps.

f. Place steering arm clamp on the left and right hand steering arms, in such a position that the clamps are butted against the outside of the chassis frame where the steering arms pass through the frame (figure 8-11).

g. Open the FORWARD STEERING Circuit Breaker on the display and control console if forward steering arms were clamped. Open the REAR STEERING Circuit Breaker if the rear steering arms were clamped.

h. Notify crew to resume operation.

8.2.8 Drive Power

The Drive Power Subsystem consists of a Control Electronics package, two Drive Controllers, and two 34 VDC Ni-Cad batteries.

8.2.8.1 Control Electronics

No periodic adjustment or maintenance is required. Inspect the exterior in accordance with Section 8.1.1

8.2.8.2 Drive Controller

No periodic adjustment or maintenance is required. Inspect the exterior in accordance with Section 8.1.1.

Mission ___J___   Basic Date __12/4/70__ Change Date __4/19/71__ Page __8-21__

LS006-002-2H
LUNAR ROVING VEHICLE
OPERATIONS HANDBOOK

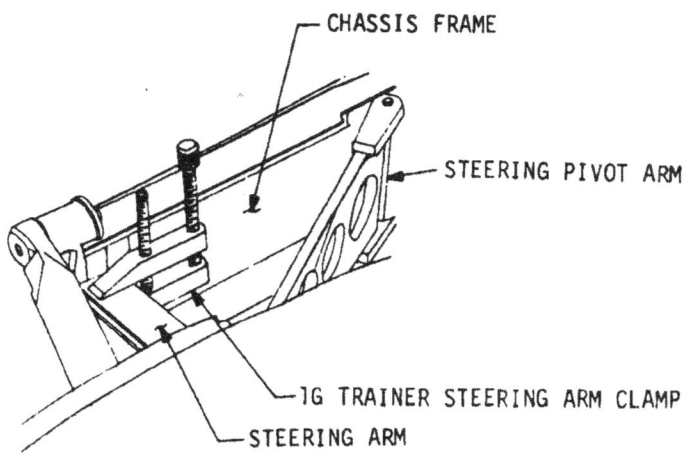

VIEW A-A
TYPICAL FOR BOTH STEERING ARMS
FORWARD AND REAR

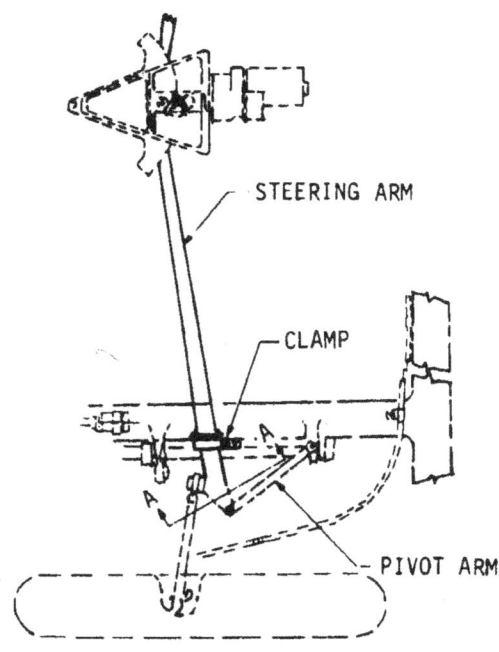

FIGURE 8-11   1G TRAINER STEERING ARM CLAMPING
TO SIMULATE STEERING DECOUPLING

LS006-002-2H
LUNAR ROVING VEHICLE
OPERATIONS HANDBOOK

8.2.8.3 Battery Change-Out

Changeout of 1G Trainer batteries can occur in two different operational conditions: (1) when the navigation system is to remain on during changeout, and (2) when the navigation system may be off during changeout. The following procedures define both cases.

a. When the navigation system is to remain on:

1) DRIVE POWER Switches (4) - OFF.
2) STEERING Switches (2) - OFF.
3) ± 15 VDC Switch - OFF.

NOTE

1G Trainer Battery #1 is on the right side, Battery #2 on the left side.

4) If changing out Battery #1, open BAT 1 BUS A and BAT 1 BUS B circuit breakers. Do not open the BAT 2 circuit breaker or navigation power will be lost.

5) If changing out Battery #2, open BAT 2 BUS C and BAT 2 BUS D circuit breakers. Do not open the BAT 1 circuit breakers or navigation power will be lost.

6) Place power selector switch (figure 8-12) for the battery to be changed out in the OFF position. Do not operate the selector switch for the other battery.

7) Disconnect the battery connector from the battery to be changed out (figure 8-12).

8) Remove the four screws from the battery mounting bracket.

9) Lift the battery clear of the 1G Trainer using the handles at either end of the battery.

10) Obtain recharged replacement battery.

11) Set the replacement battery in place with the connector inboard and the holes in the mounting brackets aligned with the screw receptacles in the 1G Trainer.

12) Install the four screws, securing the battery to the 1G Trainer.

13) Connect the battery connector to the battery.

LS006-002-2H
LUNAR ROVING VEHICLE
OPERATIONS HANDBOOK

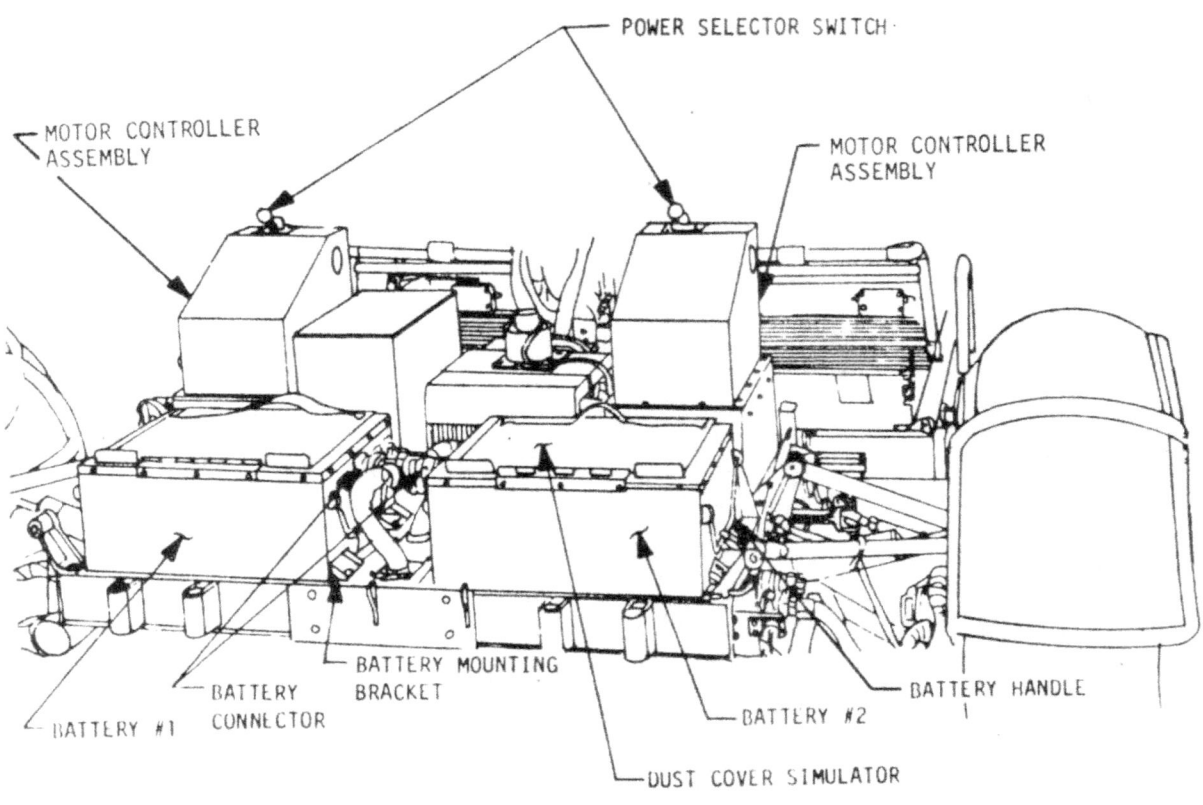

FIGURE 8-12   1G TRAINER BATTERY INSTALLATION

LS006-002-2H
LUNAR ROVING VEHICLE
OPERATIONS HANDBOOK

8.2.8.3 (Continued)

14) Place the power selector switch in the BATTERY position.

15) If Battery #1 was changed out close BAT 1 BUS A and BAT 1 BUS B circuit breakers on the control and display console.

16) If Battery #2 was changed out, close BAT 2 BUS C and BAT 2 BUS D circuit breakers on the control and display console.

17) Repeat steps 4 through 16 for the other battery.

b. When the navigation system may be off:

1) DRIVE POWER Switches - OFF.
2) STEERING Switches - OFF.
3) + 15 VDC Switch - OFF.
4) Circuit Breakers BAT 1 BUS A, BAT 1 BUS B, BAT 2 BUS C, BAT 2 BUS D - Open.
5) Place power selector switches in OFF position.
6) Perform step a7 through a14 for both batteries.
7) Close BAT 1 BUS A, BAT 1 BUS B, BAT 2 BUS C, BAT 2 BUS D Circuit Breakers.

c. If required to perform dust cover simulation, remove dust covers from one set of batteries and install on replaced batteries.

8.2.8.4 Battery Recharging

Each battery will be discharged to a condition below 29 volts when seven amps current exists.

a. Remove closure cover and all plastic fill caps.

b. Verify presence of low-pressure relief vent screw in each fill-cap.

c. Connect a charging circuit equivalent to Figure 8-13.

d. Set power supply current limiter to provide 14 amps.

e. Energize the power supply and adjust voltage to provide 14 amps out of the supply.

f. Maintain condition specified in "e" for a period of 2 hours. Adjust voltage as required for constant current charging.

g. Set power supply to provide +45.5 $\pm$ 0.2 vdc and maintain the current limiter setting of 14 amps.

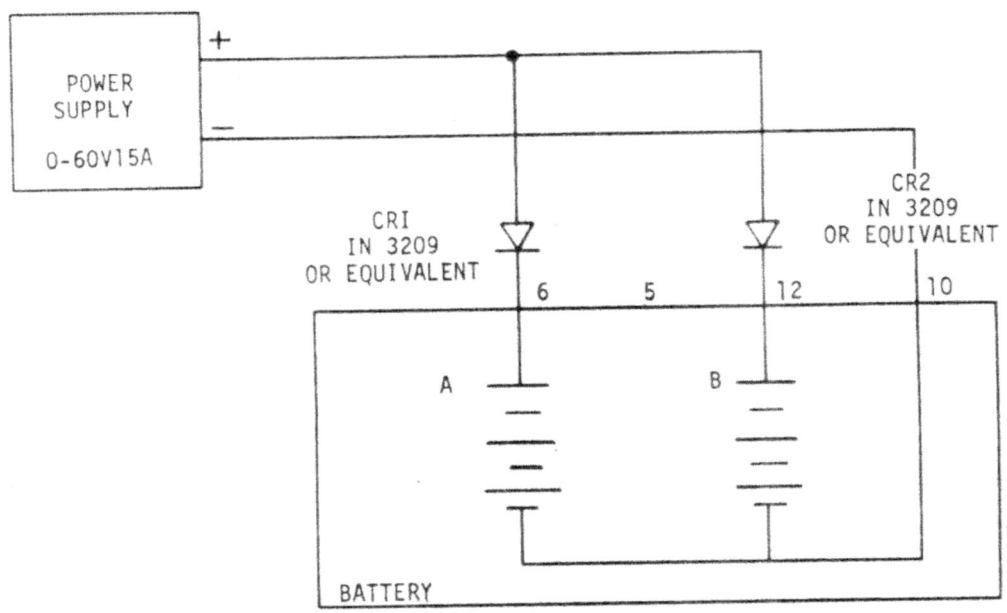

FIGURE 8-13   1G TRAINER BATTERY CHARGING CIRCUIT

LS006-002-2H
LUNAR ROVING VEHICLE
OPERATIONS HANDBOOK

8.2.8.4   (Continued)

h. Allow batteries to charge for 4 hours or until the current flow is less than 2.0 amps.

i. Remove the charging circuit from the battery.

j. Using the VOM, measure the open-circuit voltage. Voltage should be greater than +35 VDC.

k. Check and verify that the level of electrolyte in each cell is visible at the bottom of the funnel-shaped inner chamber.

l. Using a syringe, carefully add distilled water to each cell if necessary to bring electrolyte to the proper level.

m. After charging, the batteries shall be carefully rinsed with distilled water.

n. After rinsing, the batteries shall be dried.

o. Remove batteries from setup, install closure cover, and secure from test.

8.2.8.5   Precautions for Handling Batteries and Electrolyte

The electrolyte (KOH) is alkaline and corrosive. It should be handled with care, since it will cause serious burns if allowed to contact eyes or skin.

At stations where handling is done, a supply of boric acid, for eye-burns, and a solution of weak acid (5% acetic acid) for skin burns should be maintained.

Persons who work with batteries should wash their hands thoroughly after handling them.

Persons who fill batteries or otherwise handle electrolyte should wear alkali-proof aprons, gloves, and a face mask.

KOH can cause ignition between points of sufficiently high potential difference. Fire extinguishers should be available during battery operation and servicing.

Batteries should not be lifted by one man.

LS006-002-2H
LUNAR ROVING VEHICLE
OPERATIONS HANDBOOK

8.3   PREVENTIVE MAINTENANCE ASSEMBLY REMOVE AND REPLACE PROCEDURES

8.3.1   Removal and Replacement of the "Hand Controller"

> **CAUTION**
>
> During this operation do not put tape of any kind on the electrical cables. The tape will remove the silver from the protective cable cover.

Removal is accomplished in the following manner:

a. Cut brake release cable (the small cable that extends rearward from the hand controller mount).

b. Detach electrical cables from Display and Control Console. Cut cable ties as required.

c. Remove brake cables.

d. Remove the display console from its base. Remove the console base by removing the two "T" pivot handle pins and the two clevis pins at the rear of the console base.

e. Uncouple the electrical connections at the base of the hand controller.

f. Remove the four side plate mounting screws.

g. Remove the four screws from the base of the hand controller.

8.3.2   Suspension/Traction Drive - Removal and Installation Procedure

a. Remove weight from wheel by suspending vehicle at lift points, after measuring ground clearance.

b. Remove wheel disconnect simulator.

c. Remove wheel and fender.

d. Disconnect steering tie rod at steering hinge.

e. Disconnect traction drive cables.

f. Disconnect brake hoses, if traction drive is to be moved from area.

8.3.2   (Continued)

g. Loosen torsion bar adjustment screw - one only.

h. Remove torsion bar cover screws/cover at outboard suspension bracket.

i. Remove torsion bar.

   NOTE: Torsion bars are not completely interchangeable (cross corners only) and each should be marked with respect to suspension fitting to assure the same ground clearance upon re-assembly.

j. Remove top and bottom king pin screws and flat washers.

k. Remove traction drive.

l. Remove shock absorber. (May be removed without prior steps if shock absorber is only service item.)

m. Remove upper suspension attachment hardware at suspension mounting brackets.

n. Remove outboard suspension brackets. Also remove inboard suspension bracket if service is required.

o. Remove suspension arms.

p. Remove suspension to king pin links (both arms), if service of parts required.

q. To remove inboard torsion bar brackets (with adjustable retainer), remove frame mounting screws and remove from the bottom. Adjustable retainer may be removed by disassembling the torsion bar bracket.

   NOTE: Torsion bar brackets are line bored with suspension brackets and are not interchangeable.

r. The fixed torsion bar retainer may be replaced in the lower suspension by pressing out dowel pins in the flange and unscrewing fitting.

   NOTE: Fixed retainers are drilled on assembly and are not interchangeable.

LS006-002-2H
LUNAR ROVING VEHICLE
OPERATIONS HANDBOOK

8.3.2 (Continued)

Inspect all parts for wear or damage. To reassemble, reverse above procedure. Also:

s. Drill link pins on assembly if replaced.

t. Tighten only the torsion bar adjustment screw that was loosened in step 8.3.2q.

u. Realign wheels if new traction drive or steering linkage installed.

v. Bleed brakes if lines have been opened.

w. Verify proper ground clearance as recorded during disassembly.

8.3.3 Drive Power - Removal and Installation Procedure

a. Remove batteries (see Section 8.2.8.3).

b. Loosen set screw in off-on charge switch handle and remove.

c. Remove all cover screws and hold down screws on filter cover and controller assembly.

d. Disconnect wires (4) from top of filters.

e. Disconnect wires (5) from rear of switch assembly which come from controller below.

f. Tip filter and switch to one side and disconnect wires (4) from lower side of filters.

g. Remove switch and filter assembly, allowing disconnected wires to flow through holes in lower switch and filter plate.

h. Remove wires from TB1 inside controller box. (Only wires which come from inside trainer chassis should be removed.)

i. Nuts and washers should be counted as it is important not to lose any hardware inside controller box or a short circuit may result.

j. Disconnect P2 cannon connector.

k. Remove controller box by lifting and feeding wires through hole in bottom of box.

l. Remove Control Electronic mounting screws, disconnect cables routed into the box from below, and remove package.

m. Install units in reverse of removal.

8.3.4 Steering Gearbox Removal and Installation Procedure

a. Remove Drive Power units as described in Section 8.3.3.

b. Remove transverse structural support member RTV21103.

c. Disconnect cables and remove signal processor unit.

   NOTE: Thermal controls are hard wired and may require unsoldering.

d. Disconnect cables and remove gyro reference unit with mounting brackets.

   NOTE: Thermal controls are hard wired and may require unsoldering.

e. Detach cables from the front cover panel by removing four screws, three places for the two battery cables and the signal processor cable.

f. Detach cable from signal processor mounting bracket, RTV-21105, and remove bracket.

g. Remove simulated hinges, front and rear.

h. Remove front and rear cover panels.

i. Disconnect the steering tie rods from the quadrant gear.

j. Disconnect the electrical plugs. Detach the two wires from the diode mounting block terminal strip.

k. Remove the two gearboxes to isolation, mount bolts.

l. Remove the two bolts and two screws from the large end of the gearbox.

   NOTE: On the rear of the front unit the nuts on the two bottom vertical screws have to be removed. The ends of the screws are slotted for this purpose.

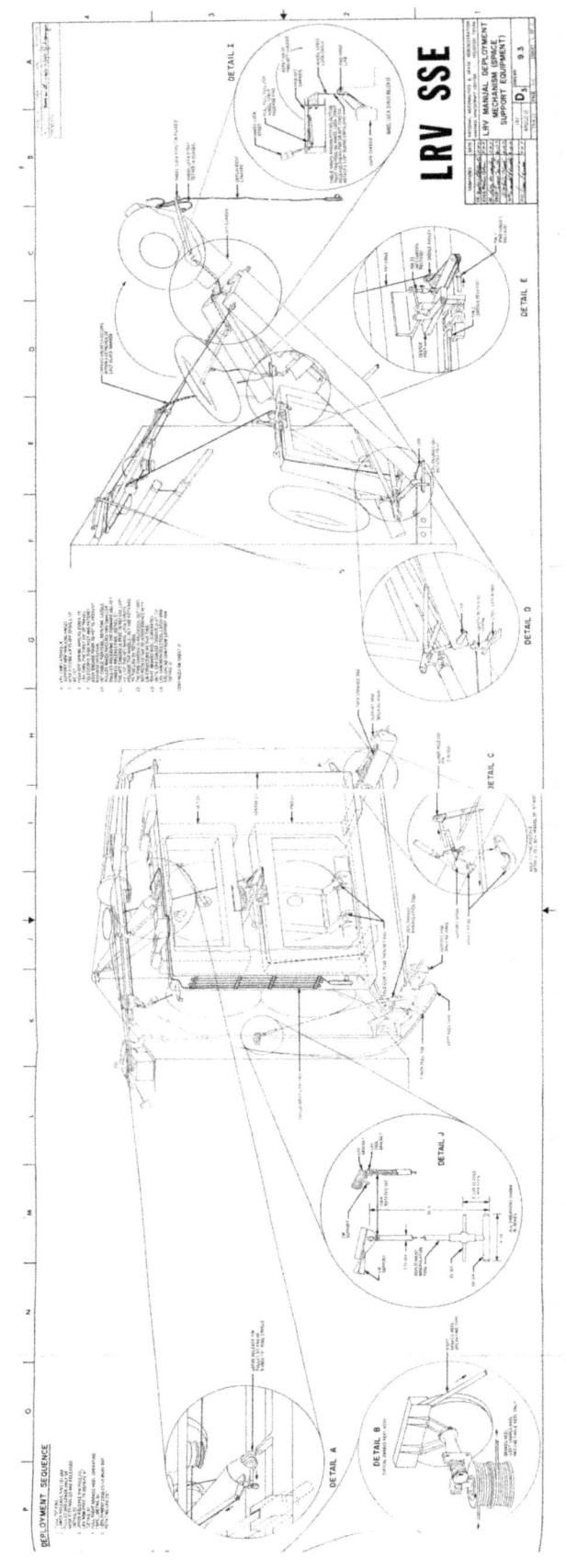

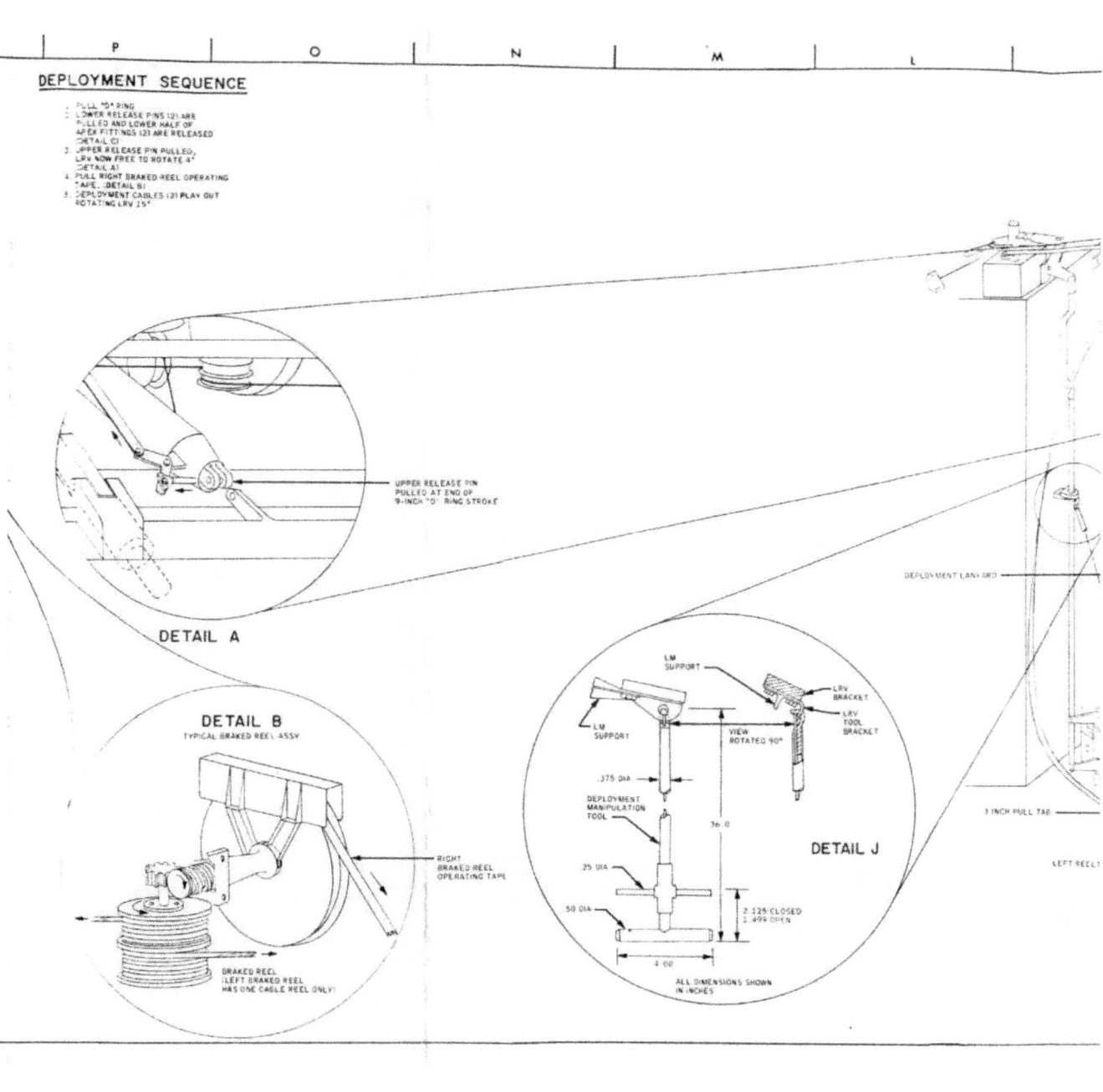

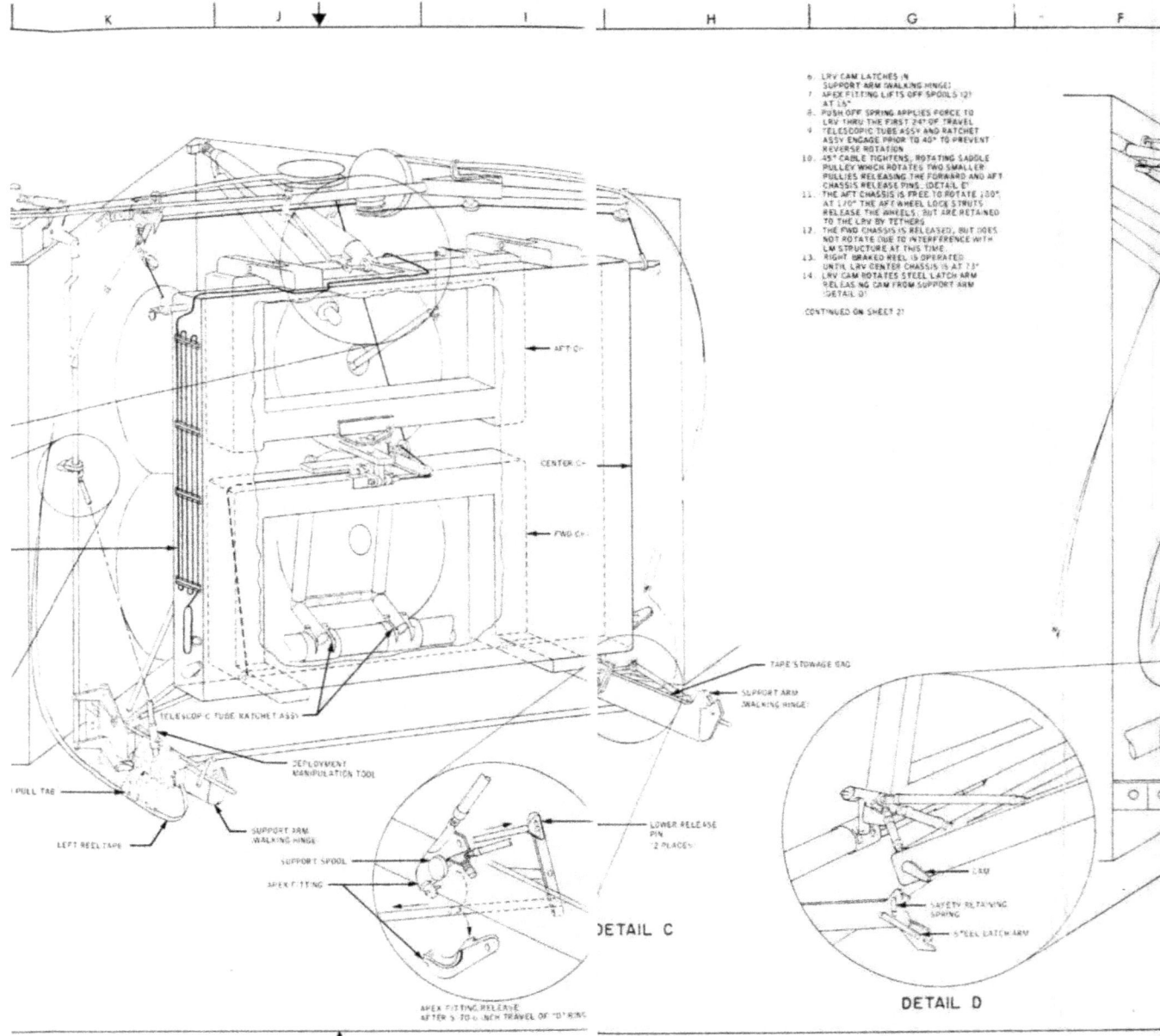

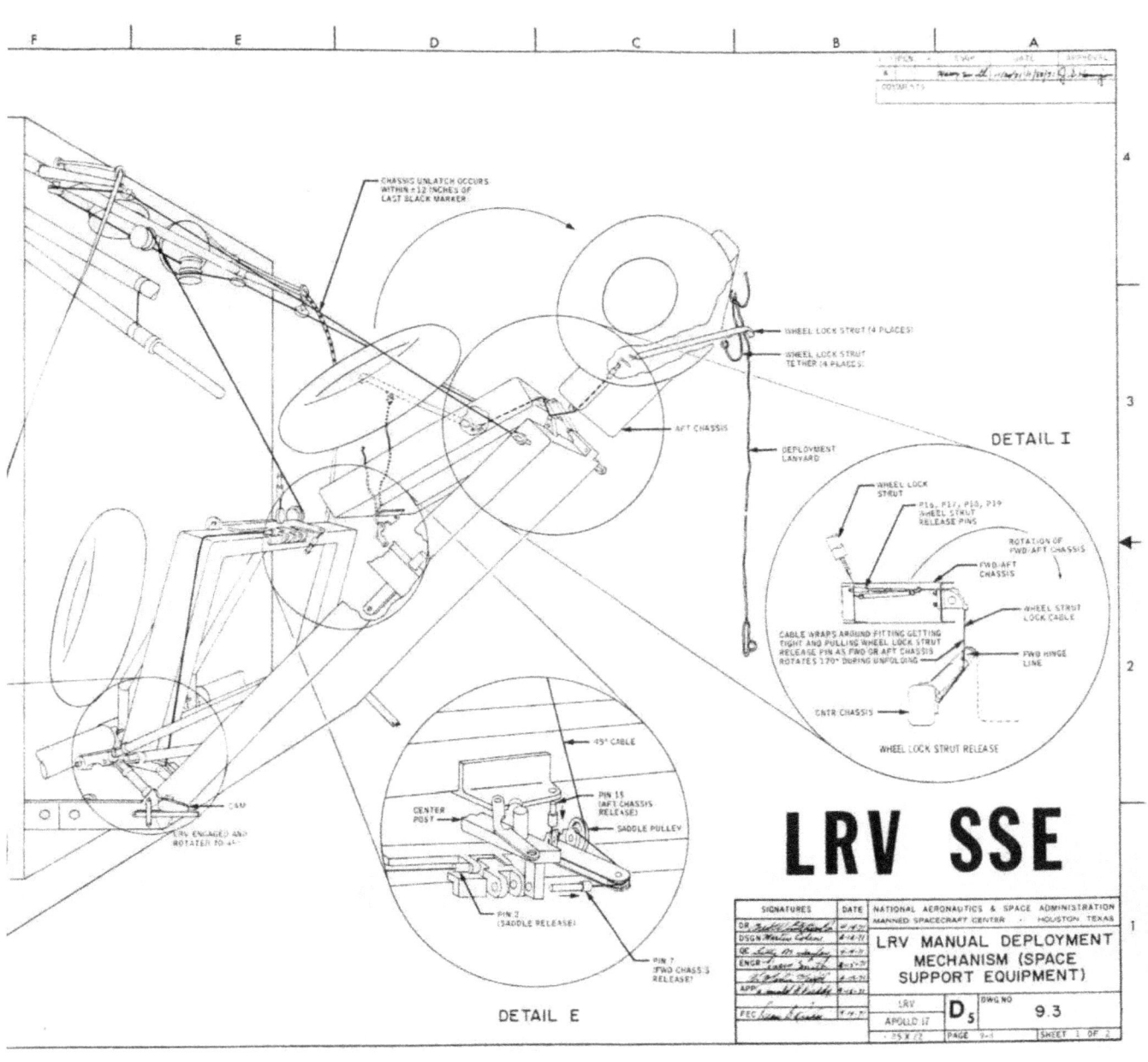

LS006-002-2H

NATIONAL AERONAUTICS AND SPACE ADMINISTRATION

LRV OPERATIONS HANDBOOK APPENDIX A (PERFORMANCE DATA)

REVISION 1

APRIL 19, 1971

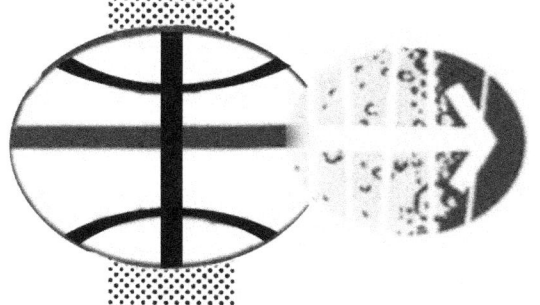

MANNED SPACECRAFT CENTER
HOUSTON, TEXAS

PREFACE

This document is the first revision issue of the Appendix A (Performance Data) to the LRV Operations Handbook. This appendix will be maintained on a controlled basis by the Systems Engineering Division of the Apollo Spacecraft Program Office, with change pages issued as required.

It is requested that any comments to the data content, requirements for additional data, and distribution changes be sent to PD4/ Mr. J. W. Mistrot, extension 4667.

REVISIONS

REV LTR	AMEND NO.	DESCRIPTION	DATE	APPROVED

LS006-002-2H
LUNAR ROVING VEHICLE
OPERATIONS HANDBOOK
APPENDIX A

TABLE OF CONTENTS

SECTION		PAGE
1.0	INTRODUCTION	A-1
1.1	PURPOSE	A-1
1.2	CONTENT	A-1
1.3	AMENDMENTS	A-1
1.4	SELECTED ABBREVIATIONS AND ACRONYMS	A-1
2.0	CONFIGURATION	A-2
2.1	CHASSIS CONFIGURATION	A-2
2.2	SUSPENSION SYSTEM CONFIGURATION	A-2
2.3	STEERING SYSTEM CONFIGURATION	A-2
2.4	TRACTION DRIVE CONFIGURATION	A-2
2.5	WHEEL CONFIGURATION	A-2
2.6	CREW STATION CONFIGURATION	A-2
2.7	POWER SYSTEM CONFIGURATION	A-3
2.8	NAVIGATION SYSTEM CONFIGURATION	A-3
2.9	THERMAL CONTROL CONFIGURATION	A-3
2.10	PAYLOAD INTERFACES	A-3
3.0	CONSTRAINTS AND OPERATIONAL LIMITATIONS	A-23
3.1	CHASSIS SUBSYSTEM CONSTRAINTS	A-23
3.2	SUSPENSION SUBSYSTEM CONSTRAINTS	A-23
3.3	MOBILITY SUBSYSTEM CONSTRAINTS	A-23
3.4	ELECTRICAL POWER SUBSYSTEM CONSTRAINTS	A-23
3.5	NAVIGATION SUBSYSTEM CONSTRAINTS	A-25
3.6	DISPLAY AND CONTROLS SUBSYSTEM CONSTRAINTS	A-25
3.7	CREW STATION SUBSYSTEM CONSTRAINTS	A-26
3.8	VEHICLE DYNAMIC OPERATION CONSTRAINTS	A-26
3.9	PARKING CONSTRAINTS	A-26
3.10	THERMAL CONSTRAINTS	A-26
4.0	SUBSYSTEM PERFORMANCE DATA	A-46
4.1	MOBILITY SUBSYSTEM PERFORMANCE	A-46
4.2	ELECTRICAL SUBSYSTEM PERFORMANCE	A-59
4.3	NAVIGATION SUBSYSTEM PERFORMANCE	A-67
4.4	DISPLAYS AND CONTROLS SUBSYSTEM PERFORMANCE	A-75
5.0	VEHICLE PERFORMANCE DATA	A-79
5.1	VEHICLE DYNAMIC RESPONSE	A-79
5.2	RANGE, SPEED, LURAIN CAPABILITY	A-82
5.3	THERMAL PERFORMANCE	A-83
5.4	CONTROLLABILITY	A-83
6.0	SPECIFIC VEHICLE DATA	A-93
6.1	LRV-1	A-93

Mission J Basic Date 2/5/71 Change Date 4/19/71

LS006-002-2H
LUNAR ROVING VEHICLE
OPERATIONS HANDBOOK
APPENDIX A

LIST OF FIGURES

FIGURE NO.		PAGE
2-1	LRV CONFIGURATION WITHOUT STOWED PAYLOAD	A-4
2-2	LRV DIMENSIONS	A-5
2-3	LRV CHASSIS	A-6
2-4	LRV SUSPENSION SYSTEM	A-7
2-5	LRV STEERING SYSTEM	A-8
2-6	LRV TRACTION DRIVE	A-9
2-7	LRV WHEEL	A-10
2-8	LRV WHEEL DEFLECTION VS LOAD	A-11
2-9	LRV DRIVE CONTROL	A-12
2-10	LRV CREW STATION COMPONENTS	A-14
2-11	LRV BATTERY AND DUST COVER ASSEMBLY	A-16
2-12	LRV NAVIGATION COMPONENTS	A-17
2-13	LRV THERMAL CONTROL	A-18
2-14	PAYLOAD INTERFACES	A-19
3-1	CREW STATION LIMIT LOADS (HANDHOLDS & TOEHOLDS)	A-27
3-2	CREW STATION LIMIT LOADS (SEATS & FOOTRESTS)	A-28
3-3	GROSS WEIGHT ALLOWABLE C.G. ENVELOPE	A-29
3-4	VEHICLE STATIC STABILITY	A-31
3-5	LATERAL LIMIT VELOCITY FOR OVERTURNING FROM COLLISION WITH IMMOVABLE OBJECT	A-32
3-6	LRV SLIDING LIMIT VELOCITY AS A FUNCTION OF TURNING RADIUS AND SLOPE ANGLE FOR A 0.8 COEFFICIENT OF FRICTION	A-33
3-7	LRV SLIDING LIMIT VELOCITY AS A FUNCTION OF TURNING RADIUS AND SLOPE ANGLE FOR A 0.6 COEFFICIENT OF FRICTION	A-34
3-8	LRV OVERTURNING LIMIT VELOCITY AS A FUNCTION OF TURNING RADIUS AND SLOPE ANGLE	A-35
3-9	MAXIMUM BUMP HEIGHTS WHICH CAN BE ENCOUNTERED WITHOUT CAUSING OVERTURN AS A FUNCTION OF VELOCITY AND SLOPE	A-36
3-10	TURNING RADIUS REQUIRED FOR LOCATING C.G. OUTSIDE NORMAL POSITION FOR PREVENTION OF OVERTURN ON 0° SLOPE	A-37
3-11	TURNING RADIUS REQUIRED FOR LOCATING C.G. OUTSIDE NORMAL POSITION FOR PREVENTION OF OVERTURN ON 10° SLOPE	A-38
3-12	TURNING RADIUS REQUIRED FOR LOCATING C.G. OUTSIDE NORMAL POSITION FOR PREVENTION OF OVERTURN ON 20° SLOPE	A-39
3-13	TURNING RADIUS REQUIRED FOR LOCATING C.G. ABOVE NORMAL POSITION FOR PREVENTION OF OVERTURN ON 0° SLOPE	A-40

Mission J   Basic Date 2/5/71   Change Date 4/19/71   Page A

LS006-002-2H
LUNAR ROVING VEHICLE
OPERATIONS HANDBOOK
APPENDIX A

LIST OF FIGURES
(Continued)

FIGURE NO.		PAGE
3-14	TURNING RADIUS REQUIRED FOR LOCATING C.G. ABOVE NORMAL POSITION FOR PREVENTION OF OVERTURN ON 10° SLOPE	A-41
3-15	TURNING RADIUS REQUIRED FOR LOCATING C.G. ABOVE NORMAL POSITION FOR PREVENTION OF OVERTURN ON 20° SLOPE	A-42
3-16	SAFE DRIVING CORRIDOR IN CASE OF STEERING FAILURE AS A FUNCTION OF VELOCITY, SLOPE, AND STEERING APPLICATION	A-43
3-17	PARKING ORIENTATION CONSTRAINTS	A-45
4-1	VEHICLE POWER CONSUMPTION VS SPEED - 0° SLOPE	A-47
4-2	VEHICLE POWER CONSUMPTION VS SPEED - 5° SLOPE	A-48
4-3	VEHICLE POWER CONSUMPTION VS SPEED - 10° SLOPE	A-49
4-4	LRV TRACTION DRIVE PERFORMANCE (DC MOTOR-HARMONIC DRIVE-DRIVE CONTROLLER) FULL VOLTAGE (36 V) PERFORMANCE	A-50
4-5	BATTERY CURRENT VS SPEED	A-51
4-6	TRACTION DRIVE-RATE OF TEMPERATURE INCREASE VS SOLAR ELEVATION ANGLE (FRONT WHEEL)	A-52
4-7	TRACTION DRIVE-RATE OF TEMPERATURE INCREASE VS ZENITH ANGLE (LEFT FRONT WHEEL)	A-53
4-8	TRACTION DRIVE-RATE OF TEMPERATURE INCREASE VS ZENITH ANGLE (LEFT REAR WHEEL)	A-54
4-9	TRACTION DRIVE-RATE OF TEMPERATURE INCREASE VS ZENITH ANGLE (RIGHT FRONT WHEEL)	A-55
4-10	TRACTION DRIVE-RATE OF TEMPERATURE INCREASE VS ZENITH ANGLE (RIGHT REAR WHEEL)	A-56
4-11	TRACTION DRIVE-RATE OF TEMPERATURE INCREASE VS SPEED (FRONT WHEEL)	A-57
4-12	TRACTION DRIVE-RATE OF TEMPERATURE INCREASE VS SLOPE (FRONT WHEEL)	A-58
4-13	VOLTAGE VS CURRENT DRAW PER BATTERY	A-60
4-14	BATTERY VOLTAGE VS STATE OF CHARGE	A-61
4-15	BATTERY RATE OF TEMPERATURE INCREASE WITH CURRENT DRAIN (RIGHT BATTERY; i.e., BATTERY #2)	A-62
4-16	BATTERY RATE OF TEMPERATURE INCREASE WITH CURRENT DRAIN (LEFT BATTERY; i.e., BATTERY #1)	A-63
4-17	BATTERY RATE OF TEMPERATURE CHANGE (DUST COVERS OPEN, LEFT BATTERY; i.e., BATTERY #1)	A-64

LS006-002-2H
LUNAR ROVING VEHICLE
OPERATIONS HANDBOOK

APPENDIX A

LIST OF TABLES

TABLE NO.		PAGE
3-I	PAYLOAD LIMITATIONS	A-24
3-II	COMPONENT TEMPERATURE LIMITS	A-30
3-III	SPEED RESTRICTIONS TO PREVENT EXCEEDING STRUCTURAL DESIGN LOADS	A-44
5-I	VIBRATION ENVIRONMENT AT LRV PAYLOAD INTERFACES	A-81
5-II	LRV STEADY STATE POWER CONSUMPTION	A-82
5-III	LRV SPEED CAPABILITY	A-83
5-IV	LRV CONTROLLABILITY SPEED LIMITS	A-84
6-I	LRV-1 NAVIGATION SYSTEM POWER CONSUMPTION	A-98
6-II	LRV-1 WEIGHT, C.G. AND MOMENTS OF INERTIA	A-99
6-III	LRV-1 LUNAR OPERATIONAL WEIGHT DISTRIBUTION - STATIC, LEVEL LURAIN CONDITION	A-100
6-IV	SUMMARY OF LRV-1 MOBILITY PARAMETERS	A-101

Mission   J   Basic Date 2/5/71   Change Date 4/19/71   Page   Av

LS006-002-2H
LUNAR ROVING VEHICLE
OPERATIONS HANDBOOK
APPENDIX A

1.0   INTRODUCTION

1.1   PURPOSE

This document is intended to supply LRV performance data necessary for mission planning and flight control personnel to adequately plan and control operation of the LRV on the lunar surface. The data contained herein is to be used in conjunction with the LRV Operations Handbook which contains crew operating procedures, timelines and system description.

1.2   CONTENT

This Appendix A contains data on LRV configuration and defines constraints and operational limitations of the various vehicle subsystems. In addition subsystems performance data are provided as well as overall vehicle performance data. The final section of the Appendix contains data defining constraints and performance data unique to each individual LRV.

1.3   AMENDMENTS

Amendments to data contained in the appendix will be issued as additional data becomes available from LRV test programs and/or as more refined data becomes available from updated analyses. When such amendments are issued, they will be approved and signed by the MSFC LRV Program Manager. MSFC will then transmit one printing master of each amendment to MSC. Reproduction and control of amendment distribution at MSC will be accomplished by MSC.

1.4   SELECTED ABBREVIATIONS AND ACRONYMS

The following abbreviations and acronyms are used throughout this document:

DCE	-	Drive Controller Electronics
DGU	-	Directional Gyro Unit
GCTA	-	Ground Controlled Television Assembly
IPI	-	Integrated Position Indicator
KPH	-	Kilometers per Hour
LCRU	-	Lunar Communications Relay Unit
LRV	-	Lunar Roving Vehicle
NSS	-	Navigation Subsystem
PSD	-	Power Spectral Density
PWM	-	Pulse Width Modulator
SPU	-	Signal Processing Unit
TBD	-	To be Determined

LS006-002-2H
LUNAR ROVING VEHICLE
OPERATIONS HANDBOOK
APPENDIX A

2.0 CONFIGURATION

The illustrations included in this section describe the configuration of the Lunar Roving Vehicle, details and locations of major vehicle subsystems, and important dimensions of the vehicle elements. Figure 2-1 describes the basic LRV configuration and identifies and locates the principal LRV subsystems. Stowed payload is not shown installed on the vehicle but payload mounting interfaces are identified. Important LRV overall dimensions are shown on Figure 2-2.

2.1 CHASSIS CONFIGURATION

The configuration details and principal dimensions of the LRV chassis are shown on Figure 2-3. The floor panels are shown removed to provide details on the primary and secondary structural members.

2.2 SUSPENSION SYSTEM CONFIGURATION

Major components of the LRV Suspension System are shown and identified on Figure 2-4. The torsion bars serve to unfold the wheels during the deployment operations and to provide both contour following and chassis resiliency during operation on the lurain.

2.3 STEERING SYSTEM CONFIGURATION

Major components of the LRV Steering System are shown and identified on Figure 2-5. Only one decoupling ring is provided for each steering system, fore and aft. The aft decoupling ring is located on the right side of the vehicle; the front decoupling ring is located on the left side of the vehicle.

2.4 TRACTION DRIVE CONFIGURATION

Internal configuration of the LRV Traction Drive is shown on Figure 2-6.

2.5 LRV WHEEL CONFIGURATION

The LRV Wheel construction details and critical dimensions are shown on Figure 2-7. The decoupling devices are actuated to provide free-wheeling capability in the event of drive seizure. Wheel deflection characteristics are shown in Figure 2-8.

2.6 LRV CREW STATION CONFIGURATION

The major components comprising the LRV Crew Station are identified and located on Figure 2-10A. The layout and dimensional characteristics of the control and display console portion of the Crew Station are shown on Figure 2-10B.

LS006-002-2H
LUNAR ROVING VEHICLE
OPERATIONS HANDBOOK
APPENDIX A

2.7 LRV POWER SYSTEM CONFIGURATION

The configuration of the typical LRV Battery and Dust Cover Assembly is defined in Figure 2-11. Battery dimensions are provided on this figure.

2.8 LRV NAVIGATION SYSTEM CONFIGURATION

The components comprising the LRV Navigation System are identified and located on Figure 2-12. Control and Display Console mounted items included in the Navigation System are shown on Figure 2-10B.

2.9 LRV THERMAL CONTROL CONFIGURATION

The major components of the LRV Thermal Control System are identified and located on Figure 2-13. The dust covers are raised to permit radiation of heat from the space radiators.

2.10 LRV PAYLOAD INTERFACES

The Payload Interfaces with the LRV are shown on Figures 2-14A, 2-14B, 2-14C, and 2-14D. Portions of the interface connectors which are not part of the basic LRV system but which are permanently installed on the LRV are identified on these figures.

LS006-002-2H

LUNAR ROVING VEHICLE
OPERATIONS HANDBOOK
APPENDIX A

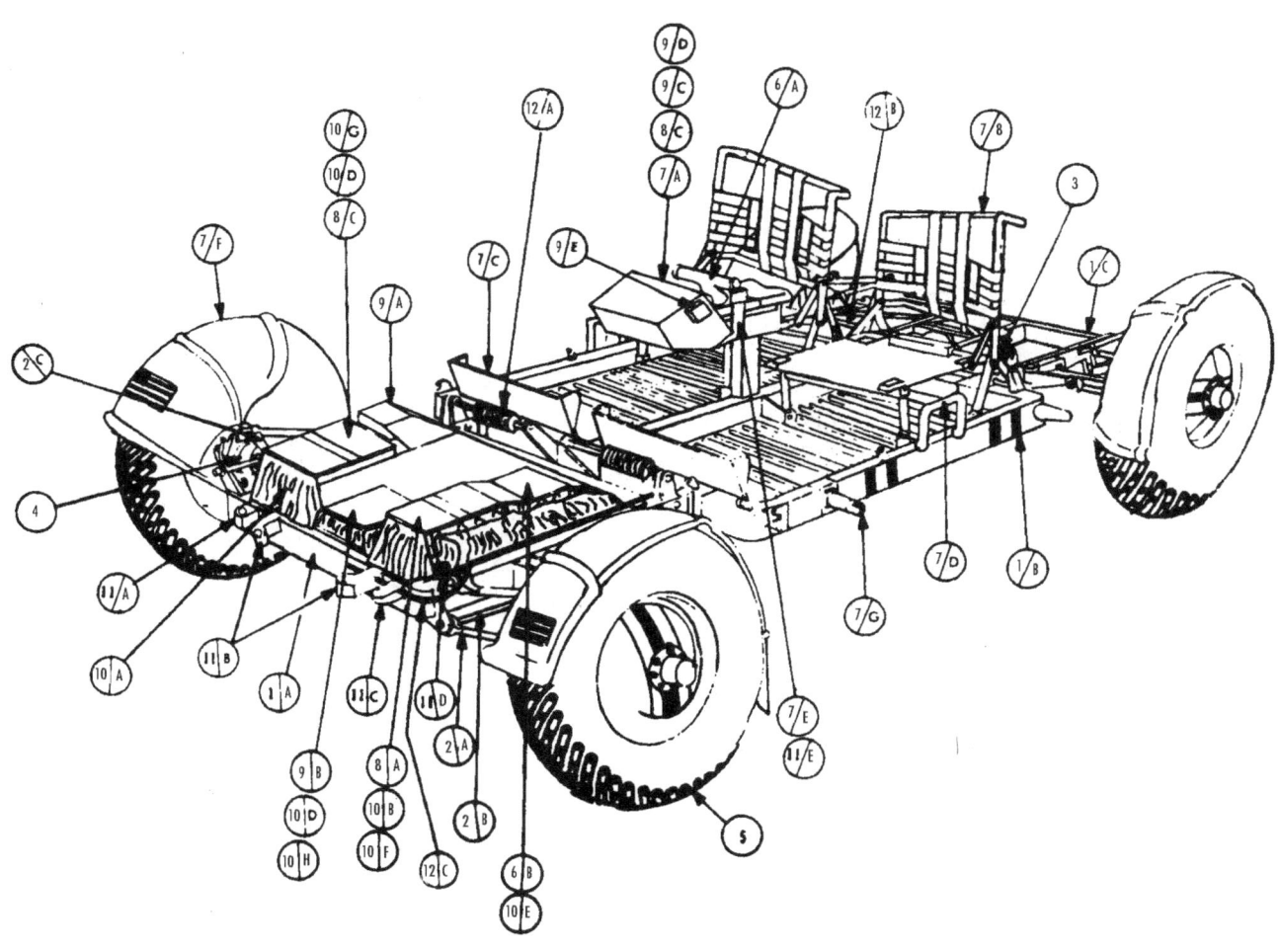

① CHASSIS
  A. FORWARD CHASSIS
  B. CENTER CHASSIS
  C. AFT CHASSIS

② SUSPENSION SYSTEM
  A. SUSPENSION ARMS (UPPER AND LOWER)
  B. TORSON BARS (UPPER AND LOWER)
  C. DAMPER

③ STEERING SYSTEM (FORWARD AND REAR)

④ TRACTION DRIVE

⑤ WHEEL

⑥ DRIVE CONTROL
  A. HAND CONTROLLER
  B. DRIVE CONTROL ELECTRONICS (DEL)

⑦ CREW STATION
  A. CONTROL AND DISPLAY CONSOLE
  B. SEAT
  C. FOOTREST
  D. OUTBOARD HANDHOLD
  E. INBOARD HANDHOLD
  F. FENDER
  G. TOEHOLD
  H. SEAT BELT

⑧ POWER SYSTEM
  A. BATTERY #1
  B. BATTERY #2
  C. INSTRUMENTATION

⑨ NAVIGATION
  A. DIRECTIONAL GYRO UNIT (DGU)
  B. SIGNAL PROCESSING UNIT (SPU)
  C. INTEGRATED POSITION INDICATOR (IPI)
  D. SUN SHADOW DEVICE
  E. VEHICLE ATTITUDE INDICATOR

⑩ THERMAL CONTROL
  A. INSULATION BLANKET
  B. BATTERY NO. 1 DUST COVER
  C. BATTERY NO. 2 DUST COVER
  D. SPU DUST COVER
  E. DCE THERMAL CONTROL UNIT
  F. BATTERY NO. 1 RADIATOR
  G. BATTERY NO. 2 RADIATOR
  H. SPU THERMAL CONTROL UNIT

⑪ PAYLOAD INTERFACE
  A. TV CAMERA RECEPTACLE
  B. LCRU RECEPTACLE
  C. HIGH GAIN ANTENNA RECEPTACLE
  D. AUXILIARY CONNECTOR
  E. LOW GAIN ANTENNA RECEPTACLE

⑫ DEPLOYMENT COMPONENTS
  A. FWD CHASSIS DEPLOYMENT TORSION SPRINGS
  B. REAR CHASSIS DEPLOYMENT TORSION BARS
  C. SADDLE RELEASE CABLE

FIGURE 2-1 LRV CONFIGURATION WITHOUT STOWED PAYLOAD

Mission J Basic Date 2/5/71 Change Date 4/19/71 Page A-4

LS006-002-2H

LUNAR ROVING VEHICLE
OPERATIONS HANDBOOK
APPENDIX A

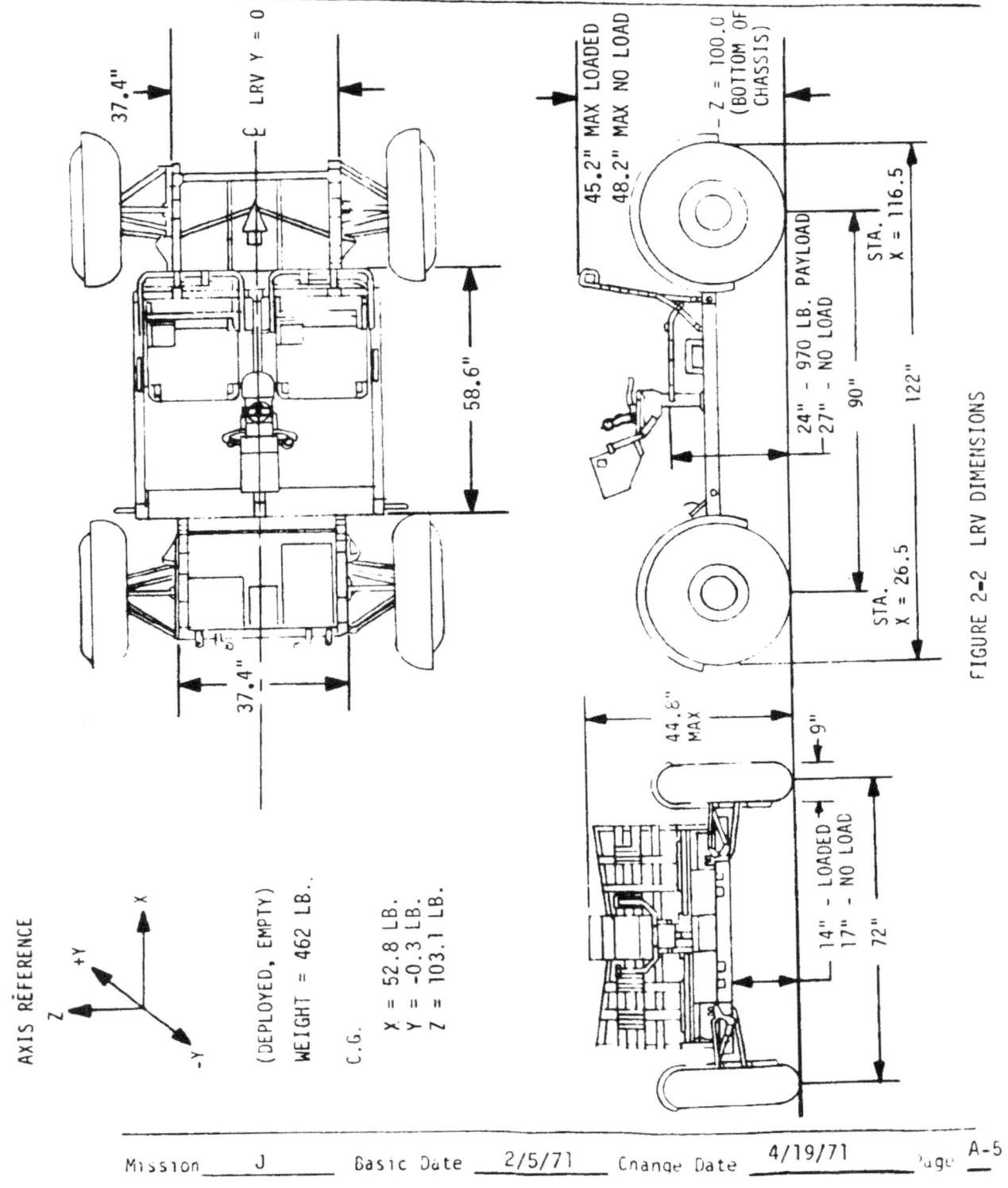

FIGURE 2-2 LRV DIMENSIONS

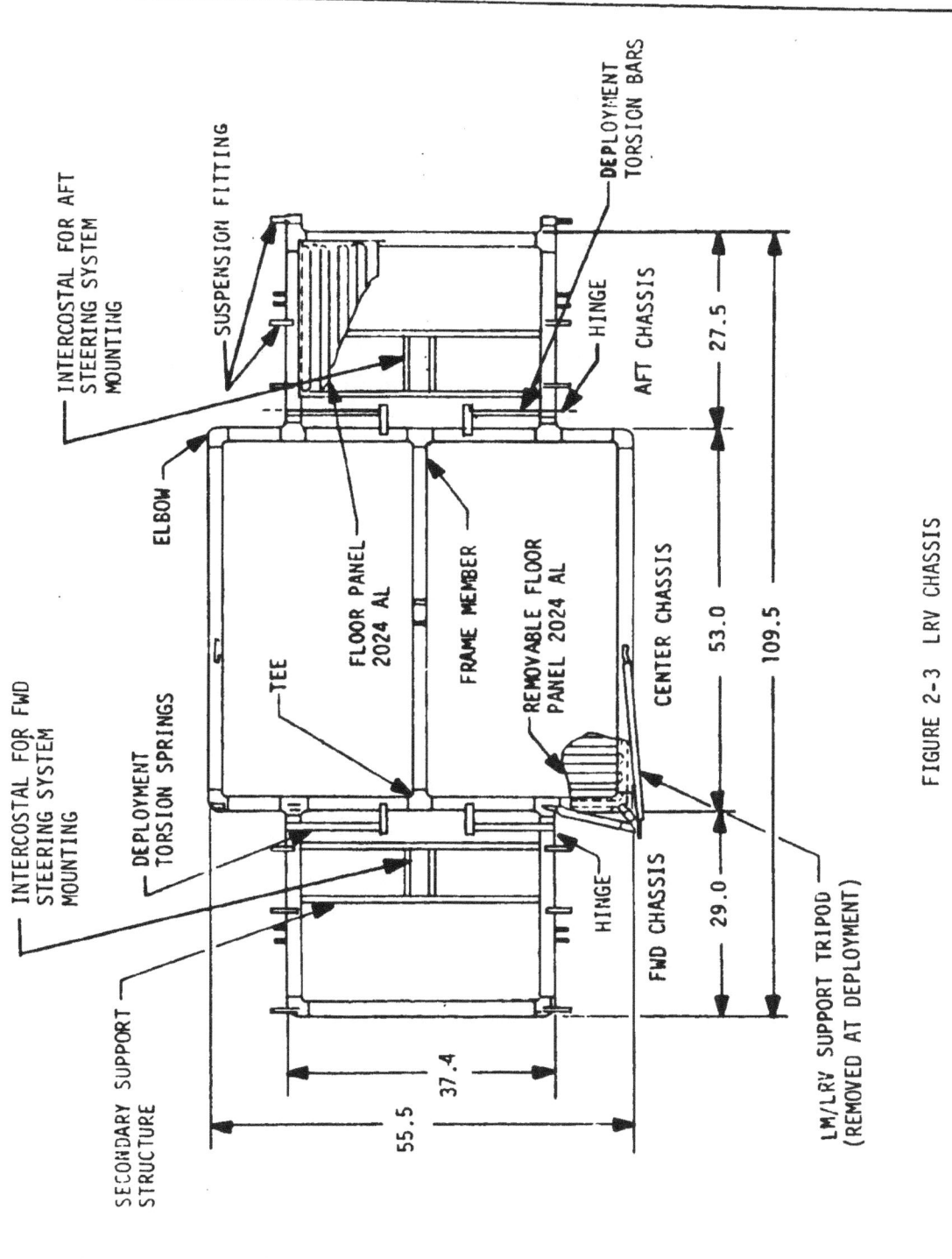

FIGURE 2-3 LRV CHASSIS

LS006-002-2H

LUNAR ROVING VEHICLE
OPERATIONS HANDBOOK
APPENDIX A

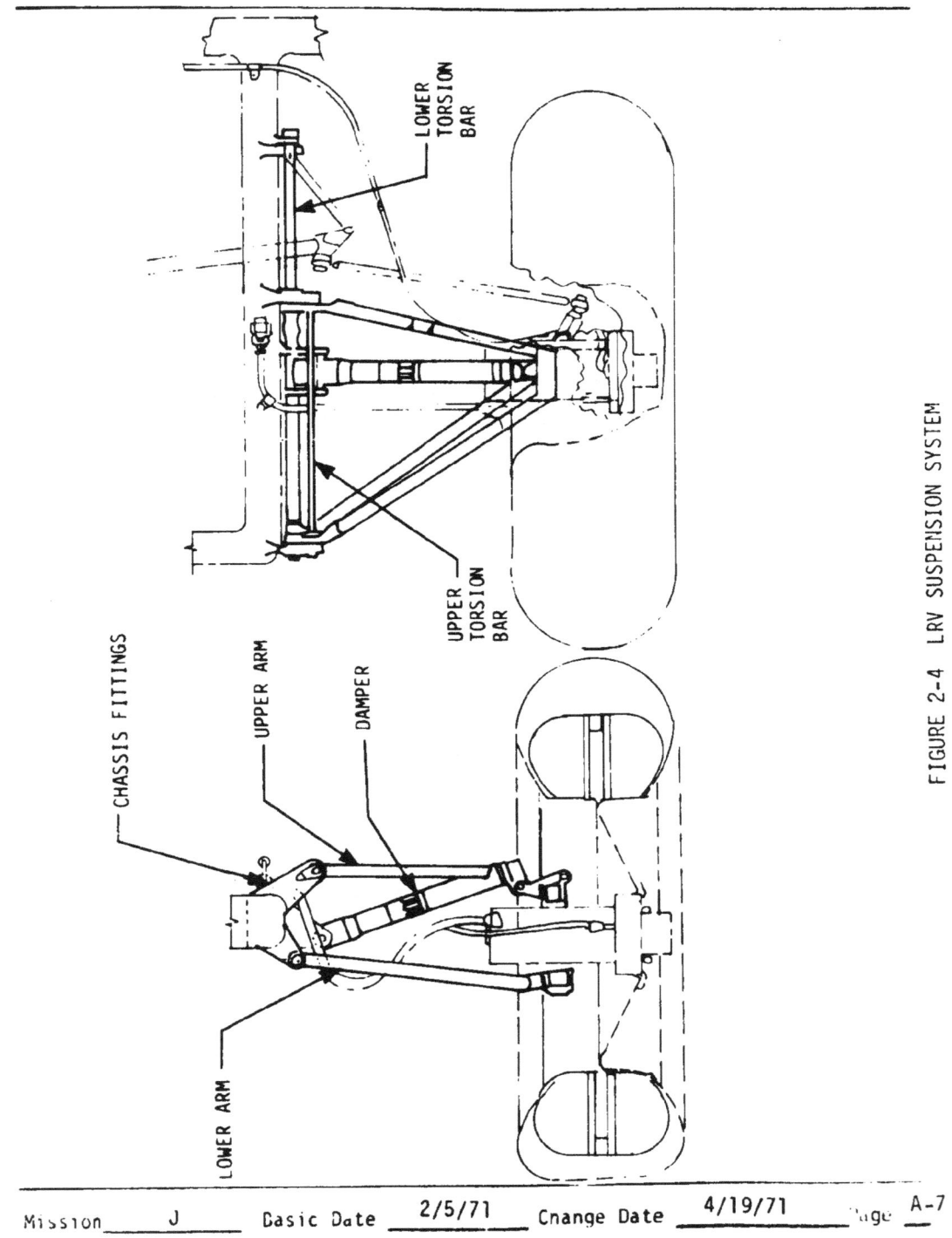

FIGURE 2-4 LRV SUSPENSION SYSTEM

LS006-002-2H
LUNAR ROVING VEHICLE
OPERATIONS HANDBOOK
APPENDIX A

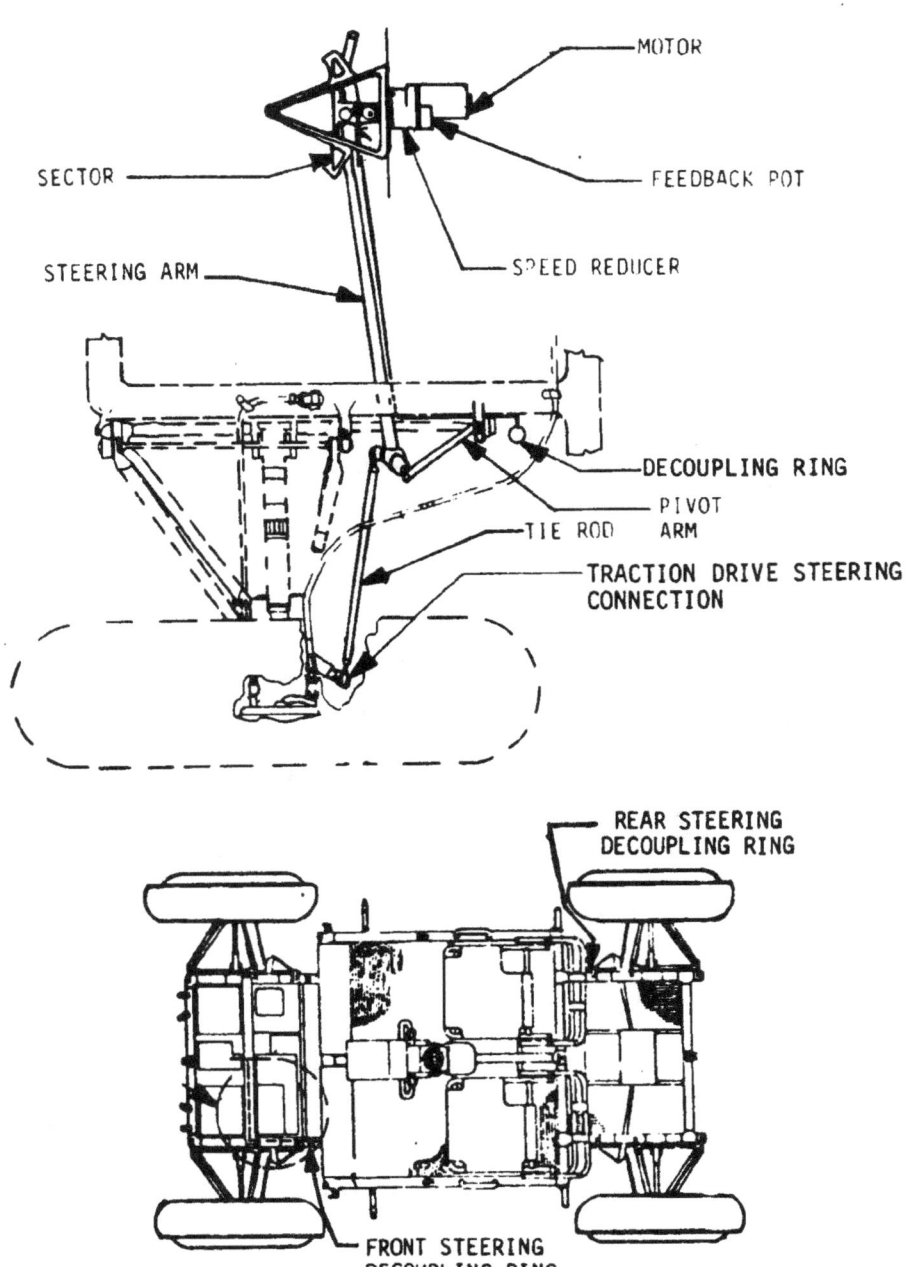

FIGURE 2-5 LRV STEERING SYSTEM

LS006-002-2H
LUNAR ROVING VEHICLE
OPERATIONS HANDBOOK
APPENDIX A

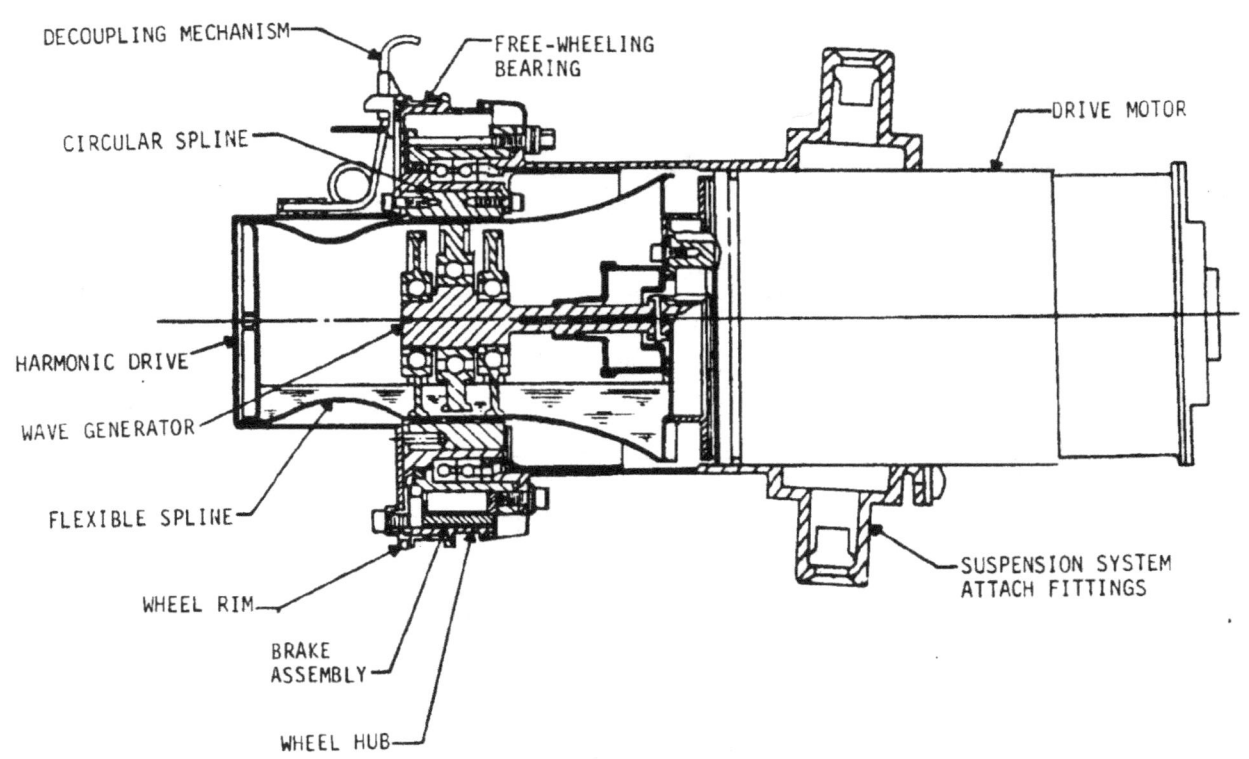

FIGURE 2-6 LRV TRACTION DRIVE

LS006-002-2H
LUNAR ROVING VEHICLE
OPERATIONS HANDBOOK
APPENDIX A

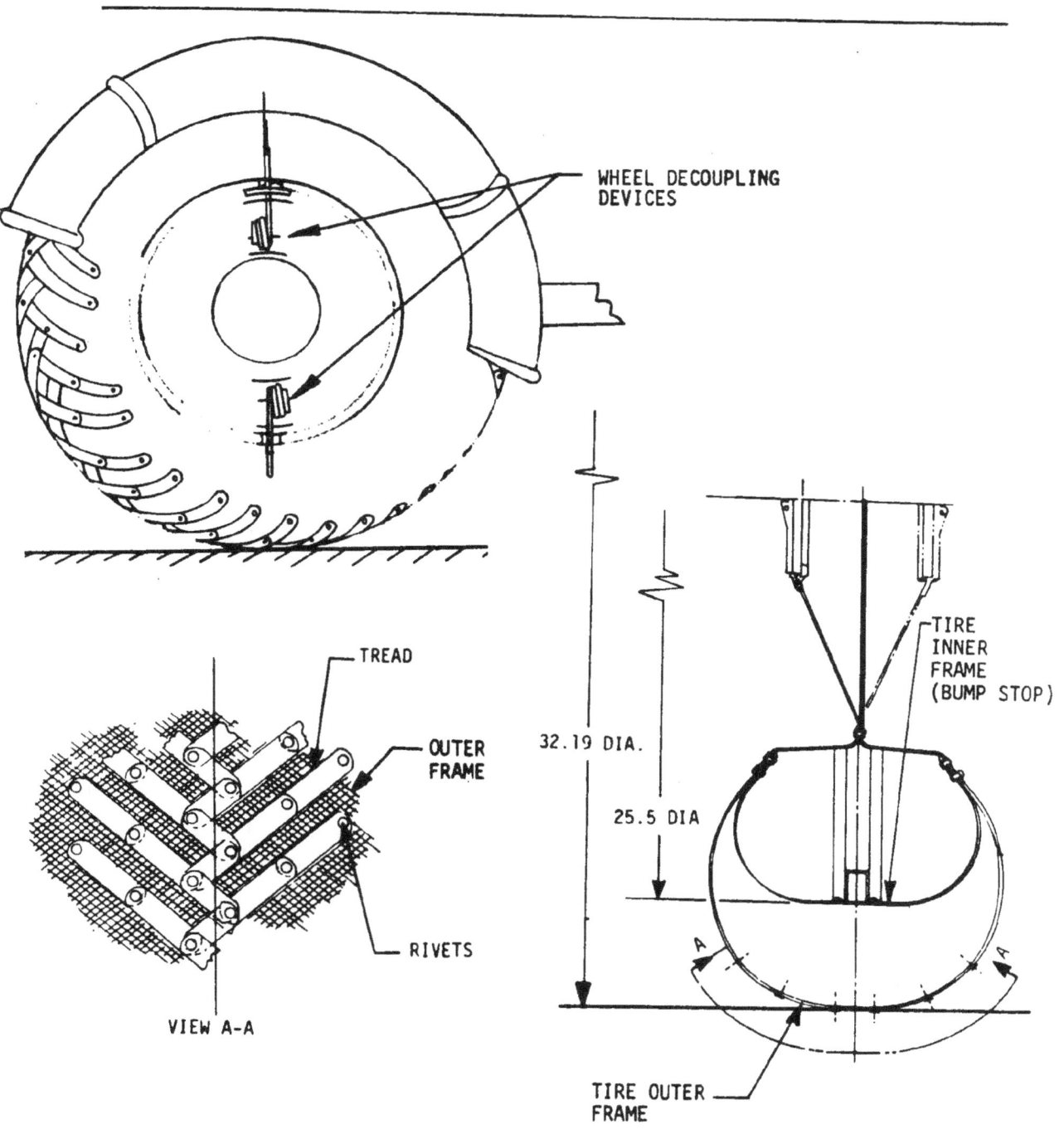

FIGURE 2-7   LRV WHEEL

Mission __J__   Basic Date __2/5/71__   Change Date __4/19/71__   Page __A-10__

LS006-002-2H
LUNAR ROVING VEHICLE
OPERATIONS HANDBOOK
APPENDIX A

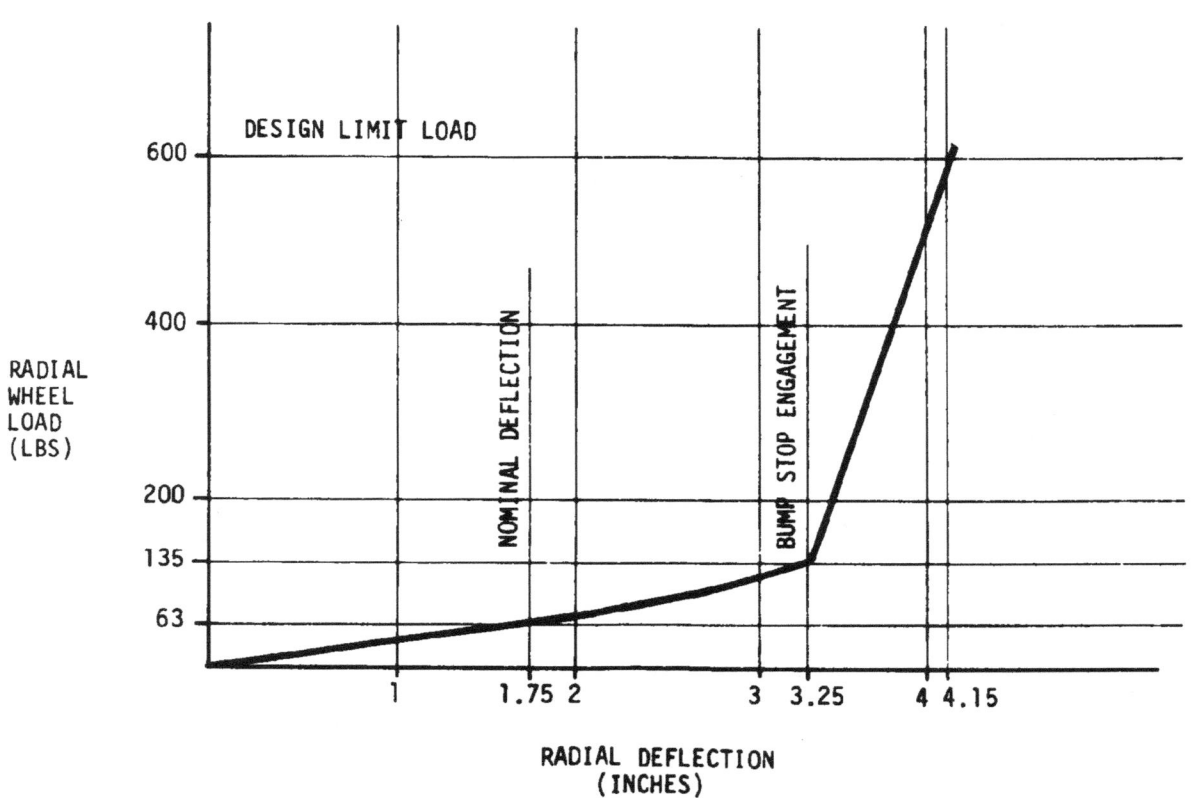

FIGURE 2-8 LRV WHEEL DEFLECTION VS LOAD

LS006-002-2H
LUNAR ROVING VEHICLE
OPERATIONS HANDBOOK
APPENDIX A

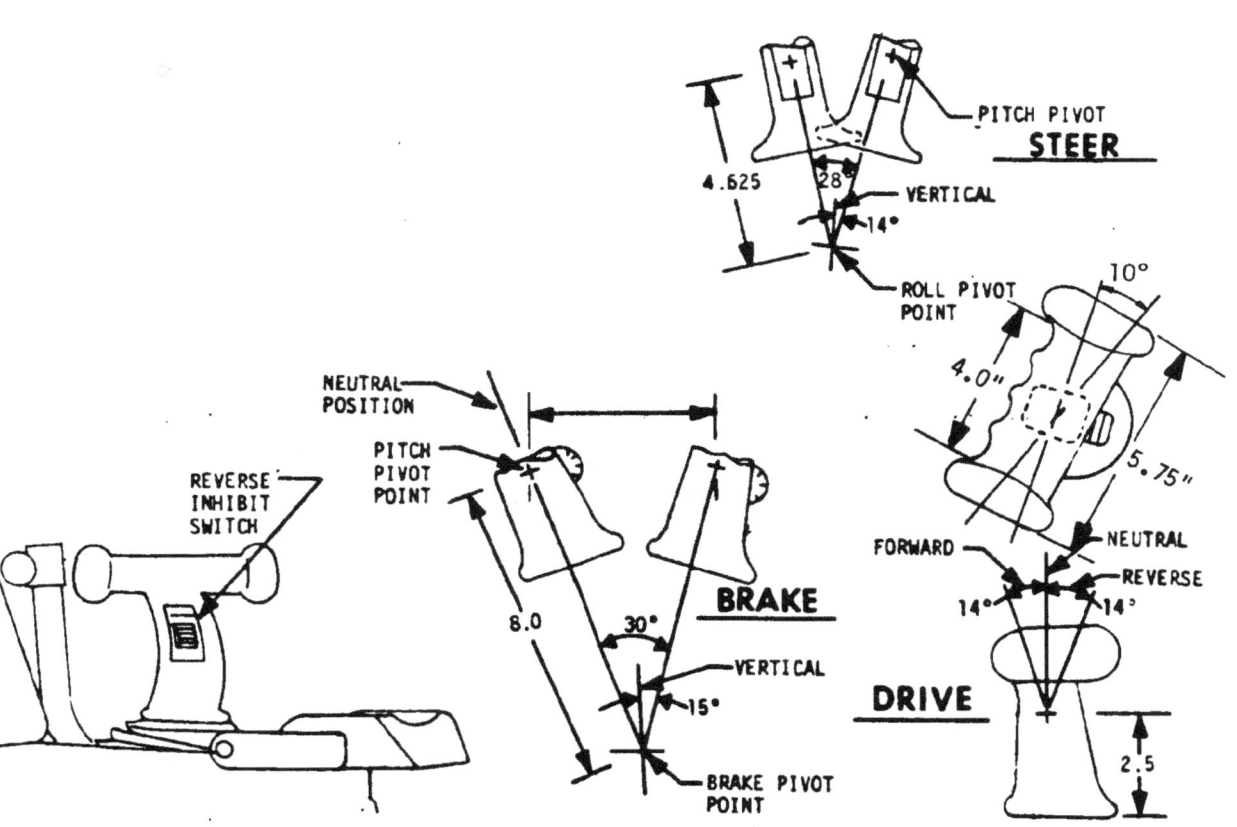

FIGURE 2-9A LRV DRIVE CONTROL - HAND CONTROLLER

LS006-002-2H

LUNAR ROVING VEHICLE
OPERATIONS HANDBOOK
APPENDIX A

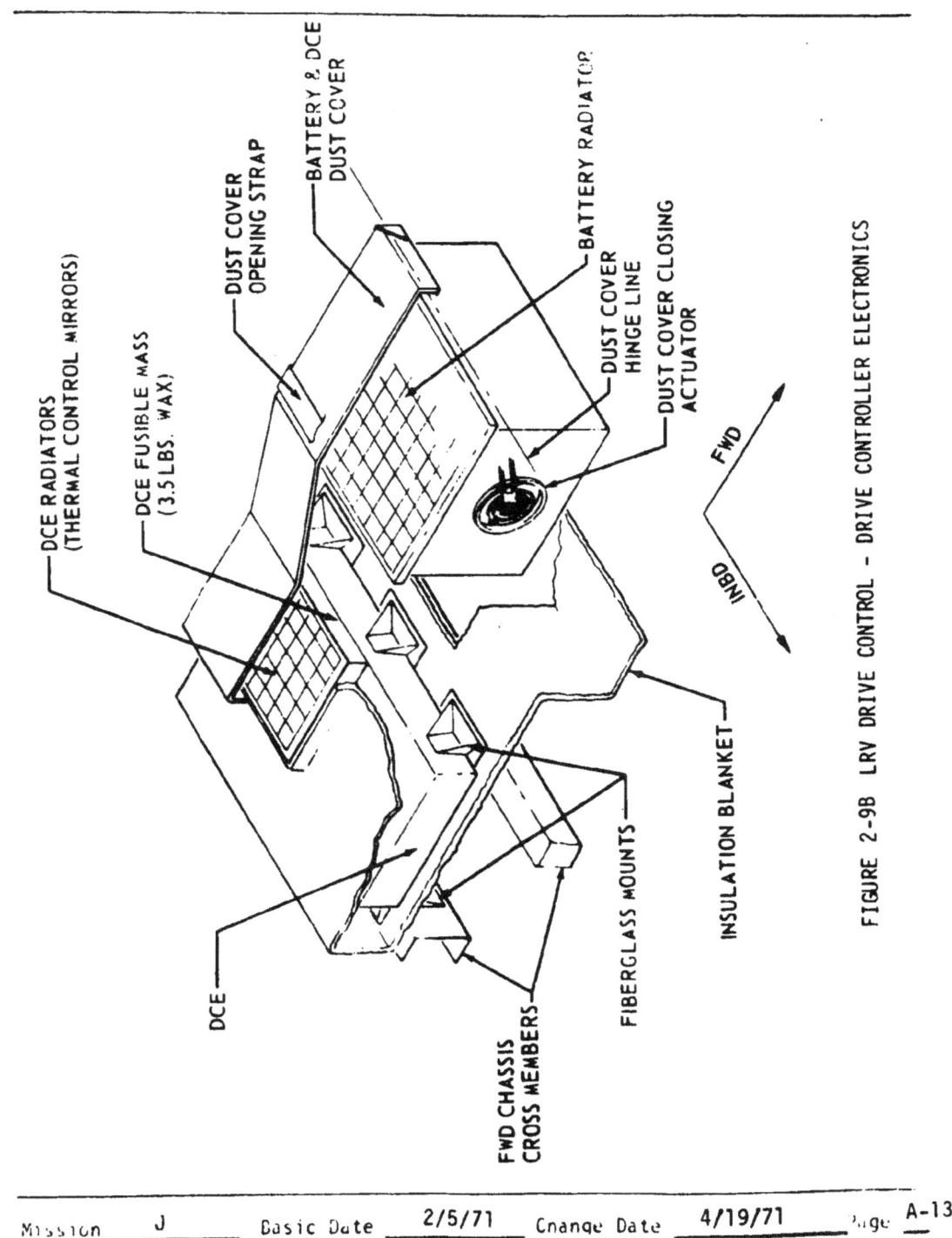

FIGURE 2-9B LRV DRIVE CONTROL – DRIVE CONTROLLER ELECTRONICS

LS006-002-2H
LUNAR ROVING VEHICLE
OPERATIONS HANDBOOK
APPENDIX A

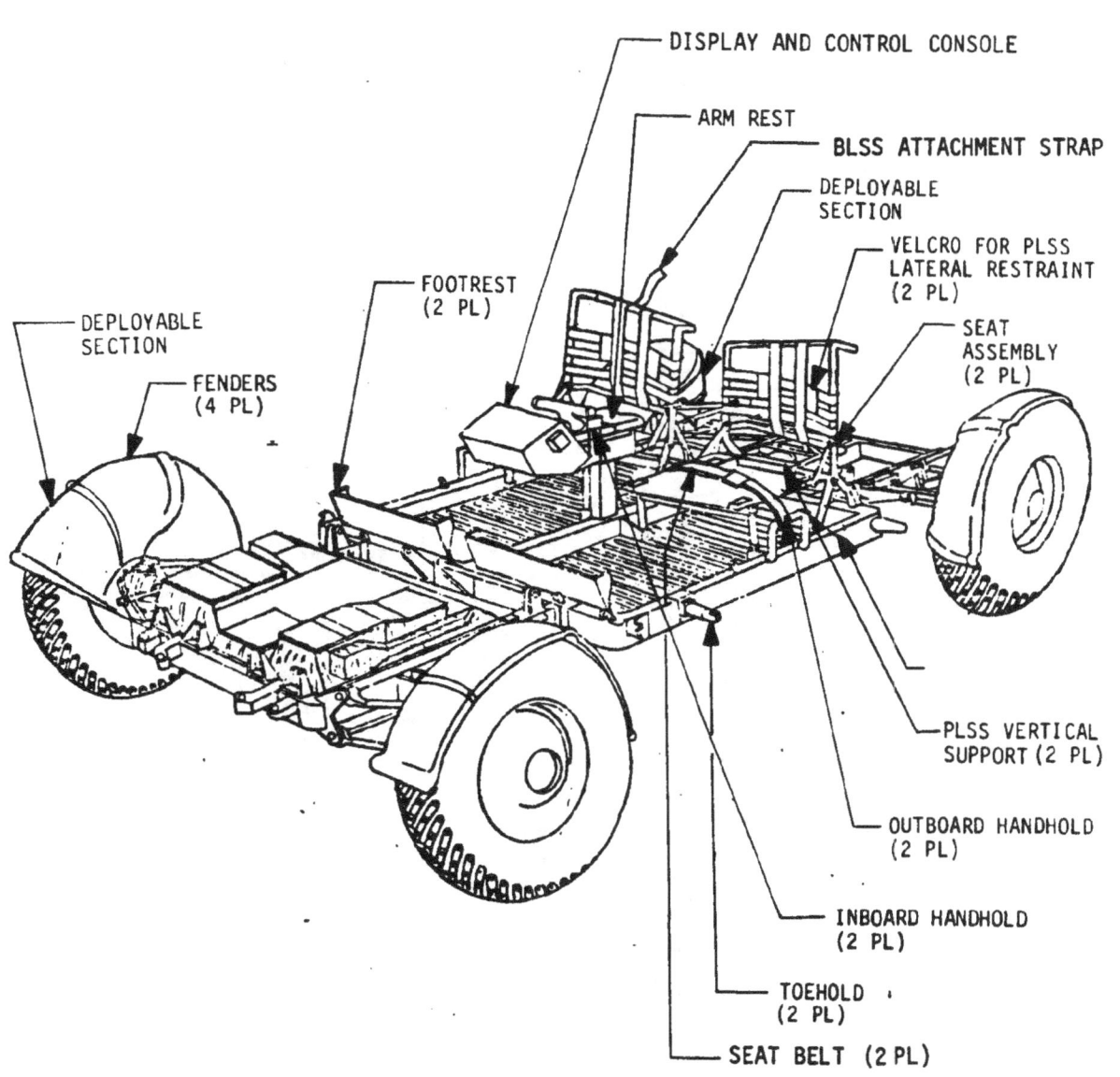

FIGURE 2-10A LRV CREW STATION COMPONENTS

LS006-002-2H
LUNAR ROVING VEHICLE
OPERATIONS HANDBOOK
APPENDIX A

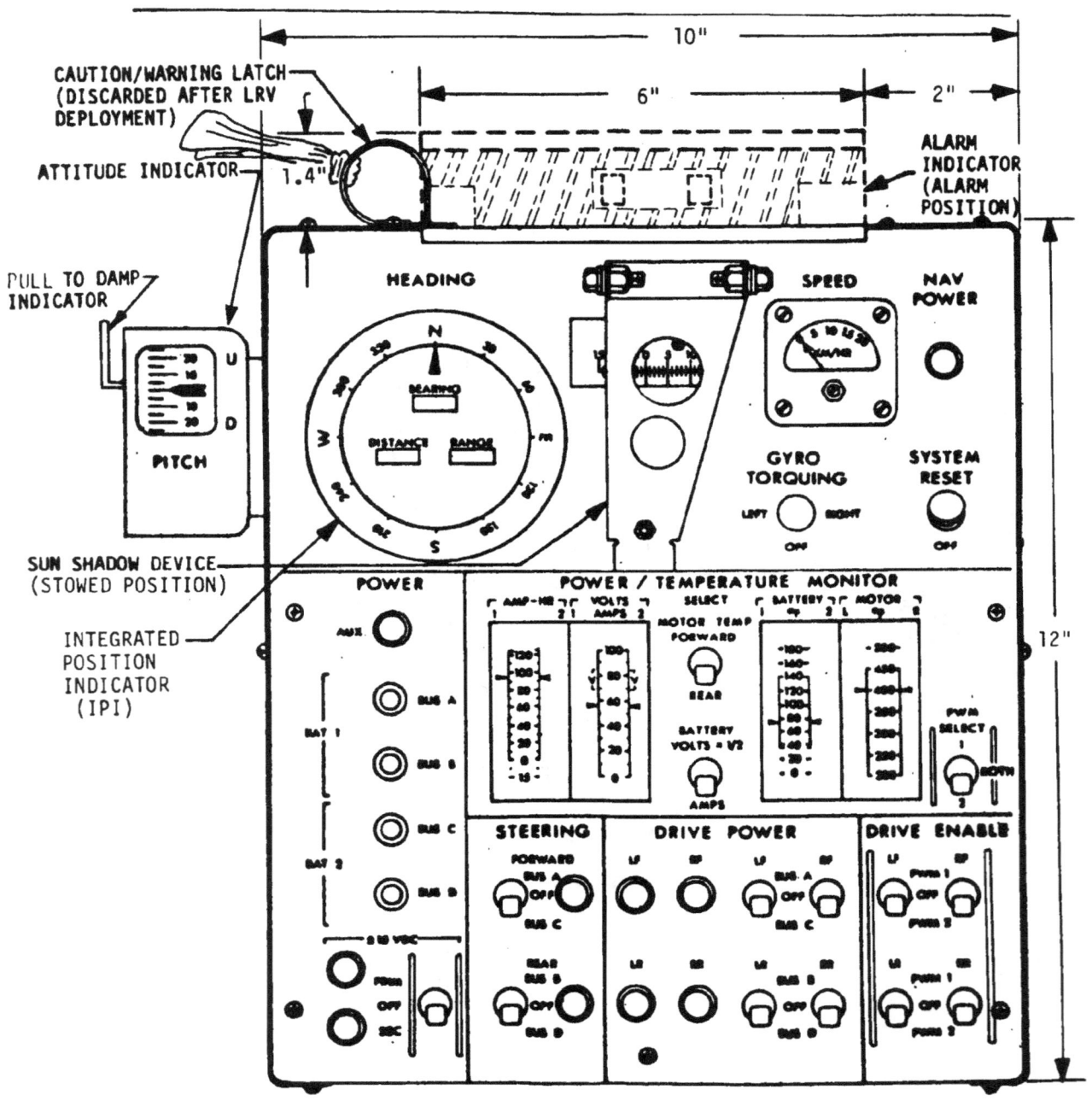

FIGURE 2-9A LRV CREW STATION COMPONENTS - CONTROL AND DISPLAY CONSOLE

Mission __J__ Basic Date __2/5/71__ Change Date __4/19/71__ Page __A-15__

LUNAR ROVING VEHICLE
OPERATIONS HANDBOOK
APPENDIX A

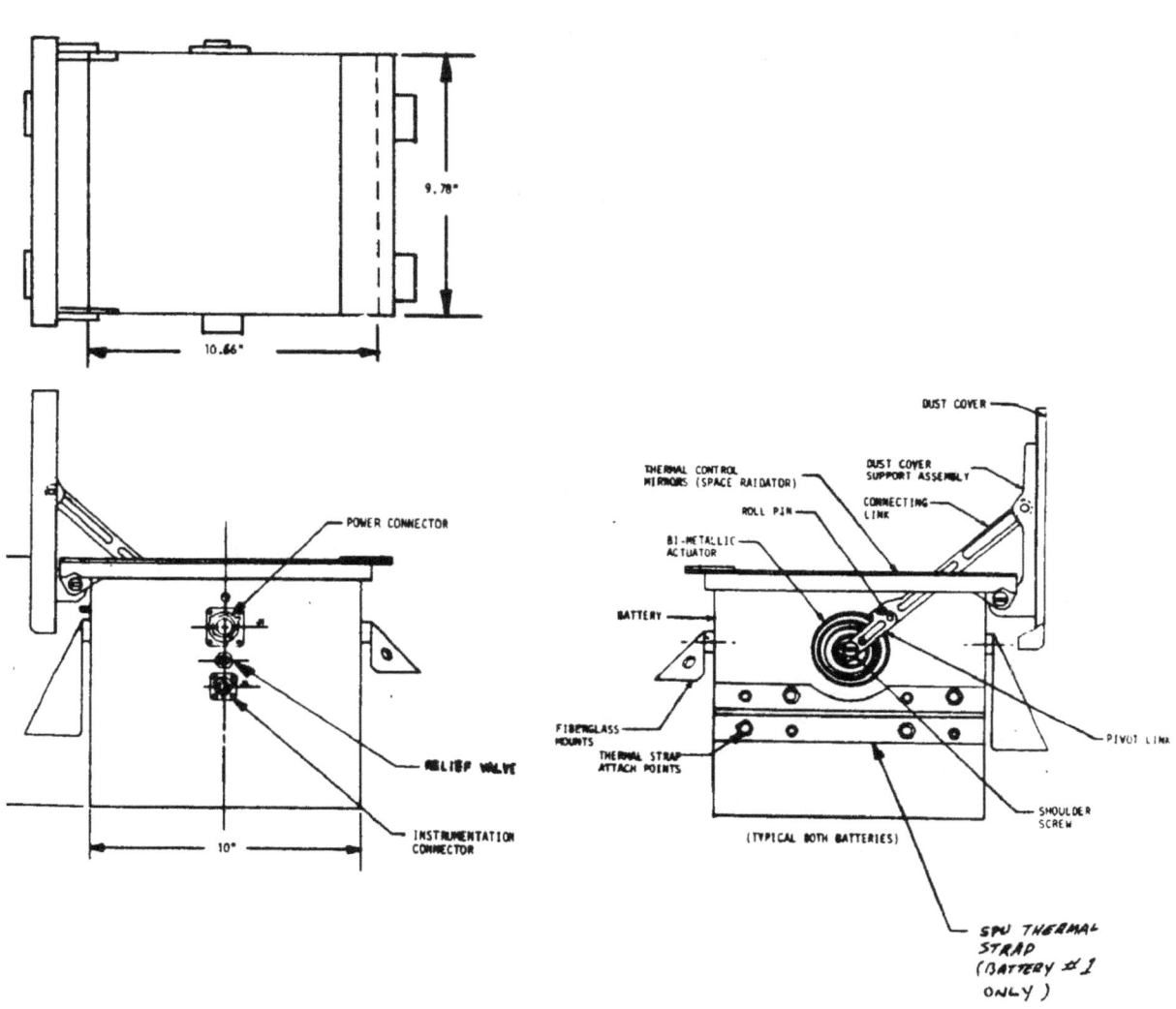

FIGURE 2-11 LRV BATTERY AND DUST COVER ASSEMBLY

LS006-002-2H
LUNAR ROVING VEHICLE
OPERATIONS HANDBOOK
APPENDIX A

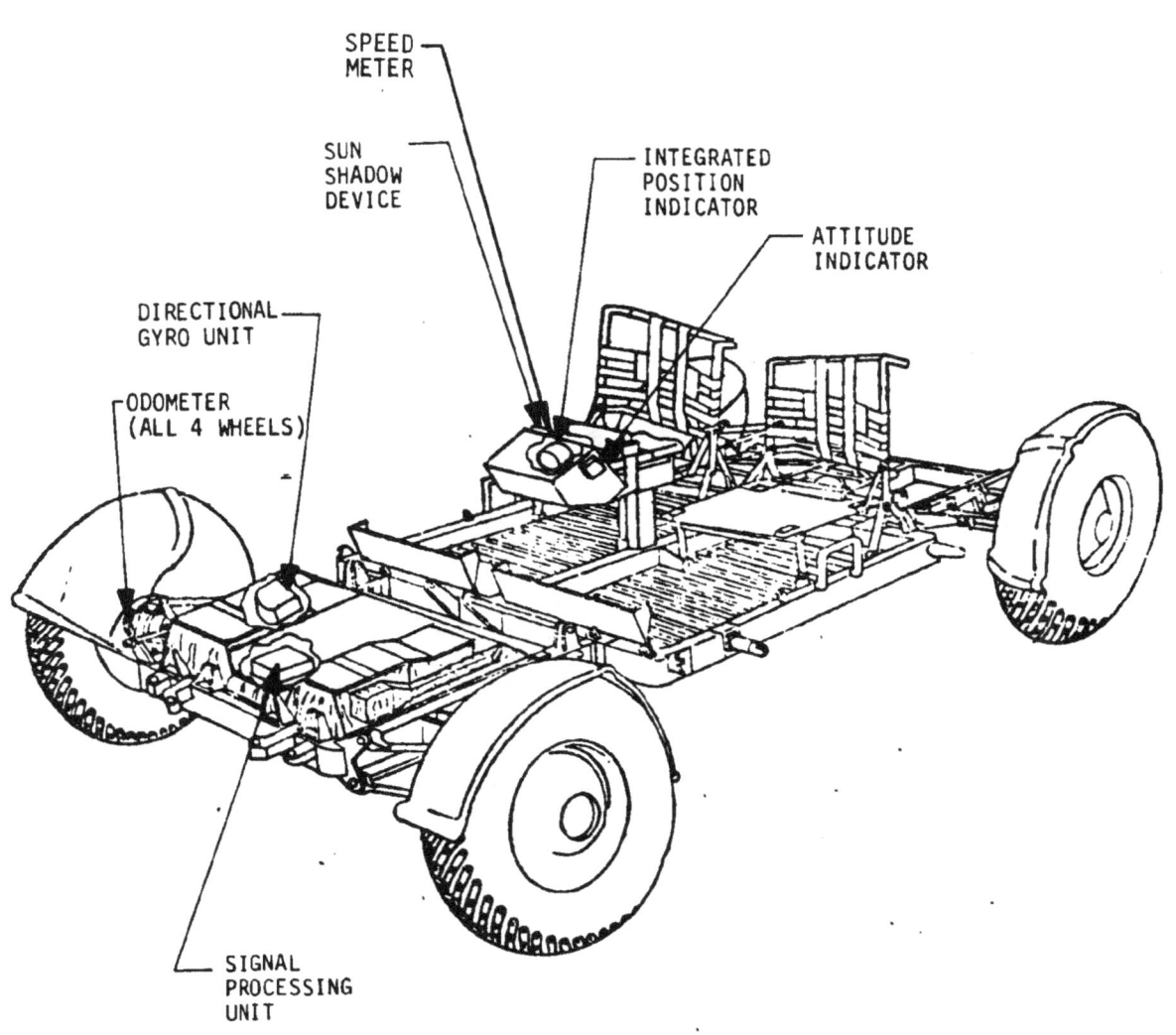

FIGURE 2-12   LRV NAVIGATION COMPONENTS

LS006-002-2H
LUNAR ROVING VEHICLE
OPERATIONS HANDBOOK
APPENDIX A

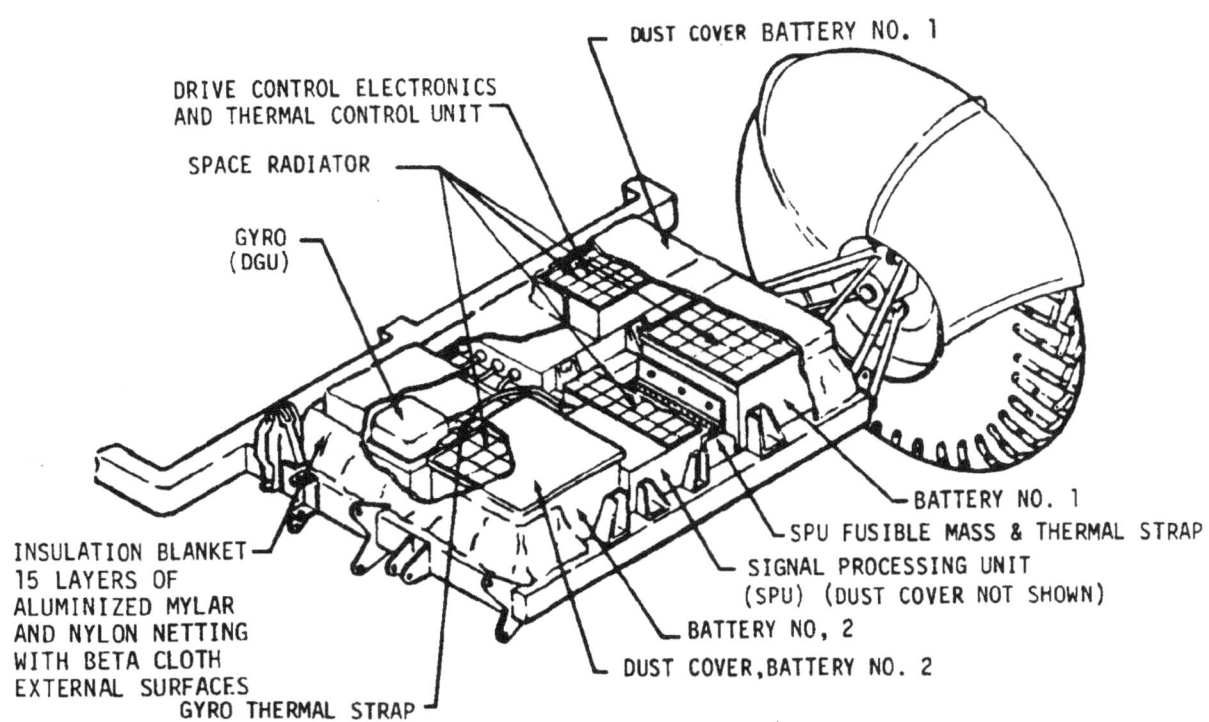

FIGURE 2-13 LRV THERMAL CONTROL

LS006-002-2H
LUNAR ROVING VEHICLE
OPERATIONS HANDBOOK
APPENDIX A

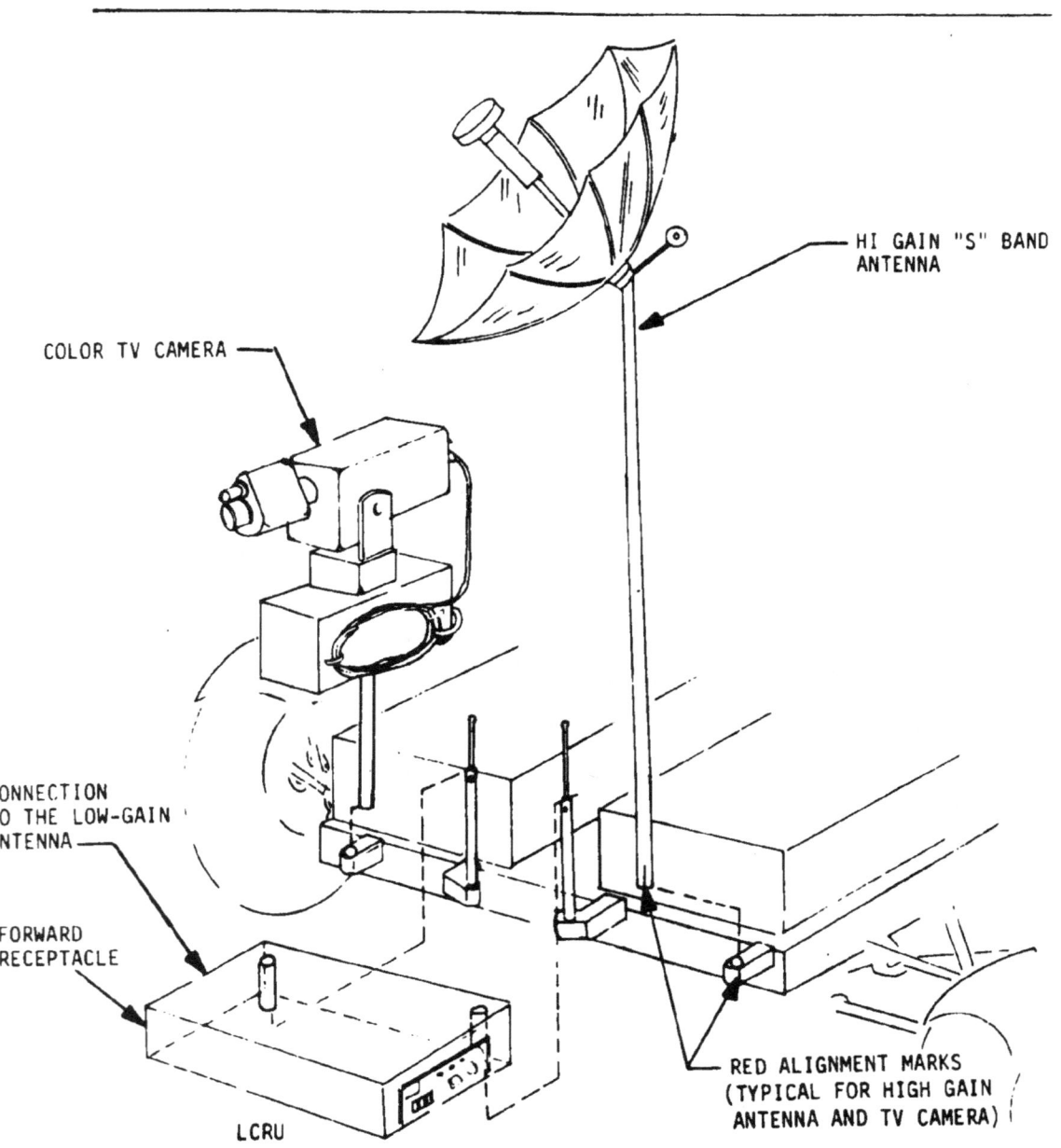

FIGURE 2-14A PAYLOAD INTERFACE-LCRU, HIGH GAIN ANTENNA, TV CAMERA INTERFACES

LS006-002-2H
LUNAR ROVING VEHICLE
OPERATIONS HANDBOOK
APPENDIX A

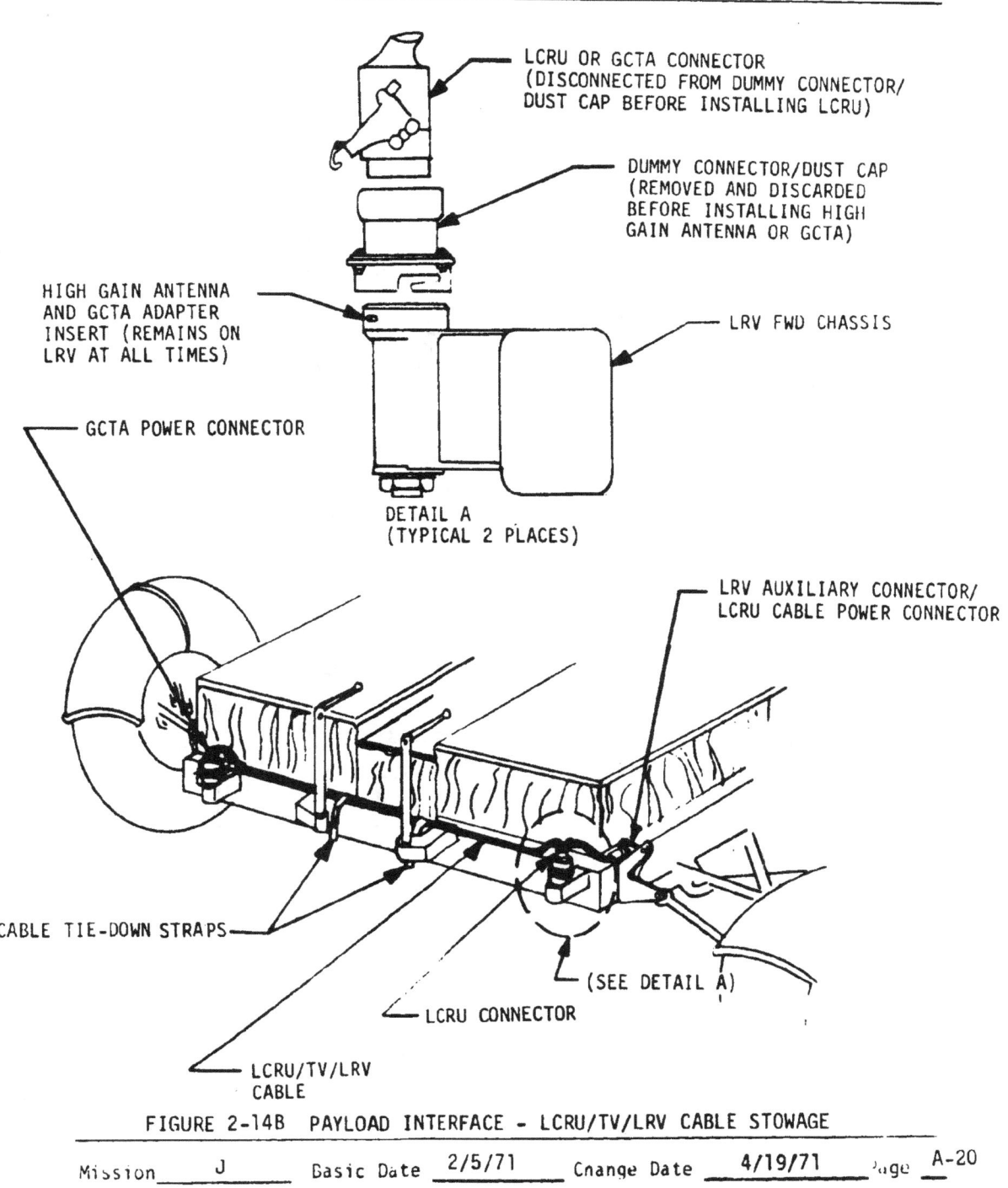

FIGURE 2-14B PAYLOAD INTERFACE - LCRU/TV/LRV CABLE STOWAGE

LS006-002-2H
LUNAR ROVING VEHICLE
OPERATIONS HANDBOOK
APPENDIX A

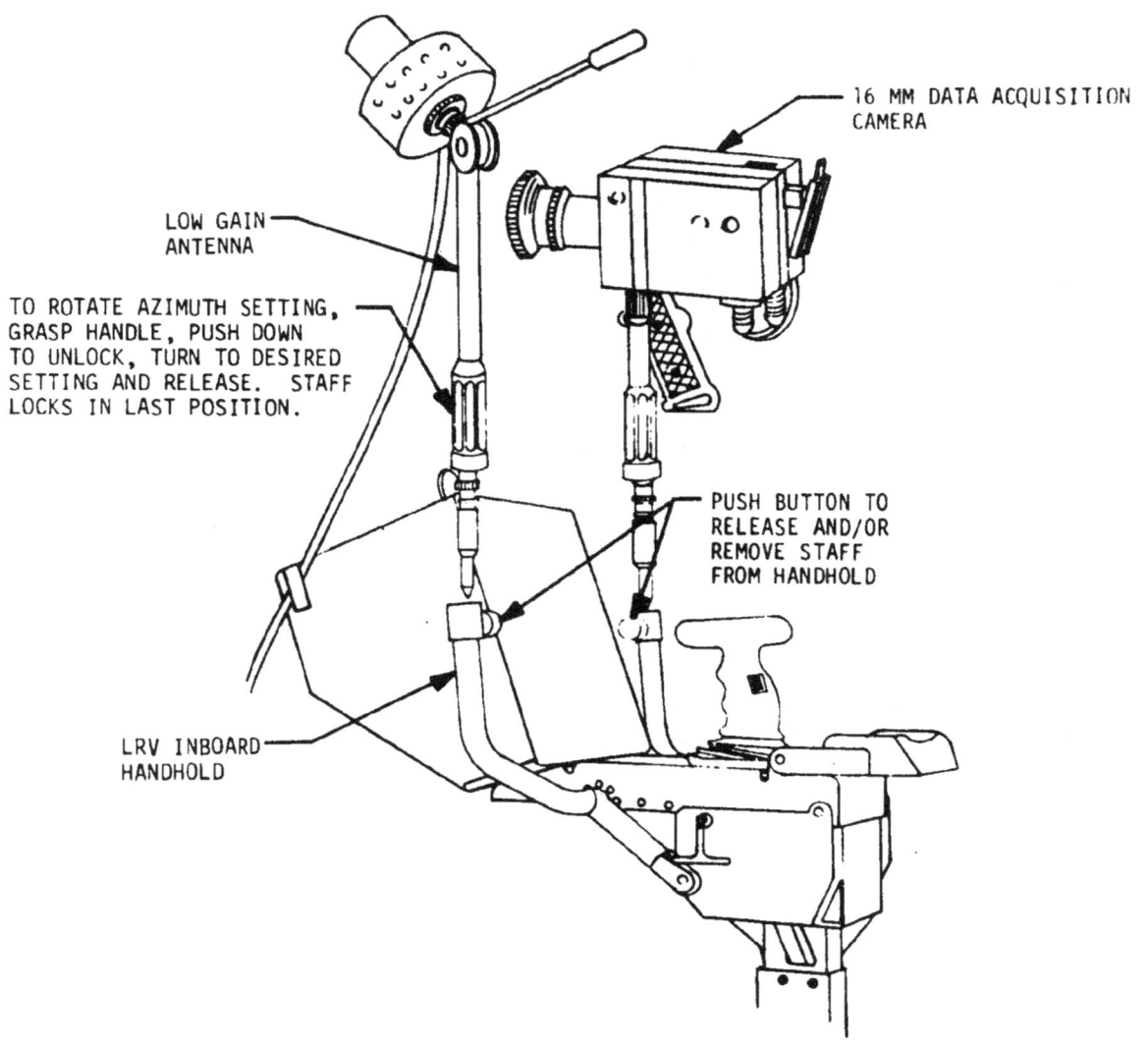

FIGURE 2-14C   PAYLOAD INTERFACE - LOW GAIN ANTENNA AND 16 MM DAC INTERFACE

LS006-002-2H
LUNAR ROVING VEHICLE
OPERATIONS HANDBOOK
APPENDIX A

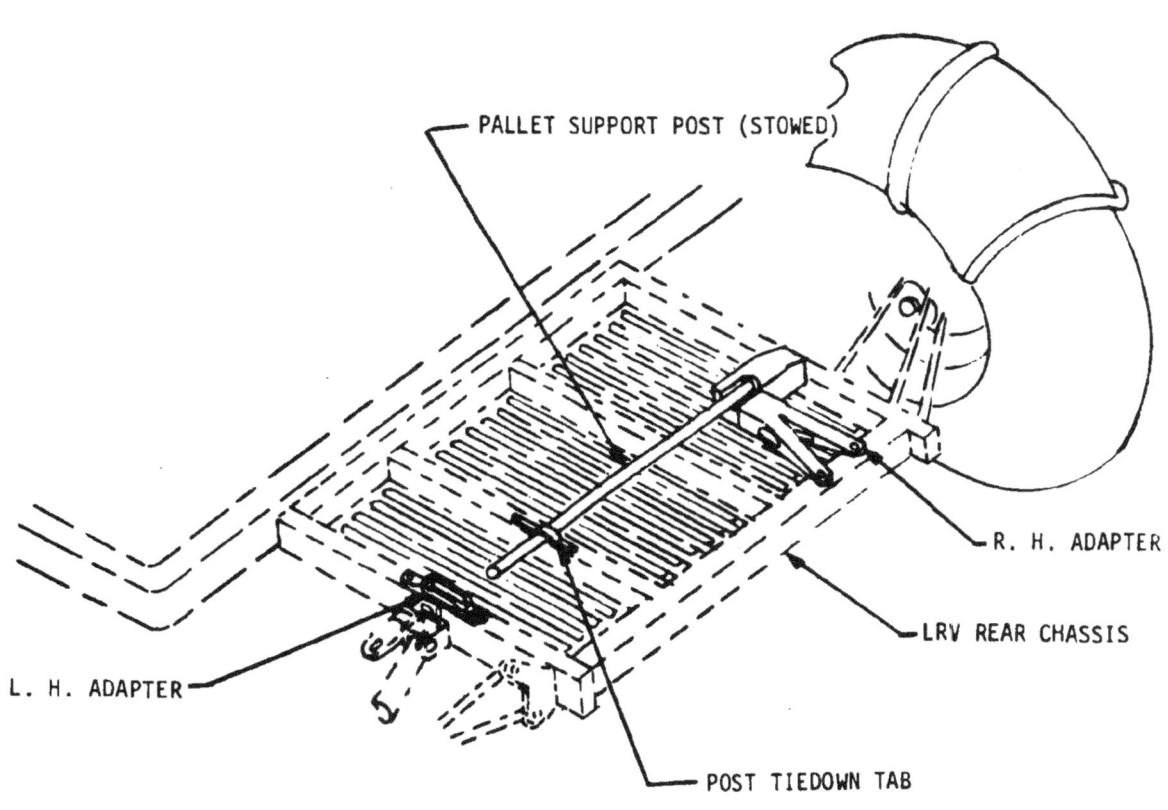

FIGURE 2-14D PAYLOAD INTERFACE - LRV REAR PAYLOAD PALLET ADAPTERS

LS006-002-2H
LUNAR ROVING VEHICLE
OPERATIONS HANDBOOK
APPENDIX A

3.0 CONSTRAINTS AND OPERATIONAL LIMITATIONS

The LRV must be operated in such a manner as not to exceed the constraints and operational limitations of the following subsystems:

3.1 CHASSIS SUBSYSTEM CONSTRAINTS

The constraints which apply to the chassis are concerned with loads imparted to the various chassis members. In order to control these loads, limitations are imposed on location of stowed payload items. To preserve the factor of safety of 1.5 the LRV shall be loaded such that the loads in the chassis payload zones do not exceed the maximum loads shown in Table 3-I. Overloading the vehicle beyond the total 970 lb. limit will also result in performance degradation as shown in Section 5.

The chassis frame is constructed of thin wall aluminum rectangular tubes which are scalloped in certain areas for weight saving. These thin wall areas are subject to damage from high-impact loads beyond the designed-for conditions. Therefore, the chassis should not be subjected to such use as supporting rock samples for breaking apart the rocks with geologic tools wherein the geologic tools could strike the chassis.

3.2 SUSPENSION SUBSYSTEM CONSTRAINTS

The suspension system is designed for a factor of safety of 1.5 based on a maximum gross weight LRV of 1500 lb. with the C.G. of the gross weight falling such that not more than 55% of the total weight is supported by the front or rear wheels. The factor of safety will be reduced if the LRV is loaded in such a manner that these limits are exceeded.

Suspension system thermal constraints are defined in paragraph 3.10.

3.3 MOBILITY SUBSYSTEM CONSTRAINTS

The only constraints concerned with the wheels, traction drives, steering and hand controller are limitations caused by component capability discussed in Section 4 and by thermal limitations discussed in paragraph 3.10.

3.4 ELECTRICAL POWER SUBSYSTEM CONSTRAINTS

3.4.1 Batteries

The power subsystem is designed to use both batteries simultaneously on an approximate-equal load basis, i.e. one front and one rear traction drive should be operated from one battery and the other front and other rear traction drives operated from the other battery. Likewise the front steering motor should be operated from one battery and the rear steering motor from the other battery. Only in a contingency mode should all loads be transferred to one battery, since battery heat rise increases with increase in battery current. This consideration is discussed in 4.2.

LS006-002-2H
LUNAR ROVING VEHICLE
OPERATIONS HANDBOOK
APPENDIX A

PAYLOAD ZONE	ZONE LOCATION ON LRV	TOTAL ALLOWABLE LOAD IN ZONE
A	AFT CHASSIS	170 LBS.
B	UNDER LEFT SEAT	30 LBS.
C	UNDER RIGHT SEAT	30 LBS.
D	CENTER CHASSIS - BETWEEN ℄ AND RIGHT SEAT	30 LBS.
E	CENTER CHASSIS - BETWEEN ℄ AND LEFT SEAT	30 LBS.
F	CENTER CHASSIS FLOOR PANEL RIGHT OF ℄. ZONE FORMED BY STOWING SEAT, CREATING CLEAR FLOOR AREA FOR ONE-ASTRONAUT DRIVING CASE.	400 LBS.
G	TOP OF INBOARD HANDHOLDS	12 LB. PER HANDHOLD, BUT TOTAL NOT TO EXCEED 15 LB. FOR BOTH HANDHOLDS COMBINED
TOTAL LRV		970 LB. MAX. INCLUDING CREW, LCRU, TV AND ALL OTHER STOWED EQUIPMENT

TABLE 3-I LRV PAYLOAD LIMITATIONS

Mission __J__   Basic Date __2/5/71__   Change Date __4/19/71__   Page __A-24__

LS006-002-2H
LUNAR ROVING VEHICLE
OPERATIONS HANDBOOK
APPENDIX A

3.0   Continued

3.4.2   Switches and Circuit Breakers

To provide maximum crew safety in event of undetected failure of drive logic components, the following restrictions should be adhered to in sequential selection of switch positions:

a. The two STEERING and four DRIVE POWER switches should be in the OFF position when changing position of the ±15VDC switch.

b. The +15VDC PRIM or +15VDC SEC circuit breaker should be closed and the ± 15 VDC switch in PRIM or SEC position, respectively, before operating the two STEERING and four DRIVE POWER switches.

Proper order for selection of all circuit breakers and switches is given in Section 2 of the basic LRV Operations Handbook.

3.5   NAVIGATION SUBSYSTEM CONSTRAINTS

a. The GYRO TORQUING switch is not to be kept in the LEFT or RIGHT position for more than two minutes. After two minutes on, the switch must be kept OFF for a minimum of five minutes to prevent damage to the torquing motor.

b. The navigation subsystem input voltage must not be allowed to be less than 30 VDC to prevent excessive computation and display errors and to prevent damage to navigation equipment if the under-voltage situation is prolonged. If the VOLTS indicator indicates less than 60 the NAV POWER circuit breaker should be opened. The VOLTS indicator should be checked periodically (at least each 15 minutes) to verify readings of not less than 60.

NOTE

Navigation readings should be reported and recorded by MCC before closing the NAV POWER circuit breaker. Reapplication of power will cause loss of display retention.

c. NAV POWER should be on for at least 3 minutes before driving the LRV to allow the directional gyro to reach operating spin speed.

d. The LRV must be level within ± 6° in pitch and roll and parked downsun within ± 3° during navigation initialization and update to achieve required system accuracy.

3.6   DISPLAY AND CONTROLS SUBSYSTEM CONSTRAINTS

Crew procedures for operating the displays and controls are contained in the basic LRV Operations Handbook.

No constraints exist other than those already listed in 3.4 and 3.5.

Mission   J   Basic Date   2/5/71   Change Date   4/19/71   Page   A-25

LS006-002-2H
LUNAR ROVING VEHICLE
OPERATIONS HANDBOOK
APPENDIX A

3.7 CREW STATION SUBSYSTEM CONSTRAINTS

No constraints to crew station operation exist when operated in accordance with the crew procedures in Section 2 of the basic LRV Operations Handbook.

These procedures are based on the crew station structural capability. To maintain the safety factor of 1.5 the loads imposed on crew station components should not exceed the values shown in Figures 3-1 and 3-2.

3.8 VEHICLE DYNAMIC OPERATION CONSTRAINTS

The LRV is designed with inherent stability characteristics of wide wheel track and low center of gravity. Static stability limits are shown in Figure 3-4. Overturn of the vehicle is a remote possibility, occurring only under severe conditions of extremely tight turns at high speeds on steep slopes or collision with immovable objects. Speeds, slopes, turning radii limits, and obstacle height to prevent overturn and sliding are shown in Figures 3-5 through 3-9. These curves are based on the C.G. of the loaded vehicle falling within the envelope shown on Figure 3-3. The required increase in turning radius for preventing overturn caused by locating the loaded LRV C.G. outside the Figure 3-3 envelope is shown in Figures 3-10 through 3-15. Maximum allowable speeds to prevent exceeding structural design loads are shown in Table 3-III. The safe driving corridor for driving with one steering assembly failed is shown in Figure 3-16.

3.9 PARKING CONSTRAINTS

3.9.1 During-Traverse Parking

There are no orientation constraints imposed on parking in sunlight during traverses. The LRV must not be parked in shadow for longer than 2 hours to prevent display and control console electronics damage.

Switch and circuit breaker positions for short-term parking are given in the basic LRV Operations Handbook.

3.9.2 Between-Traverses Parking

To prevent thermal damage to the control and display console instruments and to be compatible with LCRU thermal requirements the LRV must be parked in sunlight oriented relative to the sun azimuth in accordance with Figure 3-17 when parking the LRV between sorties.

3.10 THERMAL CONSTRAINTS

The LRV is designed for operation in the lunar day at sun elevation angles of greater than five degrees. All qualification tests have been conducted to the lunar environment temperature levels expected to be encountered by the LRV. Operations on the moon should be restricted to maintain components within these temperature limits. Temperature limits for specific components are defined in Table 3-II.

LS006-002-2H
LUNAR ROVING VEHICLE
OPERATIONS HANDBOOK
APPENDIX A

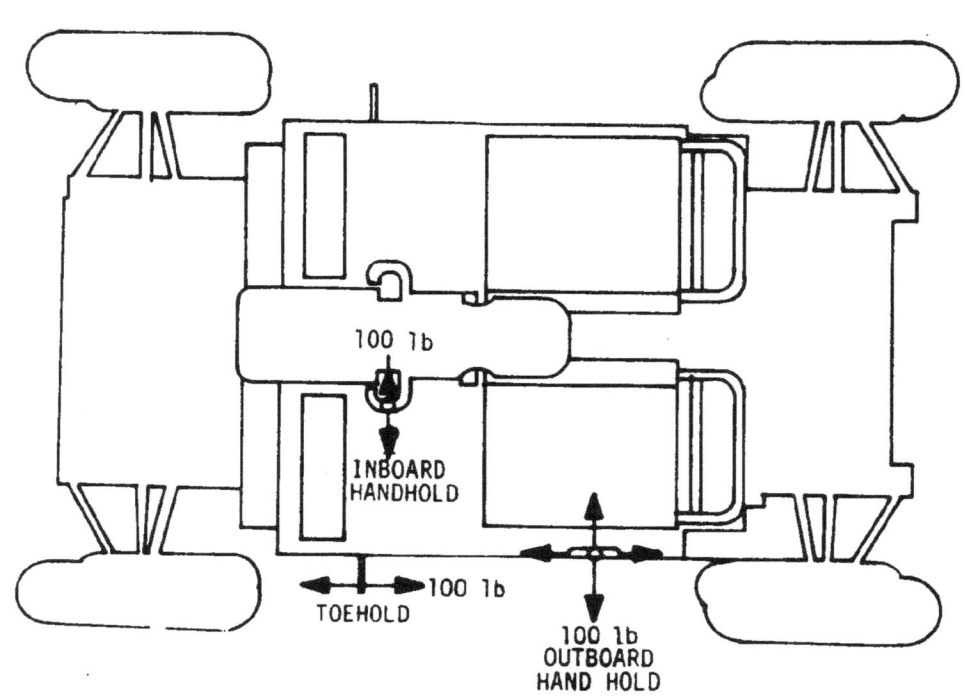

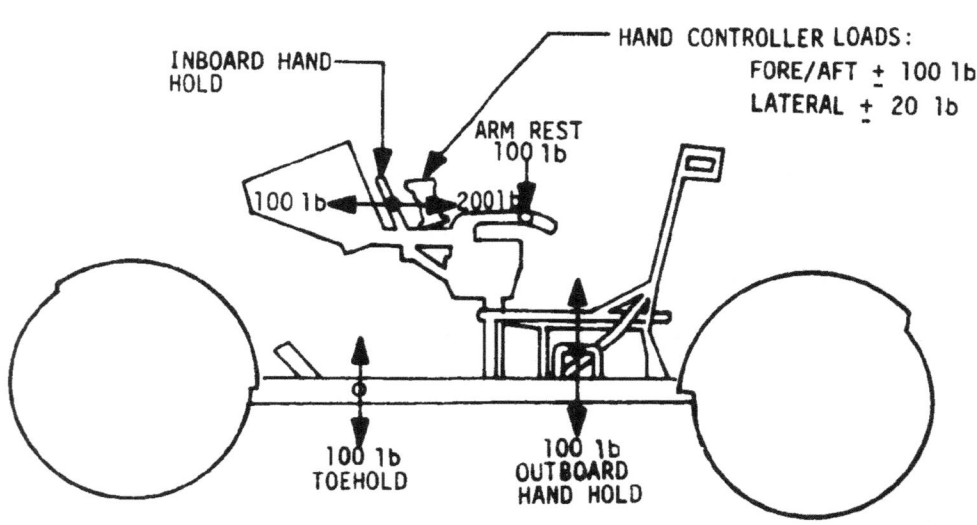

FIGURE 3-1   CREW STATION LIMIT LOADS (HANDHOLDS AND TOEHOLDS)

LS006-002-2H
LUNAR ROVING VEHICLE
OPERATIONS HANDBOOK
APPENDIX A

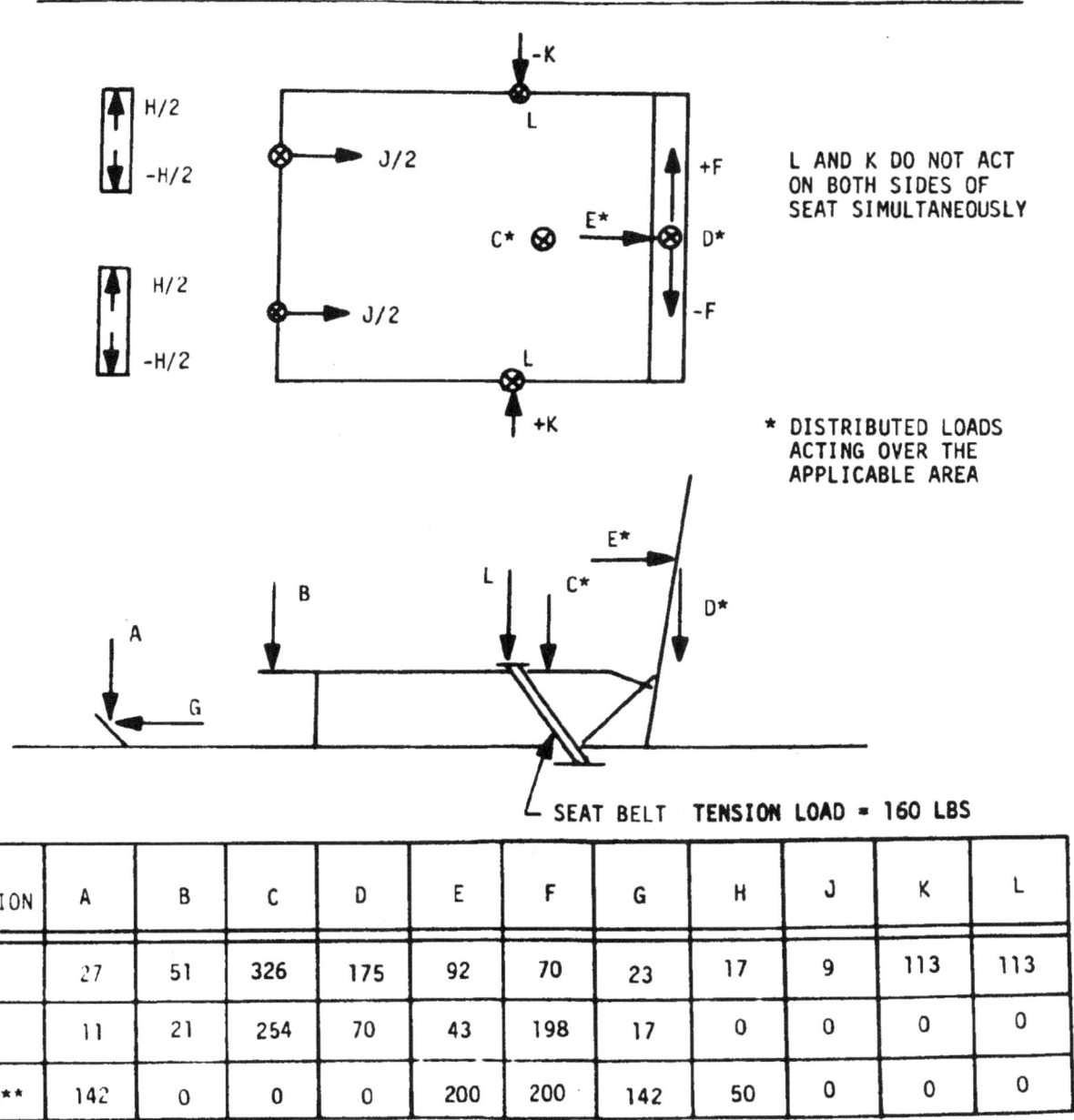

LOAD CONDITION	A	B	C	D	E	F	G	H	J	K	L
1	27	51	326	175	92	70	23	17	9	113	113
2	11	21	254	70	43	198	17	0	0	0	0
3**	142	0	0	0	200	200	142	50	0	0	0

ALL LOADS ARE LIMIT LOADS

** CONDITION 3 LOADS DO NOT ACT SIMULTANEOUSLY

FIGURE 3-2 CREW STATION LIMIT LOADS (SEATS AND FOOTRESTS)

Mission J   Basic Date 2/5/71   Change Date 4/19/71

LS006-002-2H
LUNAR ROVING VEHICLE
OPERATIONS HANDBOOK
APPENDIX A

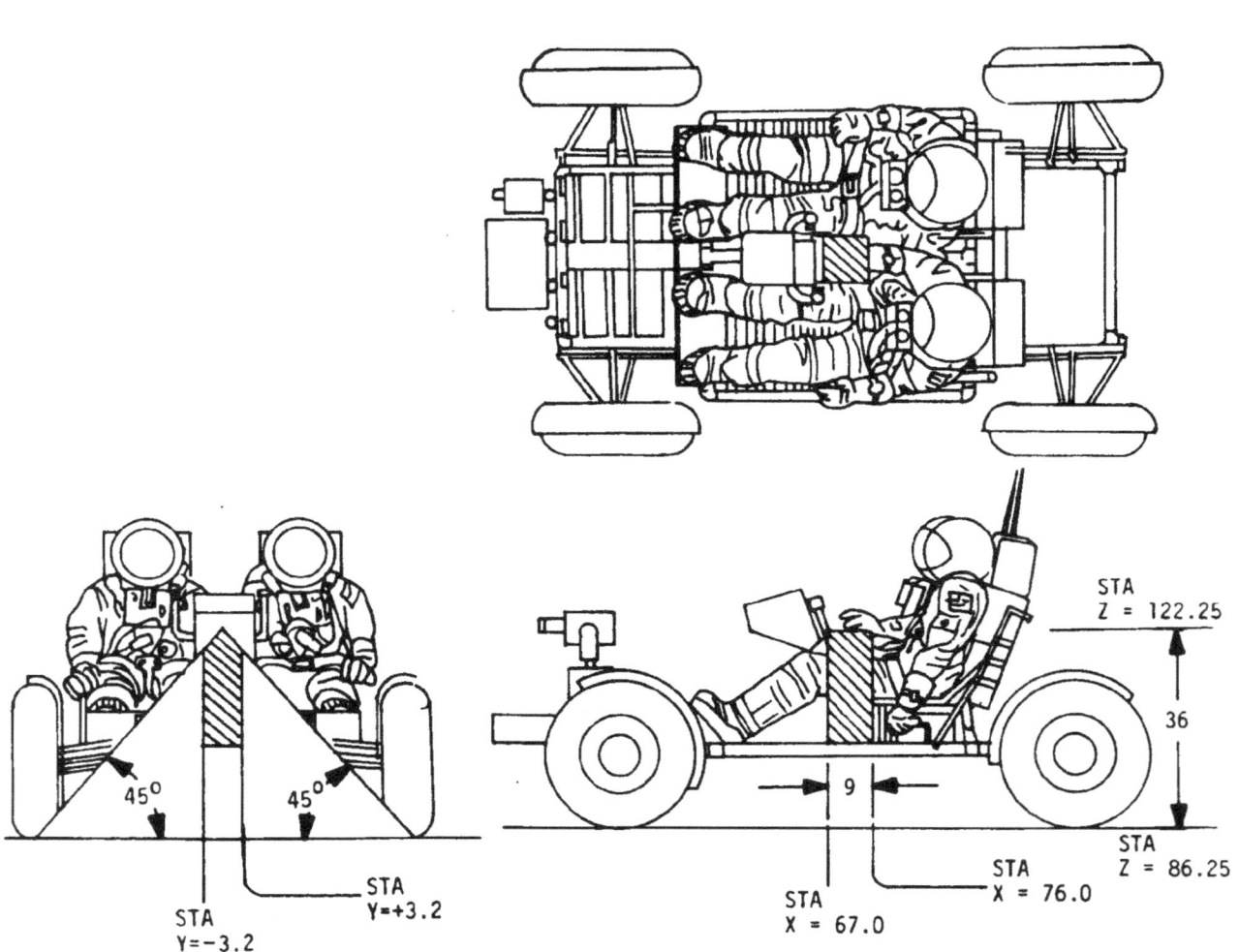

FIGURE 3-3   GROSS WEIGHT ALLOWABLE C.G. ENVELOPE

3.1.0 Continued

Component	Maximum Operating Temperature Limit °F	Survival Upper Temperature Limit °F	Minimum Operating Temperature Limit °F	Minimum Survival Temperature Limit °F
*Battery	125	140	40	-15
DCE	159	180	0	-20
*Traction Drive	400	450	-25	-50
Wheel	250	250	-200	-250
SPU	130	185	30	-65
DGU	160	200	-65	-80
IPI	185	185	-22	-65
Suspension Damper	400	450	-65	-70
Steering Motor	360	400	-25	-50

*Analog Display on LRV Panel

TABLE 3-II COMPONENT TEMPERATURE LIMITS

3.10.1 Shadow Constraints

The LRV must not be operated in shadows for longer than 30 continuous minutes to prevent damage to the wire wheels. The LRV must be exposed to 10 minutes of sunlight before re-entering shadow to allow adequate surface-to-wheel heat transfer.

The LRV must not be parked in lunar shadow for longer than two hours to prevent low temperature damage to the electronics in the control and display console.

LS006-002-2H
LUNAR ROVING VEHICLE
OPERATIONS HANDBOOK
APPENDIX A

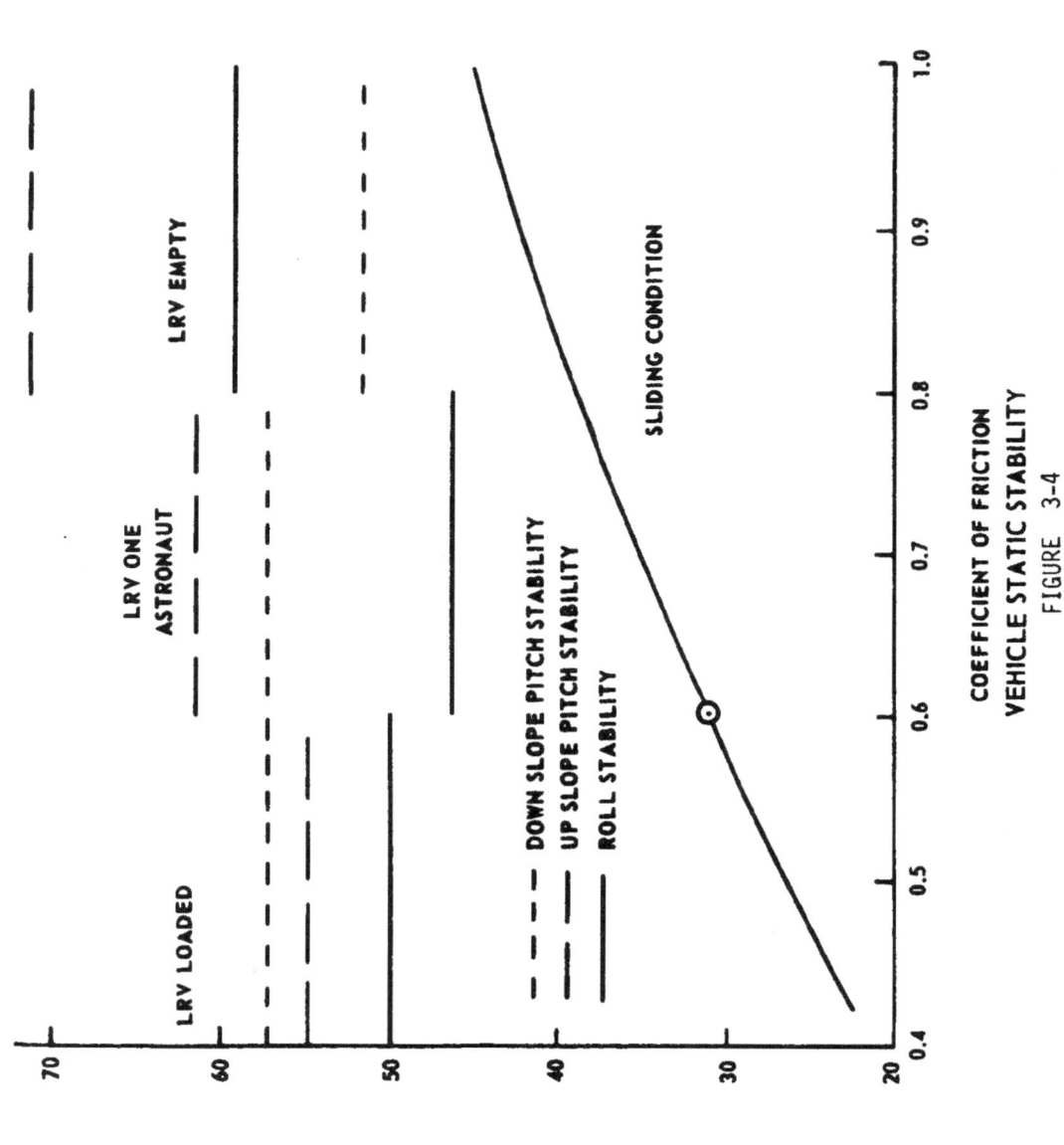

VEHICLE STATIC STABILITY
FIGURE 3-4

LS006-002-2H
LUNAR ROVING VEHICLE
OPERATIONS HANDBOOK
APPENDIX A

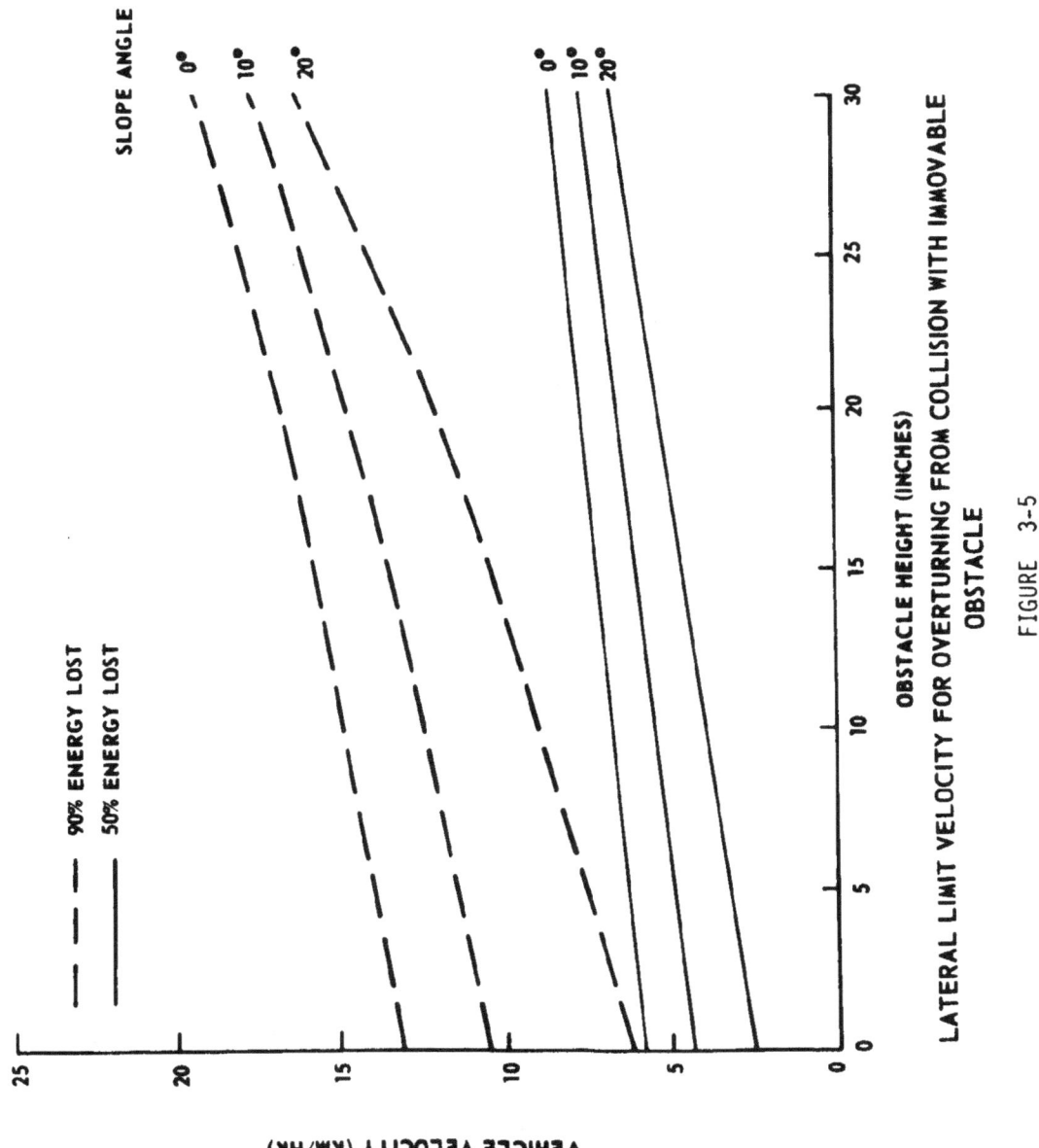

FIGURE 3-5

LS006-002-2H
LUNAR ROVING VEHICLE
OPERATIONS HANDBOOK
APPENDIX A

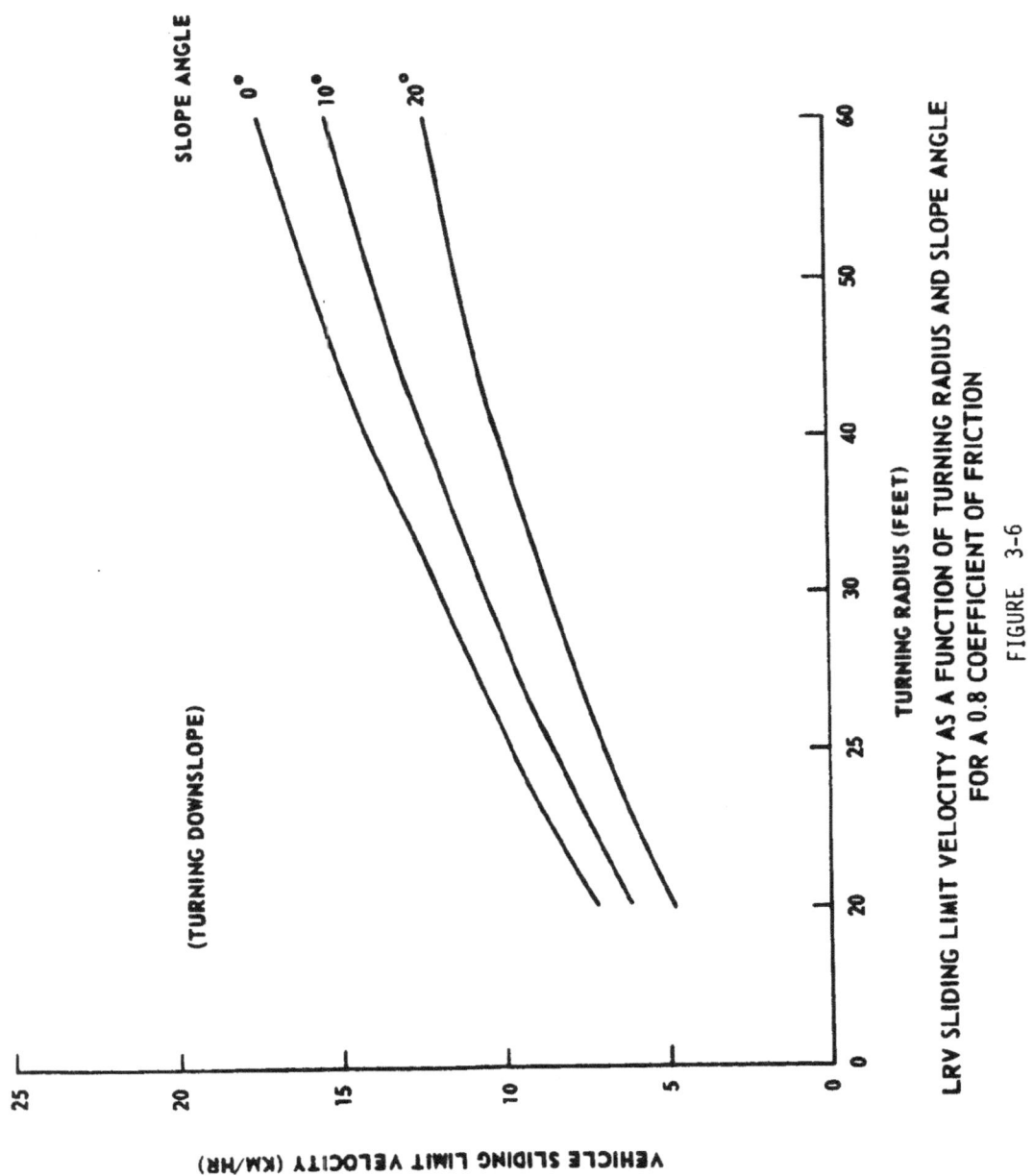

FIGURE 3-6 LRV SLIDING LIMIT VELOCITY AS A FUNCTION OF TURNING RADIUS AND SLOPE ANGLE FOR A 0.8 COEFFICIENT OF FRICTION

Mission J Basic Date 2/5/71 Change Date 4/19/71 Page A-33

LS006-002-2H
LUNAR ROVING VEHICLE
OPERATIONS HANDBOOK
APPENDIX A

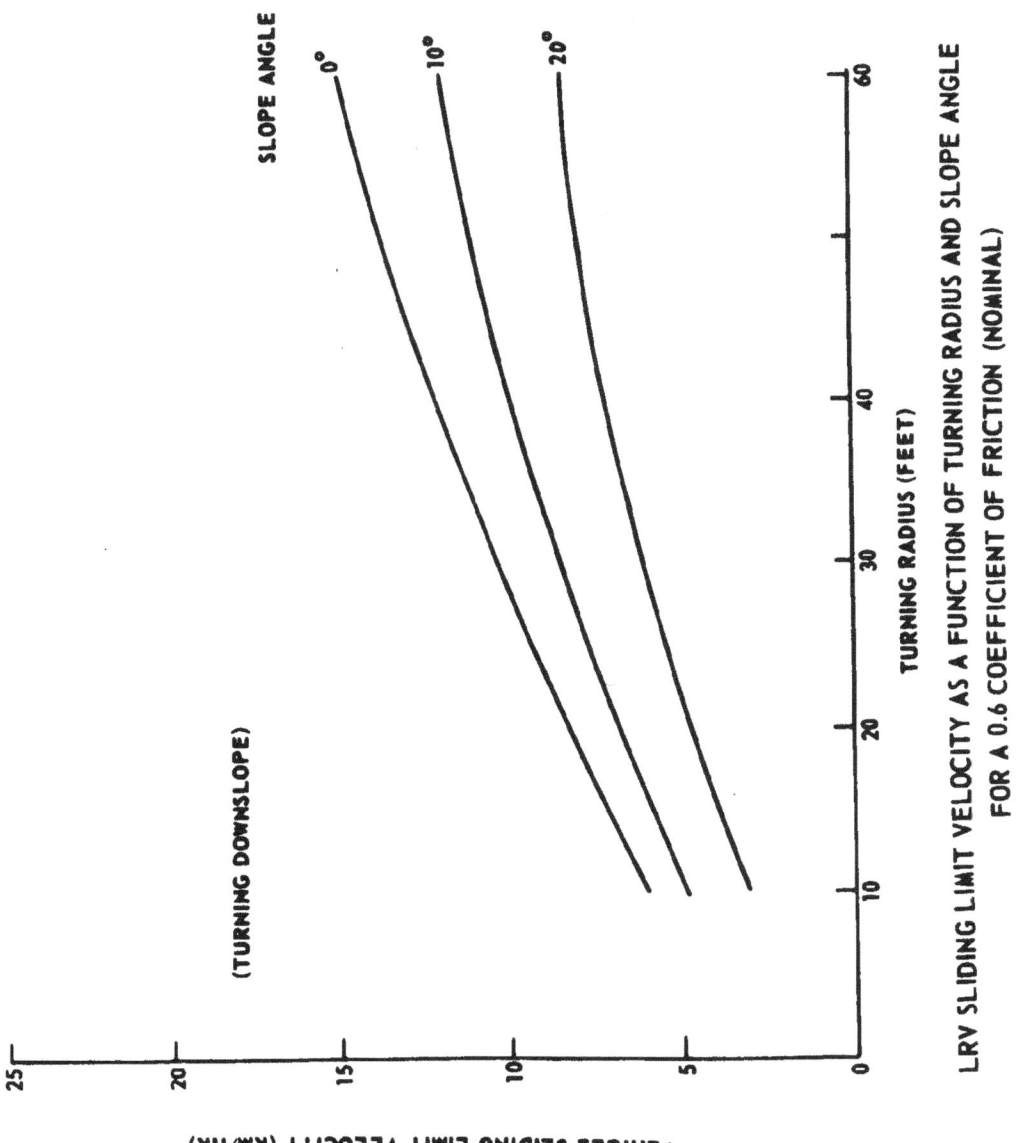

FIGURE 3-7

LRV SLIDING LIMIT VELOCITY AS A FUNCTION OF TURNING RADIUS AND SLOPE ANGLE FOR A 0.6 COEFFICIENT OF FRICTION (NOMINAL)

LS006-002-2H
LUNAR ROVING VEHICLE
OPERATIONS HANDBOOK
APPENDIX A

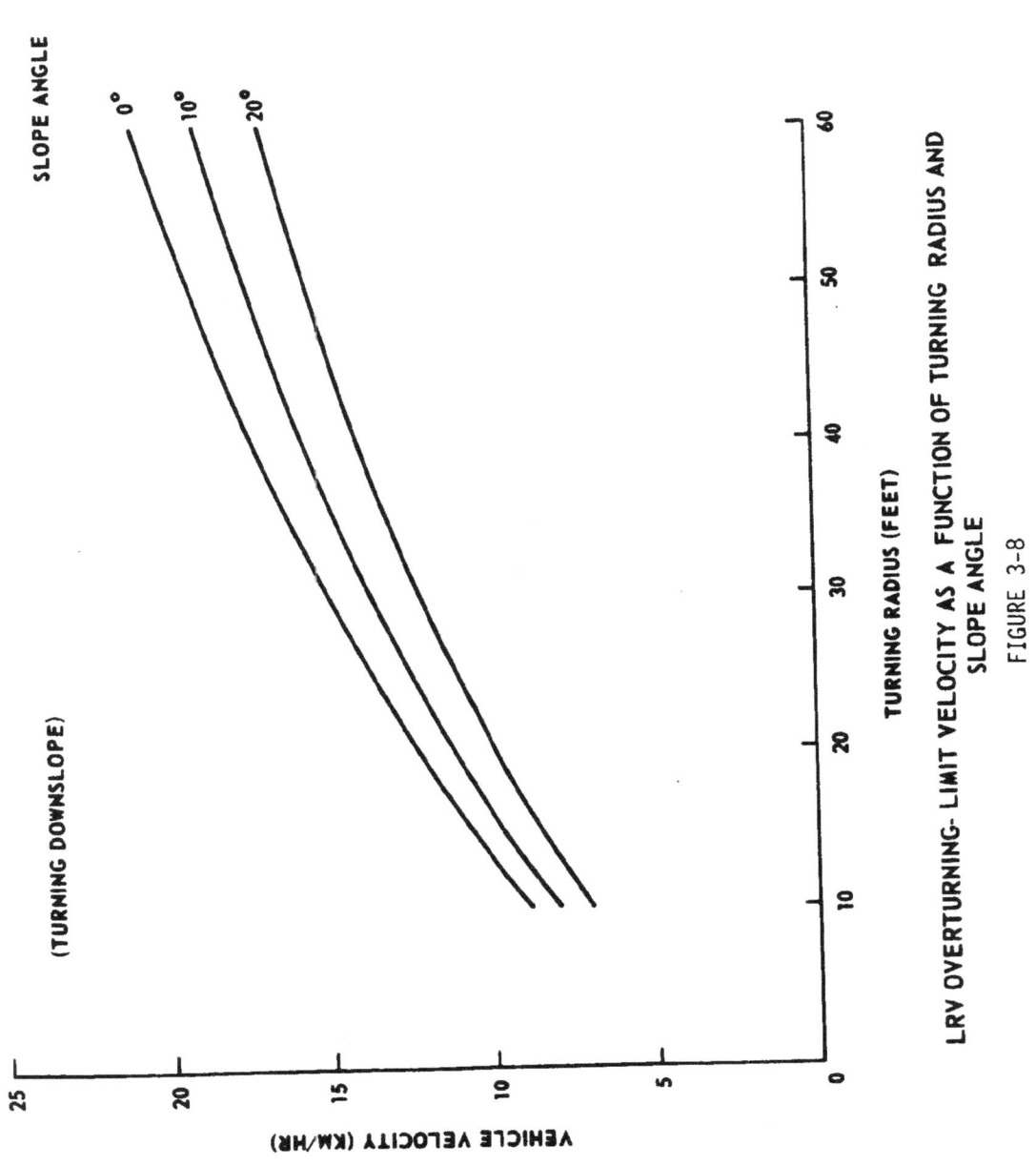

LRV OVERTURNING- LIMIT VELOCITY AS A FUNCTION OF TURNING RADIUS AND SLOPE ANGLE

FIGURE 3-8

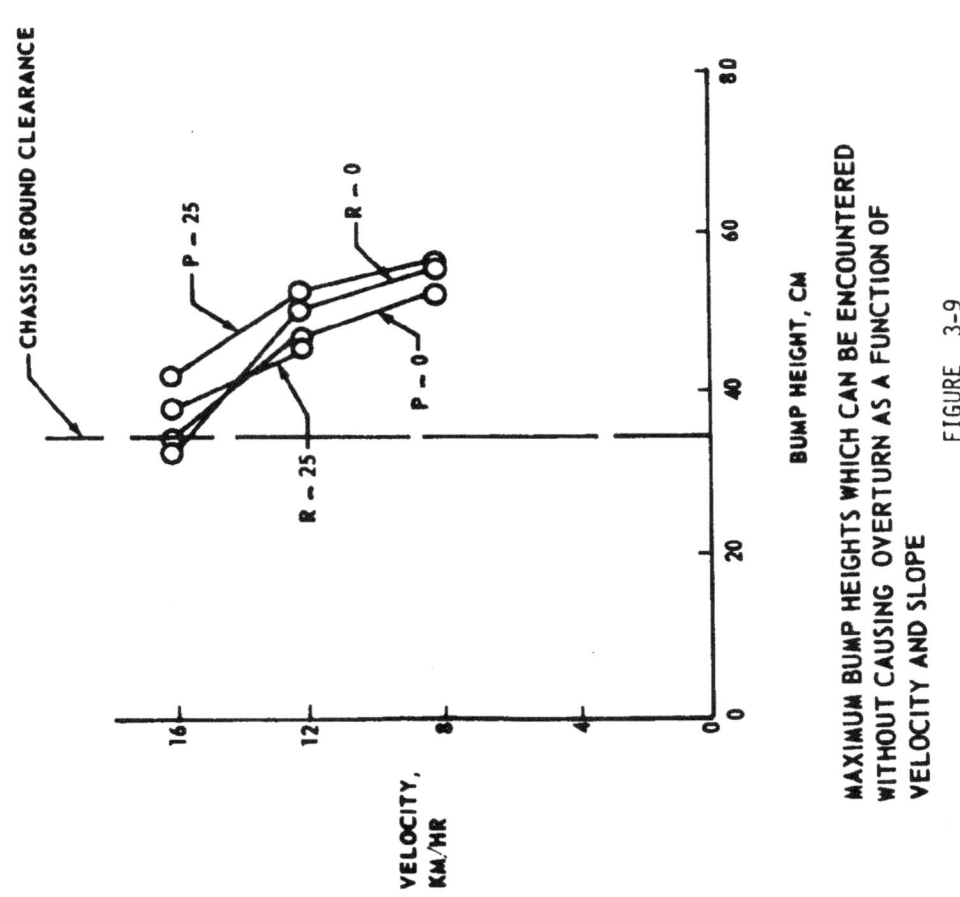

MAXIMUM BUMP HEIGHTS WHICH CAN BE ENCOUNTERED WITHOUT CAUSING OVERTURN AS A FUNCTION OF VELOCITY AND SLOPE

FIGURE 3-9

LS006-002-2H
LUNAR ROVING VEHICLE
OPERATIONS HANDBOOK
APPENDIX A

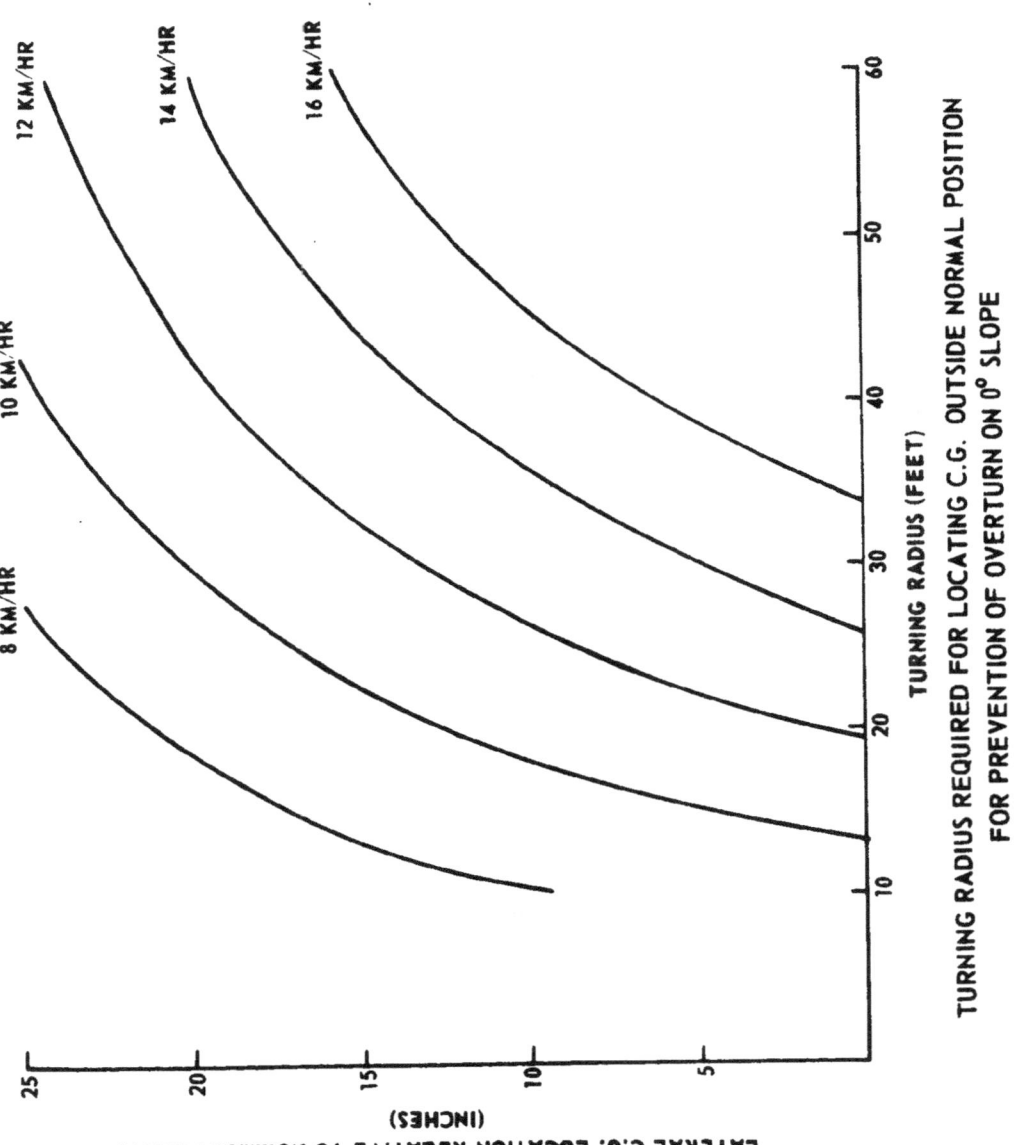

FIGURE 3-10

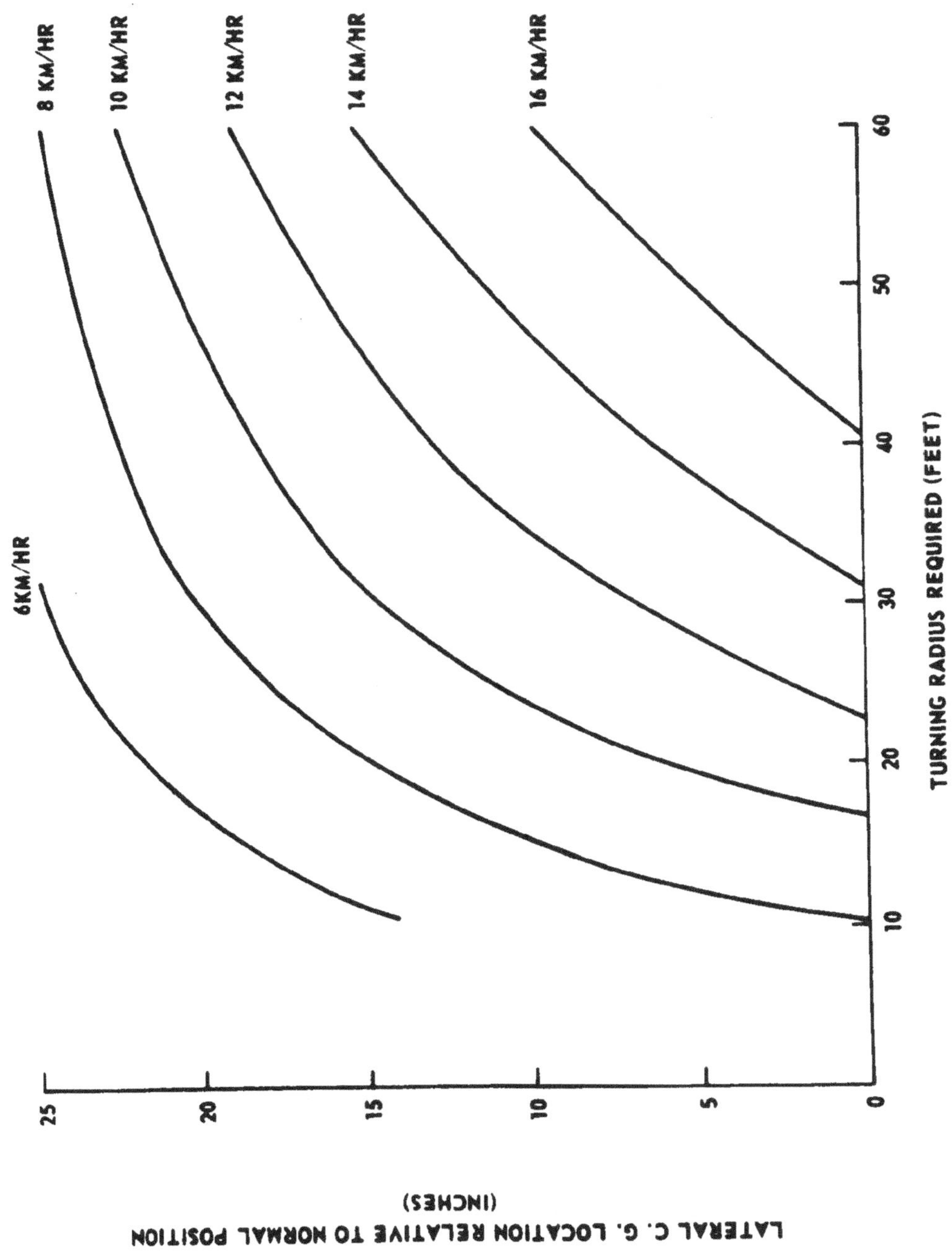

LS006-002-2H
LUNAR ROVING VEHICLE
OPERATIONS HANDBOOK
APPENDIX A

FIGURE 3-11

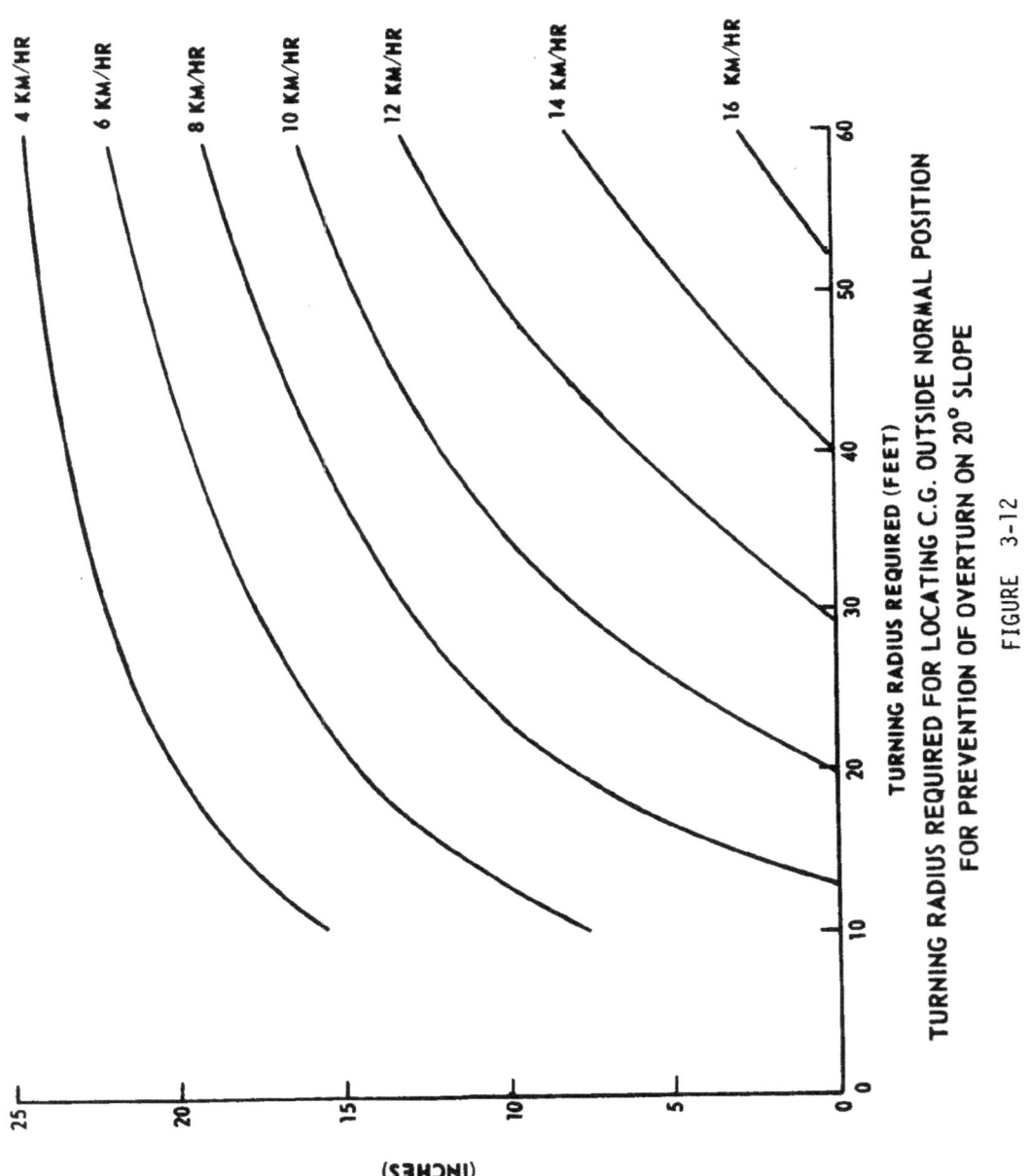

FIGURE 3-12

LS006-002-2H
LUNAR ROVING VEHICLE
OPERATIONS HANDBOOK
APPENDIX A

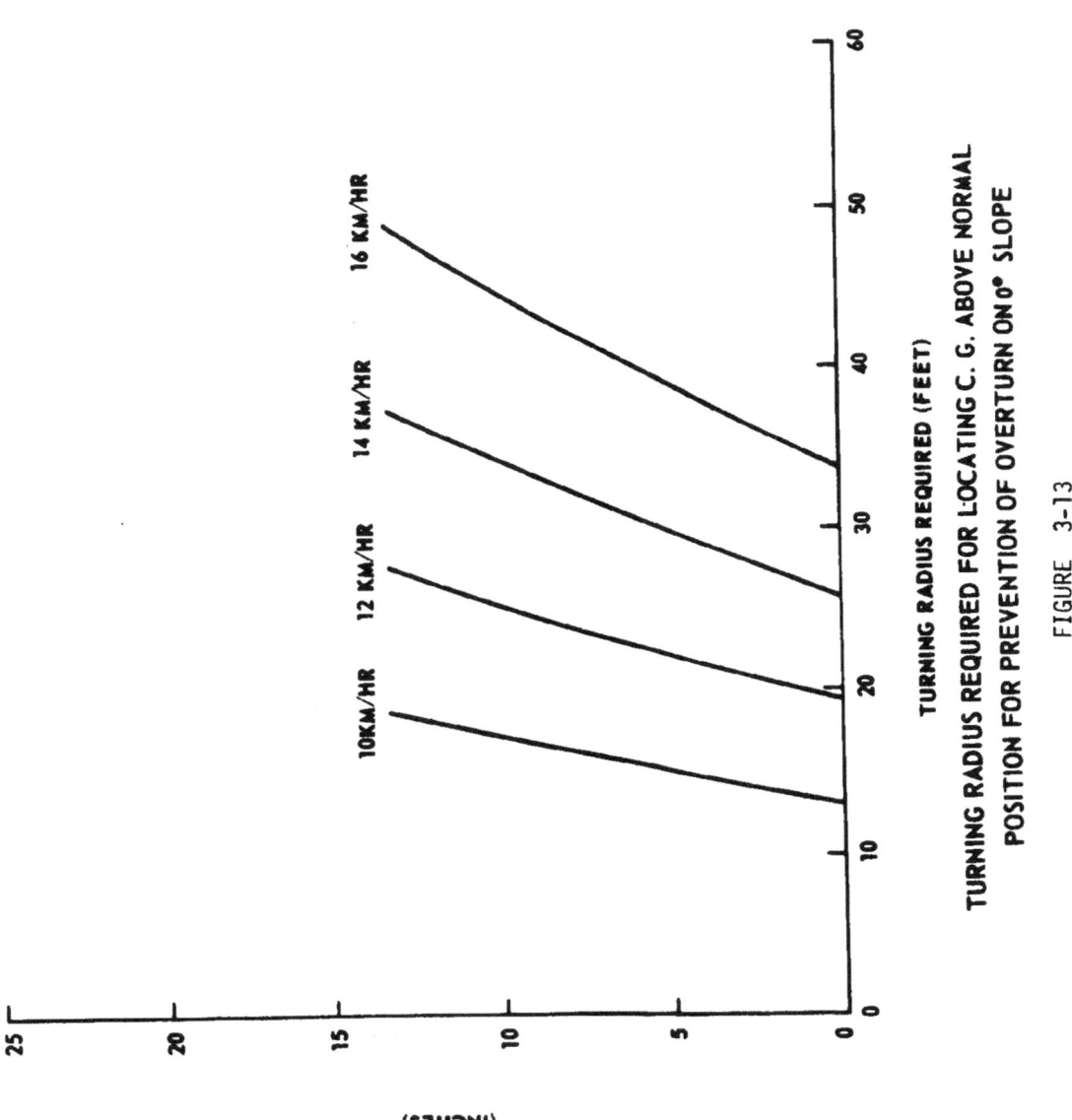

FIGURE 3-13
TURNING RADIUS REQUIRED FOR LOCATING C.G. ABOVE NORMAL POSITION FOR PREVENTION OF OVERTURN ON 0° SLOPE

LSQ06-002-2H
LUNAR ROVING VEHICLE
OPERATIONS HANDBOOK
APPENDIX A

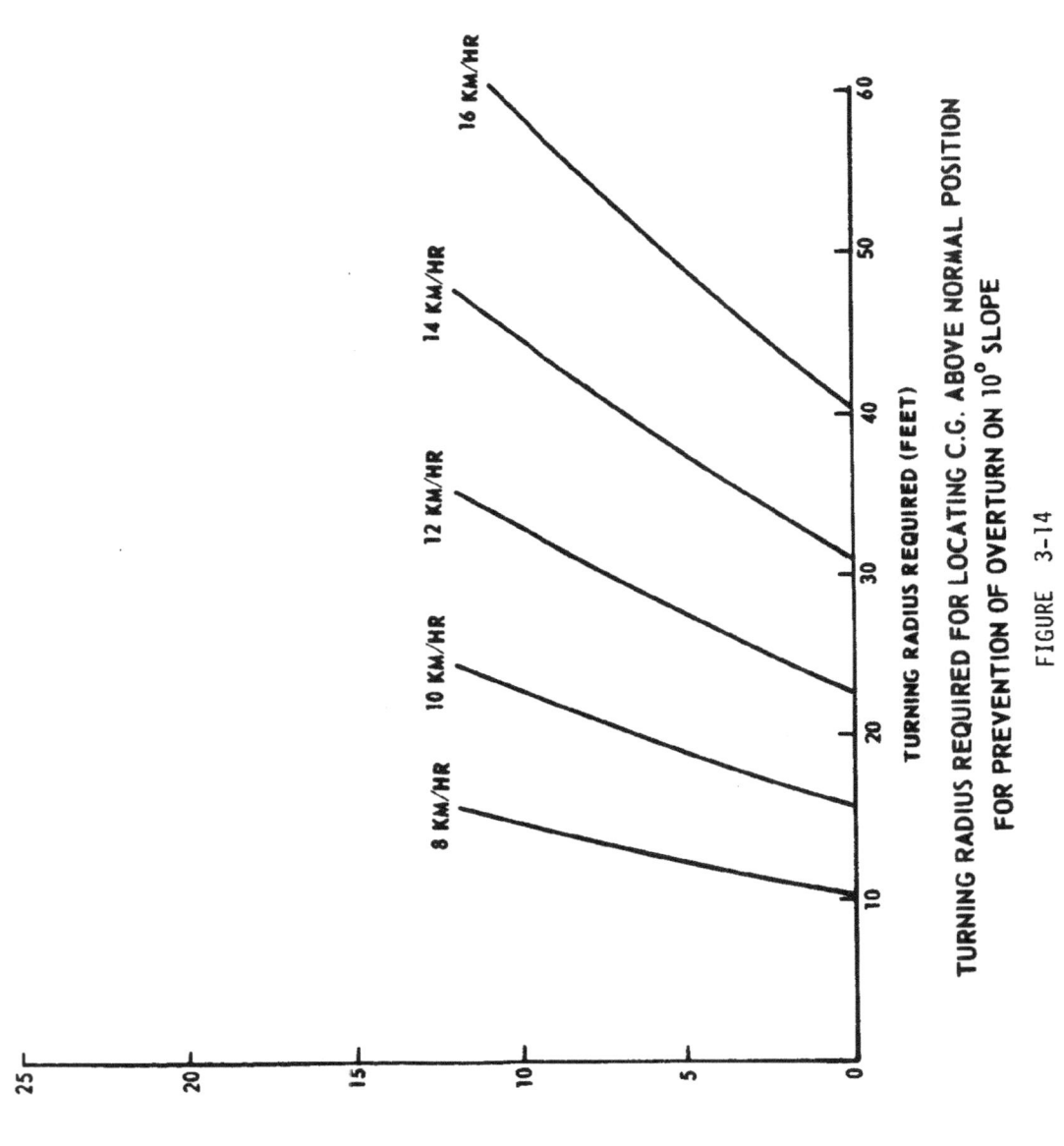

FIGURE 3-14

LS006-002-2H
LUNAR ROVING VEHICLE
OPERATIONS HANDBOOK
APPENDIX A

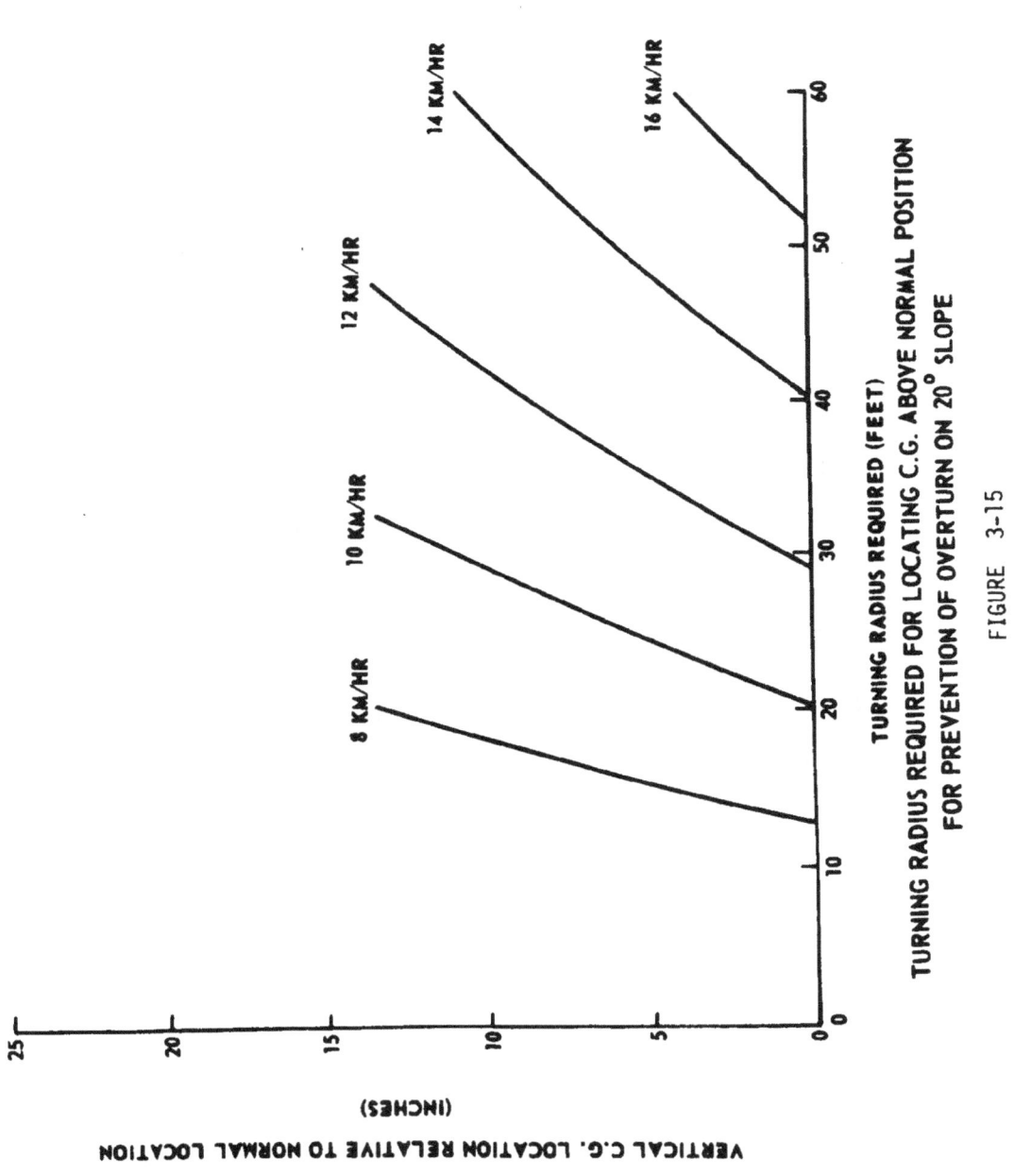

FIGURE 3-15
TURNING RADIUS REQUIRED FOR LOCATING C.G. ABOVE NORMAL POSITION FOR PREVENTION OF OVERTURN ON 20° SLOPE

Mission J   Basic Date 2/5/71   Change Date 4/19/71   Page A-42

LS006-002-2H
LUNAR ROVING VEHICLE
OPERATIONS HANDBOOK
APPENDIX A

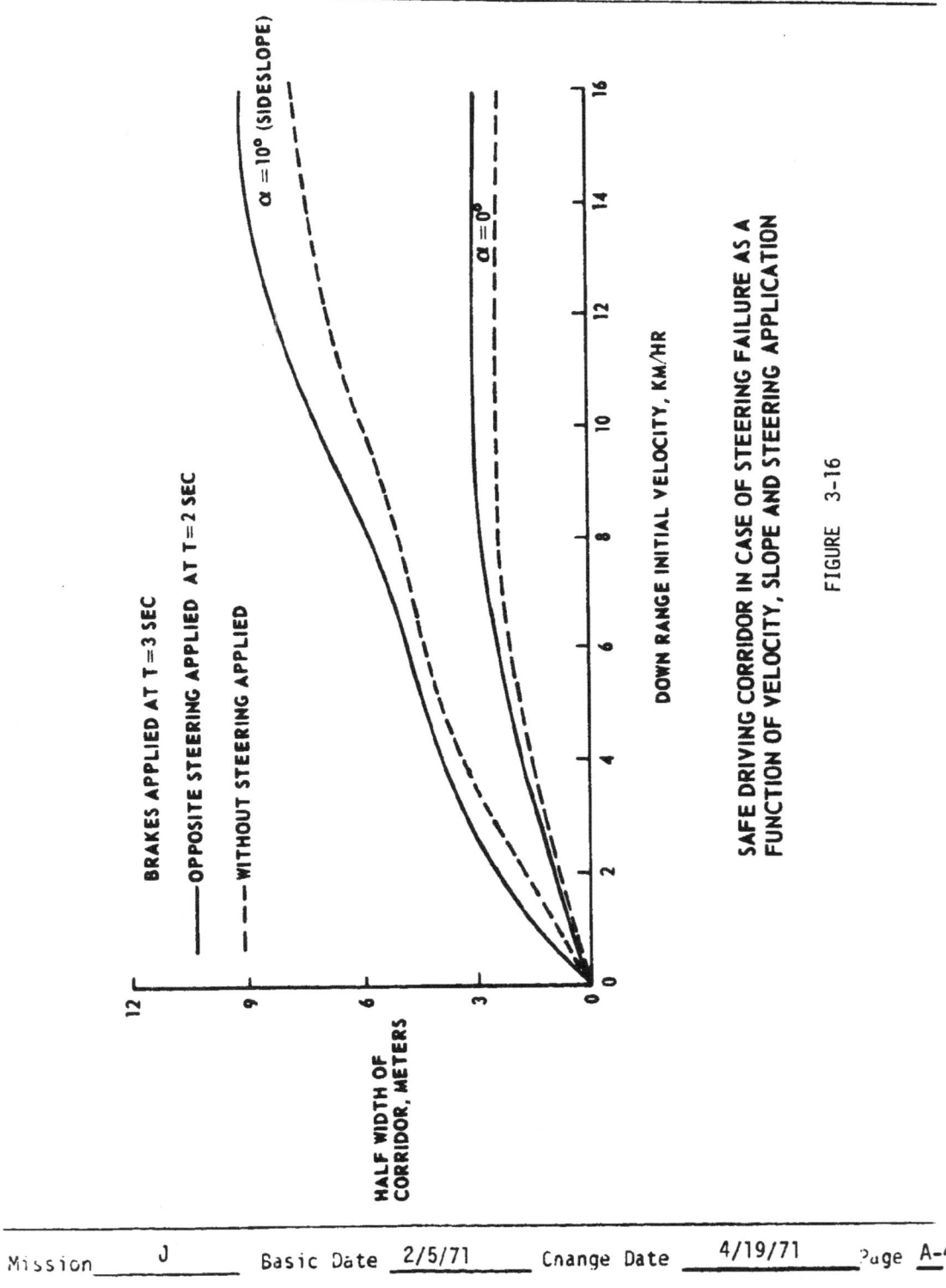

FIGURE 3-16

SAFE DRIVING CORRIDOR IN CASE OF STEERING FAILURE AS A FUNCTION OF VELOCITY, SLOPE AND STEERING APPLICATION

LS006-002-2H
LUNAR ROVING VEHICLE
OPERATIONS HANDBOOK
APPENDIX A

MAXIMUM SPEED FOR DESIGN LIMIT LOADS WITH FATIGUE CONSIDERATIONS*

LURAIN TYPE (MIDRANGE)	MAX ALLOWABLE SPEED
SMOOTH MARE	13 KM/HR
ROUGH MARE	8.5 KM/HR
HUMMOCKY UPLAND	8 KM/HR
ROUGH UPLAND	7 KM/HR

*BASED ON CEI REFERENCE MISSION

TABLE 3-III   SPEED RESTRICTIONS

LS006-002-2H
LUNAR ROVING VEHICLE
OPERATIONS HANDBOOK
APPENDIX A

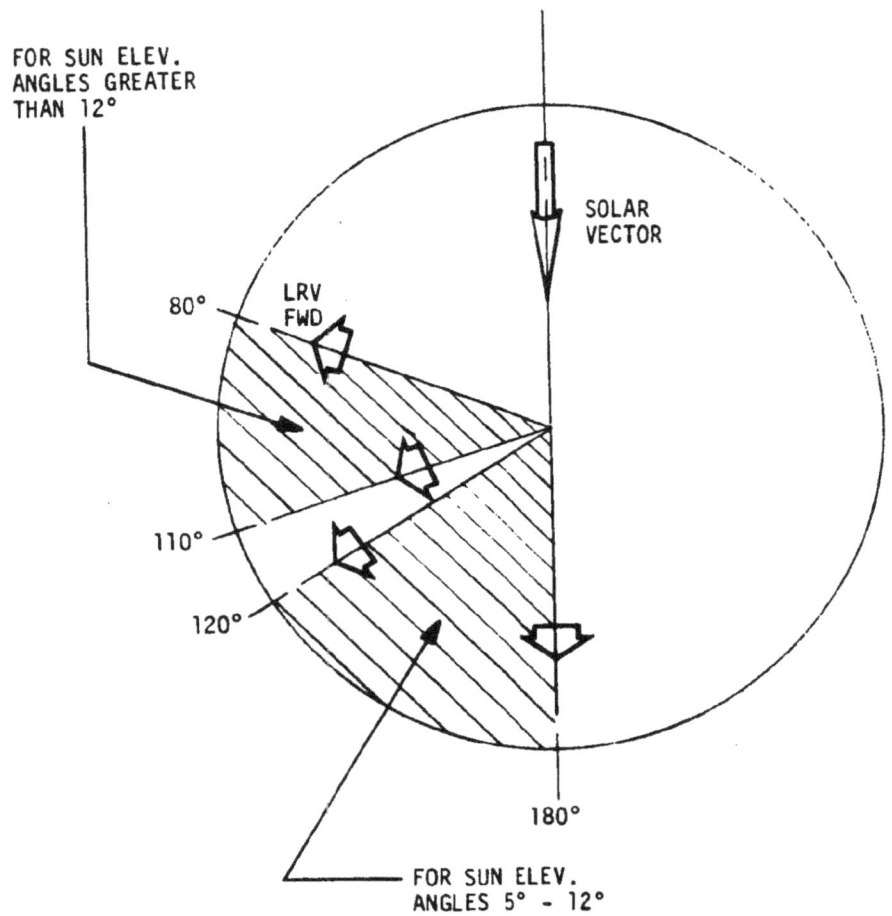

SHADED AREAS INDICATE THE PARKING ATTITUDES ACCEPTABLE
TO THE LRV AND TO THE PAYLOAD.

FIGURE 3-17 PARKING ORIENTATION CONSTRAINTS

LS006-002-2H
LUNAR ROVING VEHICLE
OPERATIONS HANDBOOK
APPENDIX A

4.0 SUBSYSTEM PERFORMANCE DATA

This section defines the performance characteristics of the various LRV subsystems. Thermal performance characteristics are included for those subsystems having thermal limitations.

4.1 MOBILITY SUBSYSTEM PERFORMANCE

LRV power consumption per kilometer of travel as functions of speed, lurain slope and soil type is defined in Figures 4-1, 4-2, and 4-3. A constant speed was considered in the preparation of these curves. The effect of vehicle acceleration through the indicated velocity was not evaluated but the curves do include wheel slippage effects. Power consumption shown on these curves is based on traction drive characteristics shown in Figure 4-4.

Battery current provided to the traction drives as a function of speed and hand controller throttle position is defined on Figure 4-5.

The rate of traction drive temperature increase as a function of solar elevation angle and drive temperature is defined on Figures 4-6. A zero degree slope was considered in the preparation of these curves.

The effects of zenith angle and traction drive temperature upon the wheel temperature increase for a solar elevation angle of 60° are shown in Figures 4-7 through 4-10.

The rate of traction drive temperature increase as a function of vehicle speed and traction drive temperature under zero degree slope and 60 degree solar angle conditions is shown in Figures 4-7 through 4-10.

The rate of traction drive temperature increase as a function of vehicle speed and traction drive temperature under zero degree slope and 60 degree solar angle conditions is shown in Figure 4-11.

The rate of traction drive temperature increase for a solar angle of 60° and as a function of lurain slope and drive temperature is defined in Figure 4-12.

LS006-002-2H
LUNAR ROVING VEHICLE
OPERATIONS HANDBOOK

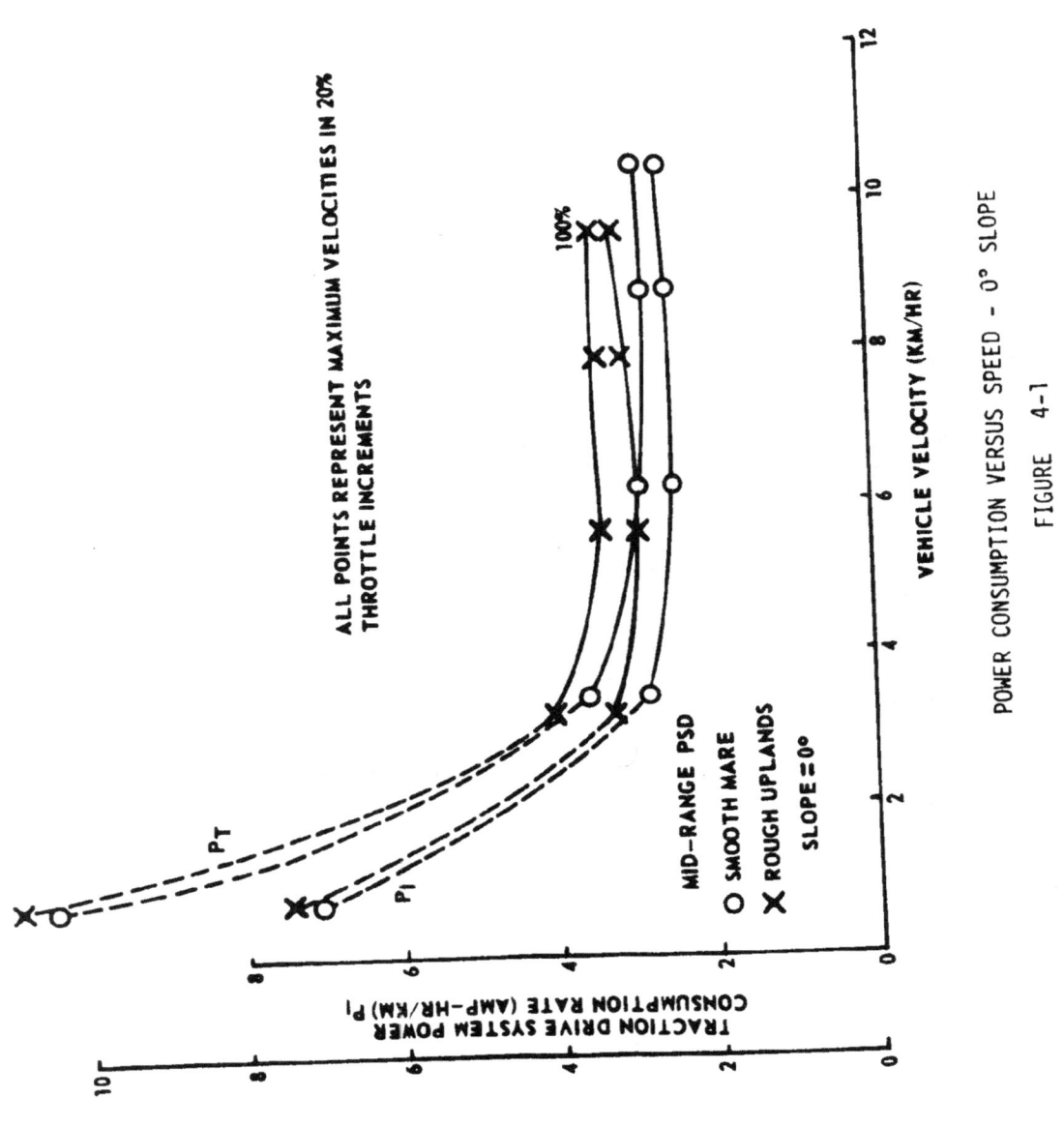

POWER CONSUMPTION VERSUS SPEED - 0° SLOPE

FIGURE 4-1

LS006-002-2H
LUNAR ROVING VEHICLE
OPERATIONS HANDBOOK
APPENDIX A

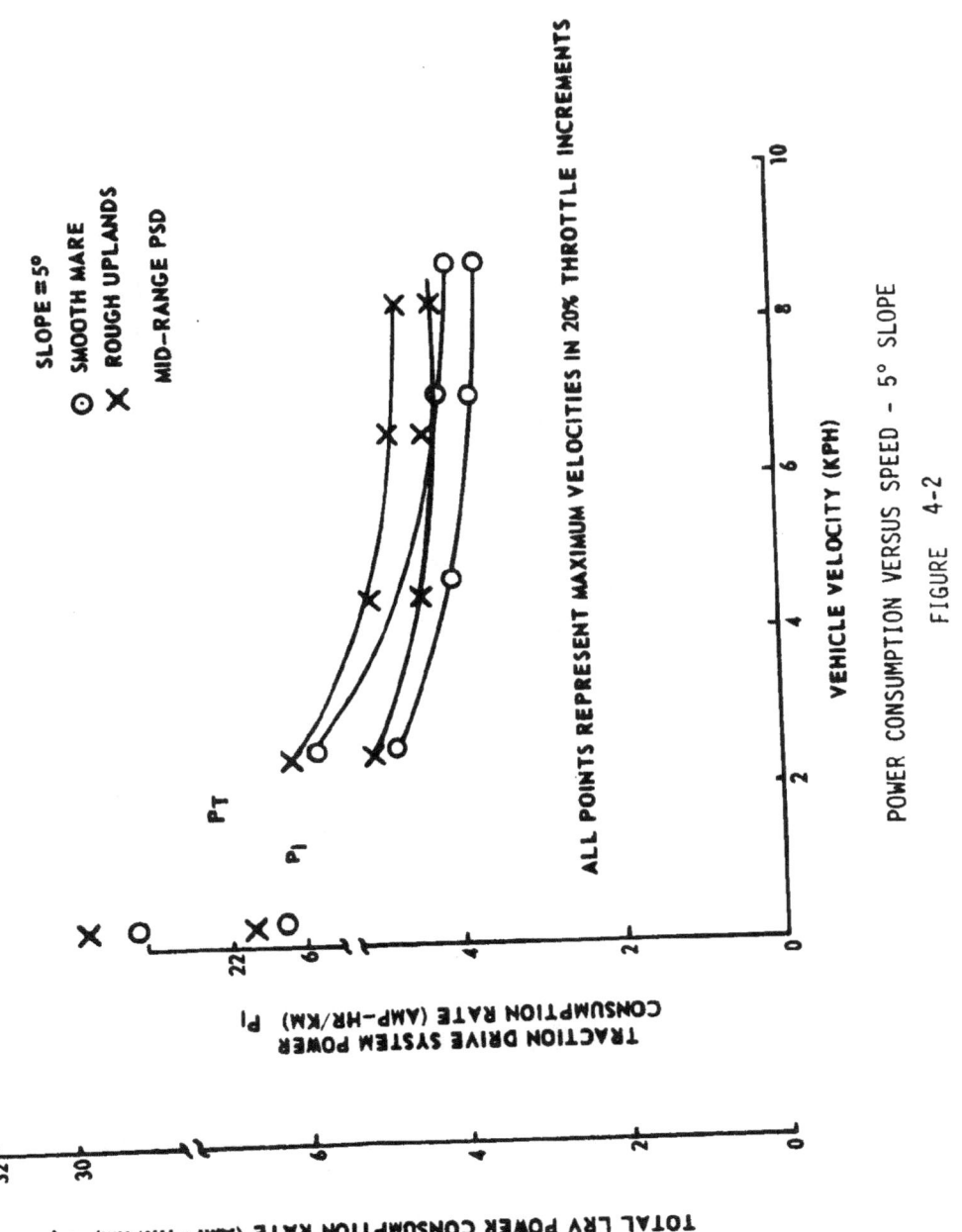

POWER CONSUMPTION VERSUS SPEED - 5° SLOPE
FIGURE 4-2

LS006-002-2H
LUNAR ROVING VEHICLE
OPERATIONS HANDBOOK
APPENDIX A

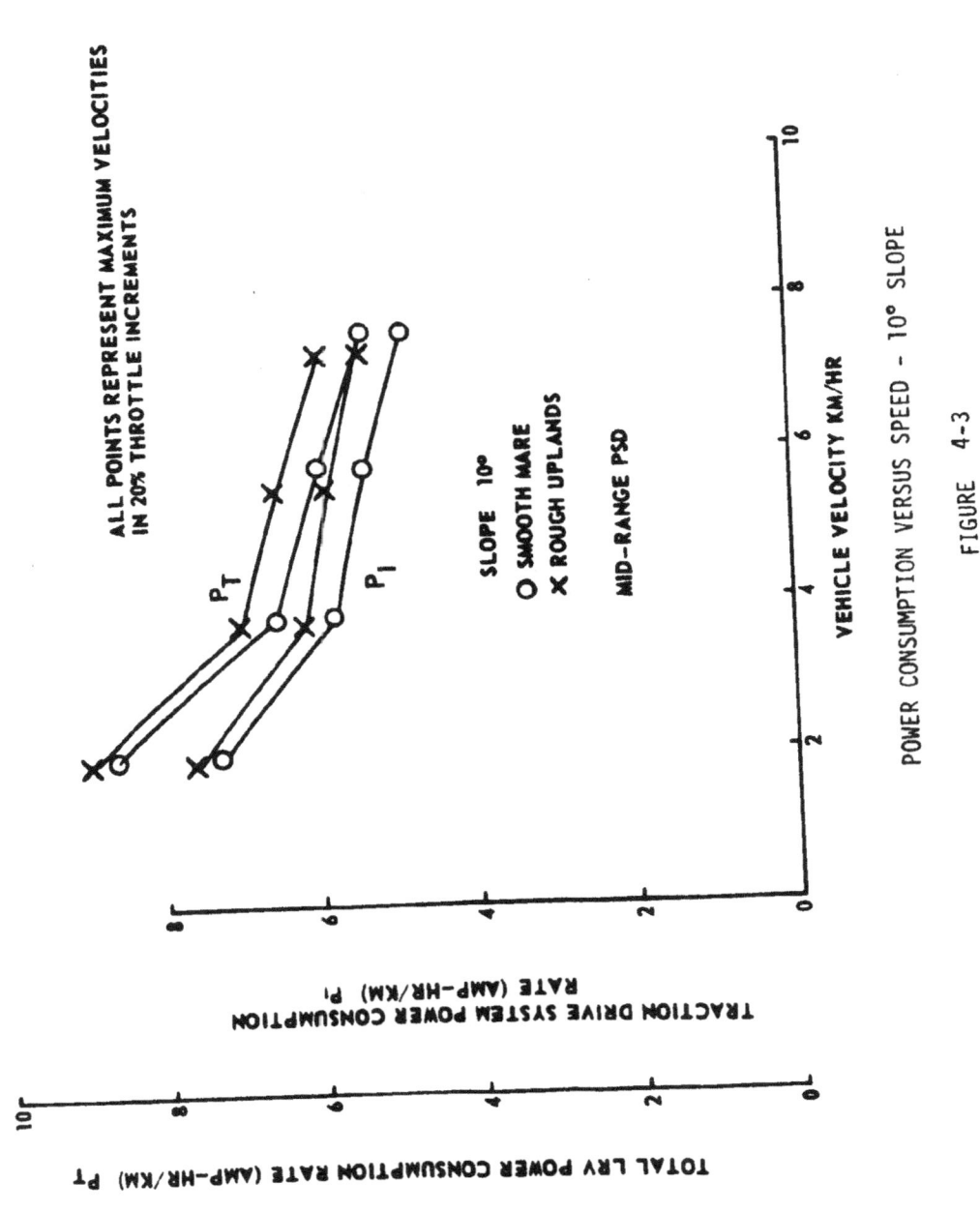

POWER CONSUMPTION VERSUS SPEED - 10° SLOPE
FIGURE 4-3

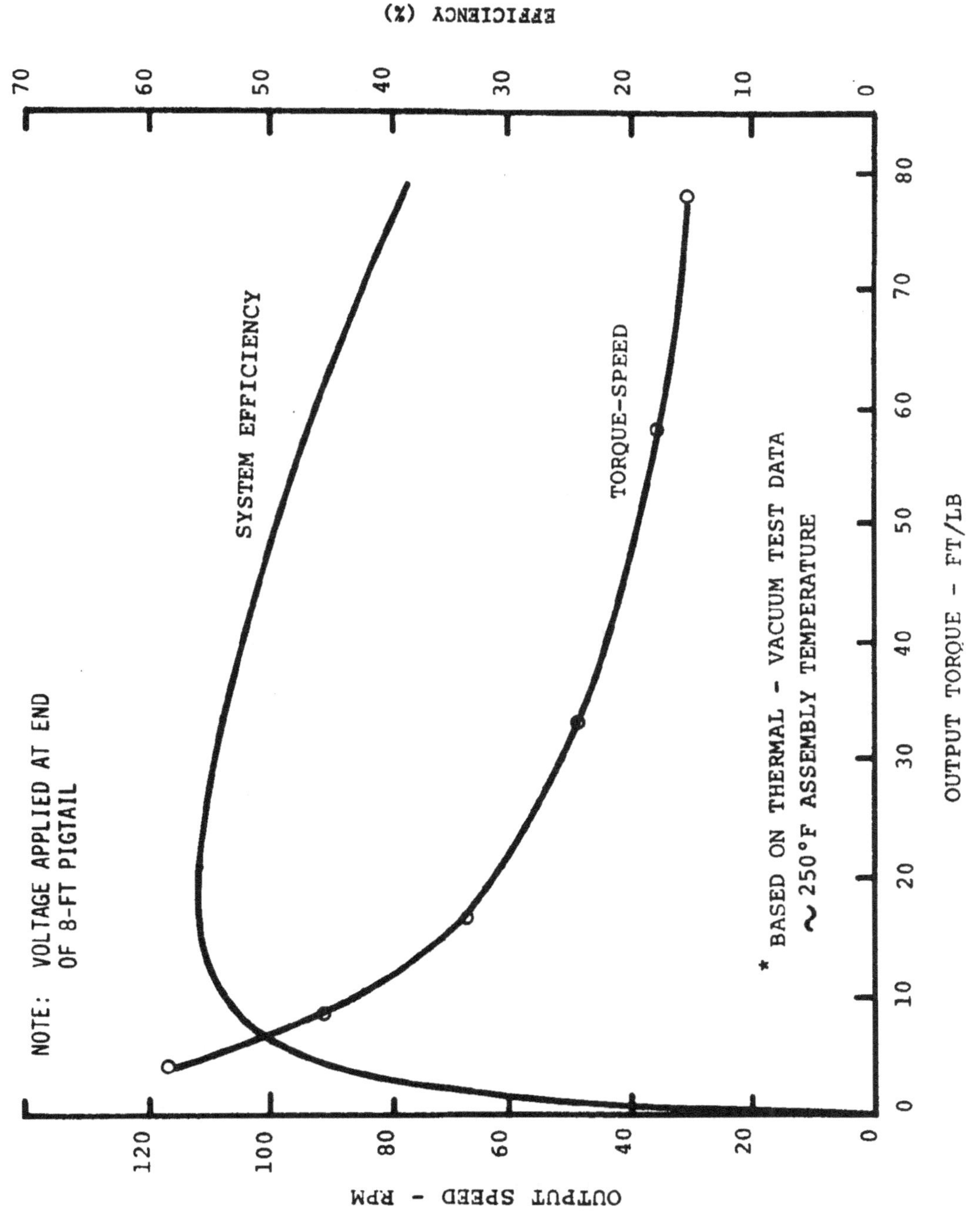

LS006-002-2H
LUNAR ROVING VEHICLE
OPERATIONS HANDBOOK
APPENDIX A

FIGURE 4-4 LRV TRACTION DRIVE PERFORMANCE (DC MOTOR - HARMONIC DRIVE - DRIVE CONTROLLER) FULL VOLTAGE (36V) PERFORMANCE

LS006-002-2H
LUNAR ROVING VEHICLE
OPERATIONS HANDBOOK
APPENDIX A

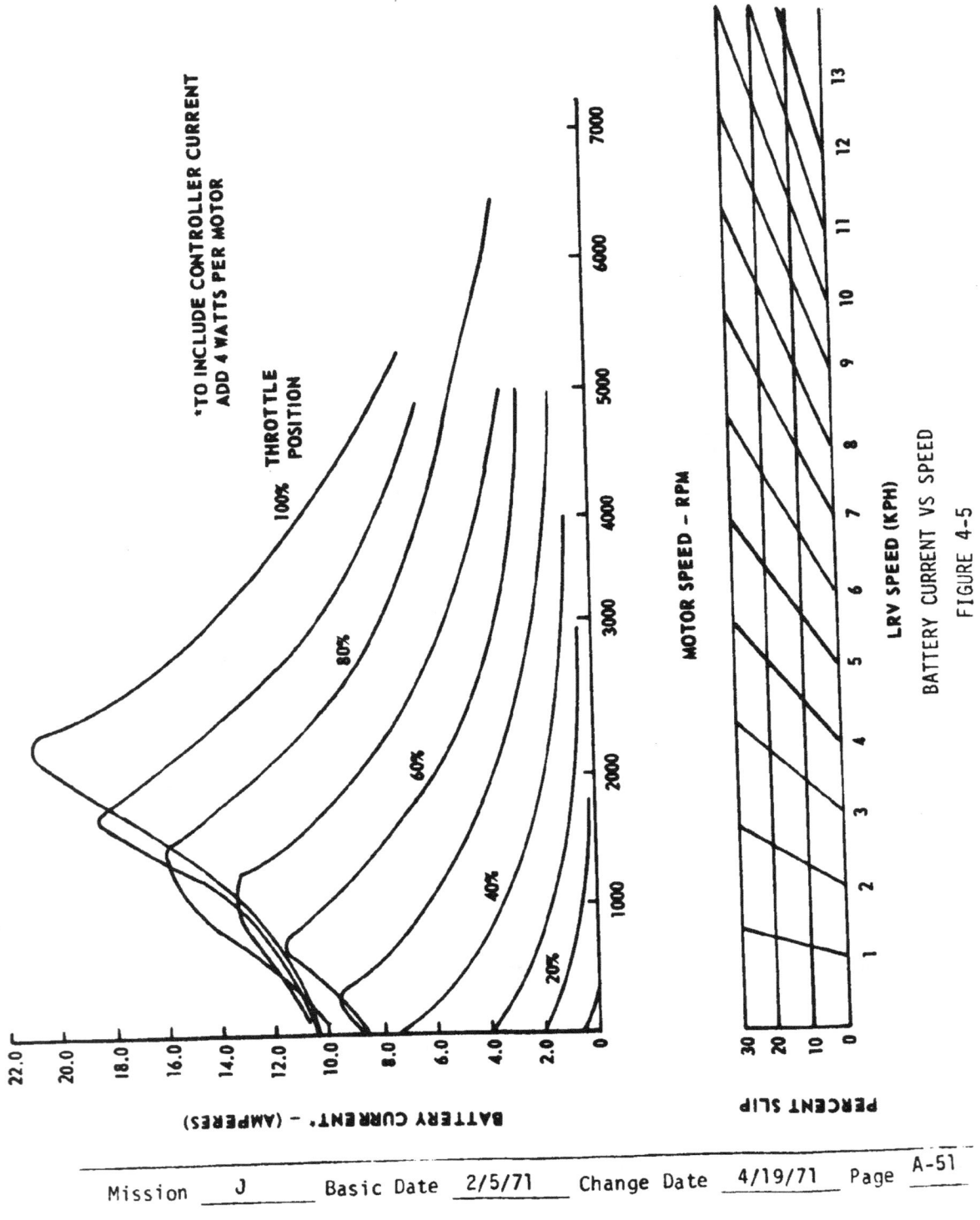

FIGURE 4-5

LS006-002-2H
LUNAR ROVING VEHICLE
OPERATIONS HANDBOOK
APPENDIX A

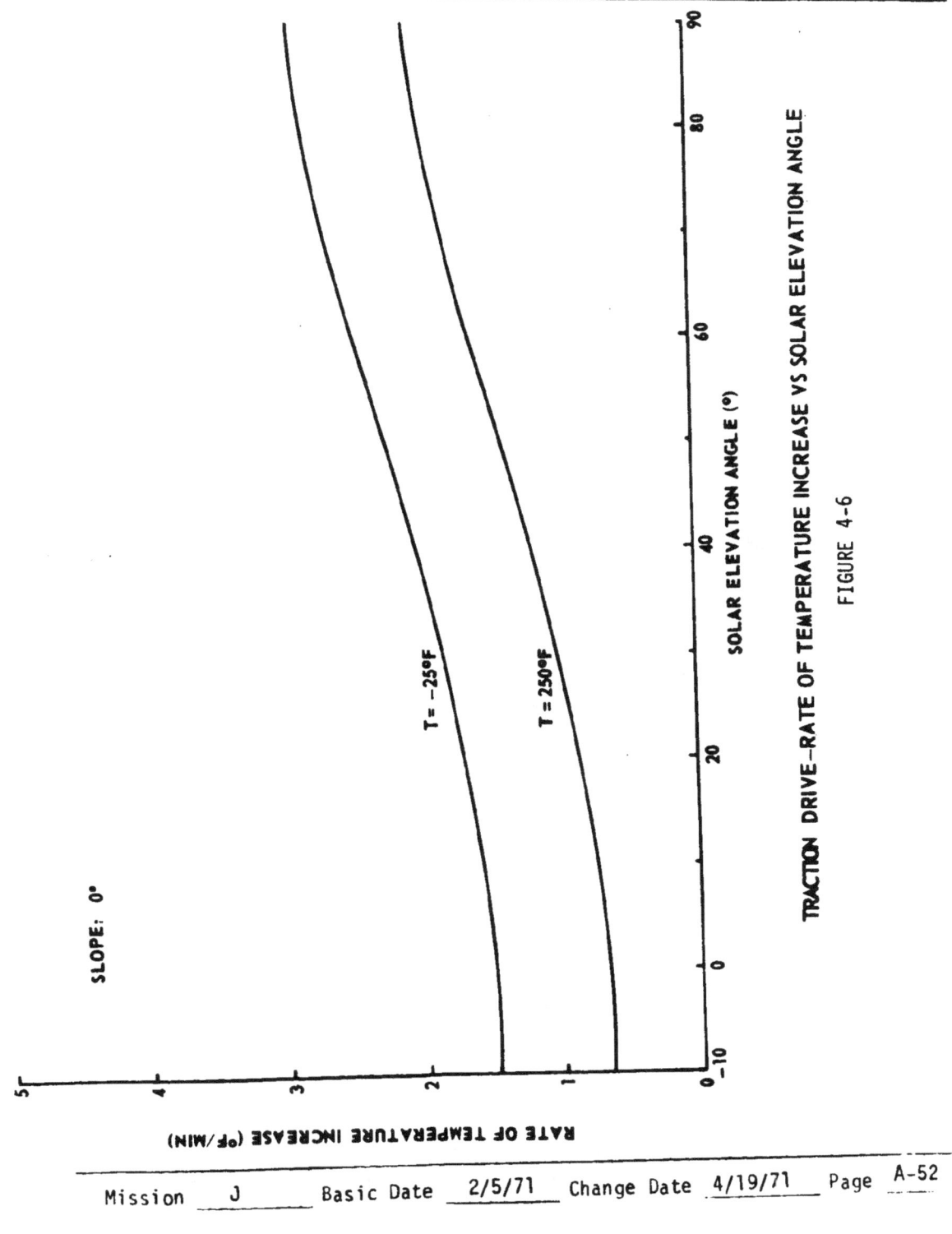

TRACTION DRIVE–RATE OF TEMPERATURE INCREASE VS SOLAR ELEVATION ANGLE

FIGURE 4-6

LS006-002-2H
LUNAR ROVING VEHICLE
OPERATIONS HANDBOOK
APPENDIX A

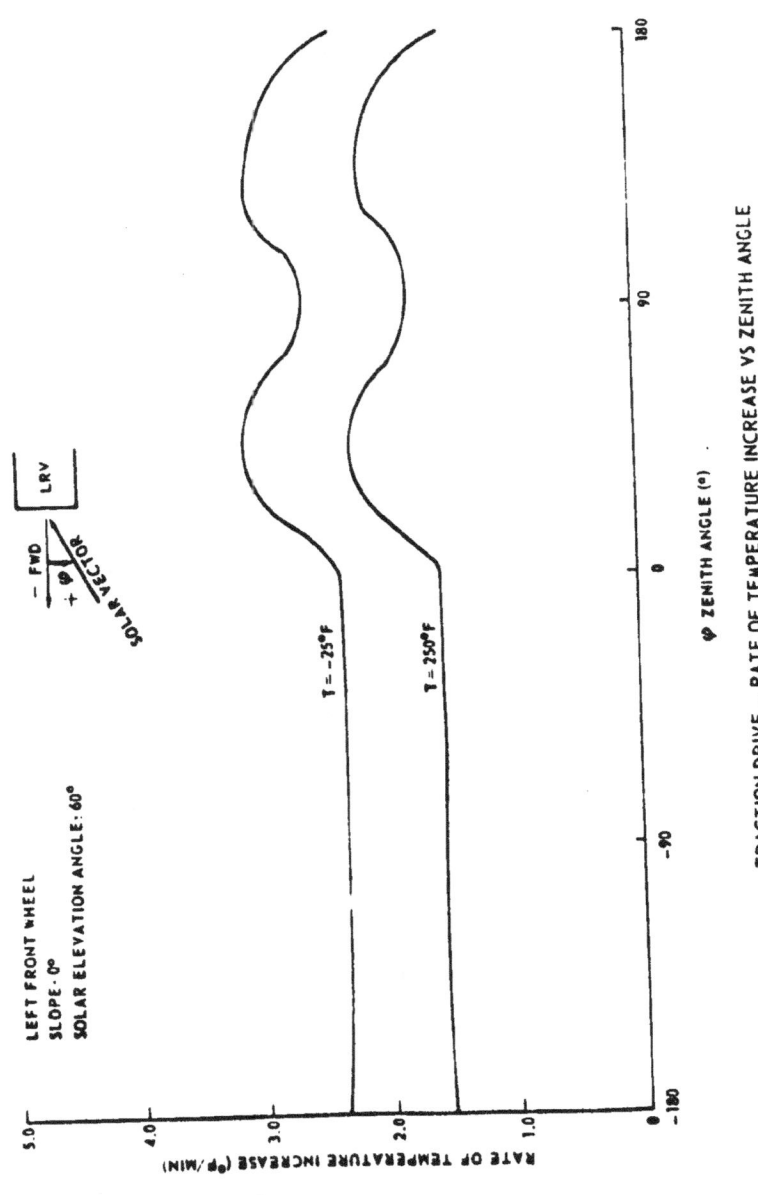

FIGURE 4-7
TRACTION DRIVE - RATE OF TEMPERATURE INCREASE VS ZENITH ANGLE

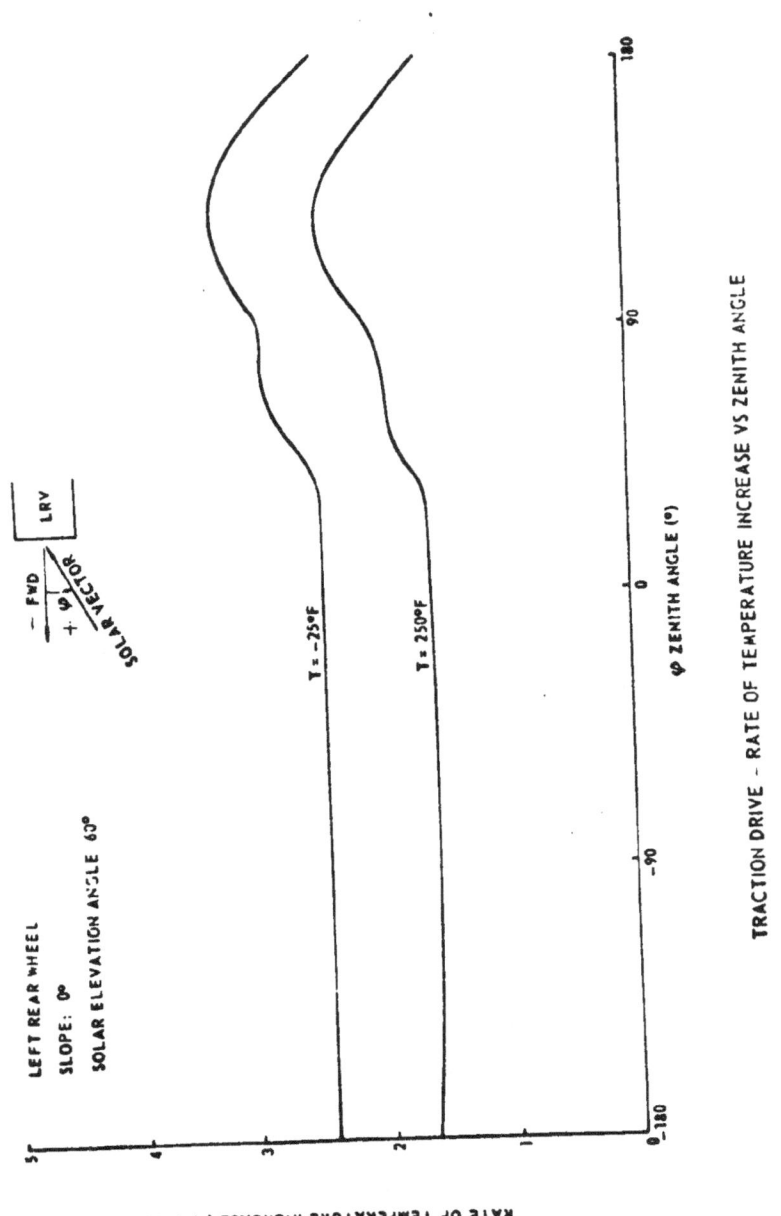

FIGURE 4-8

TRACTION DRIVE - RATE OF TEMPERATURE INCREASE VS ZENITH ANGLE

LS006-002-2H
LUNAR ROVING VEHICLE
OPERATIONS HANDBOOK
APPENDIX A

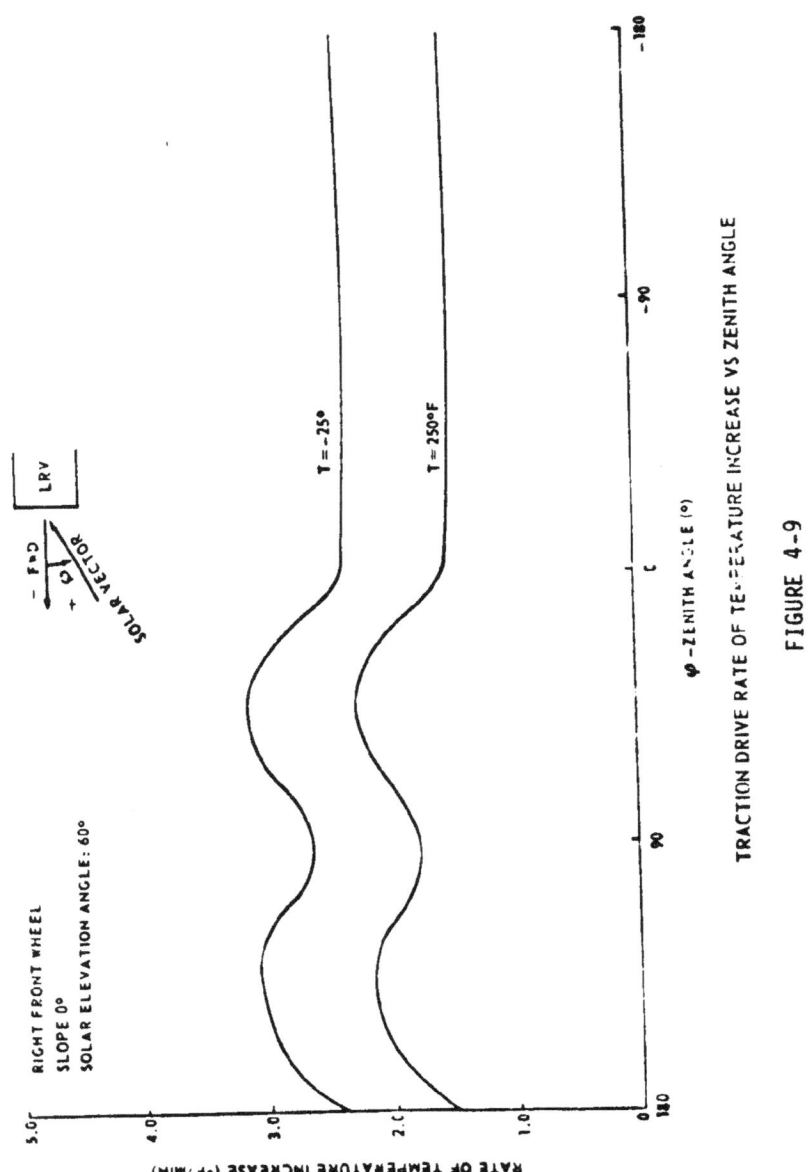

TRACTION DRIVE RATE OF TEMPERATURE INCREASE VS ZENITH ANGLE
FIGURE 4-9

LS006-002-2H
LUNAR ROVING VEHICLE
OPERATIONS HANDBOOK
APPENDIX A

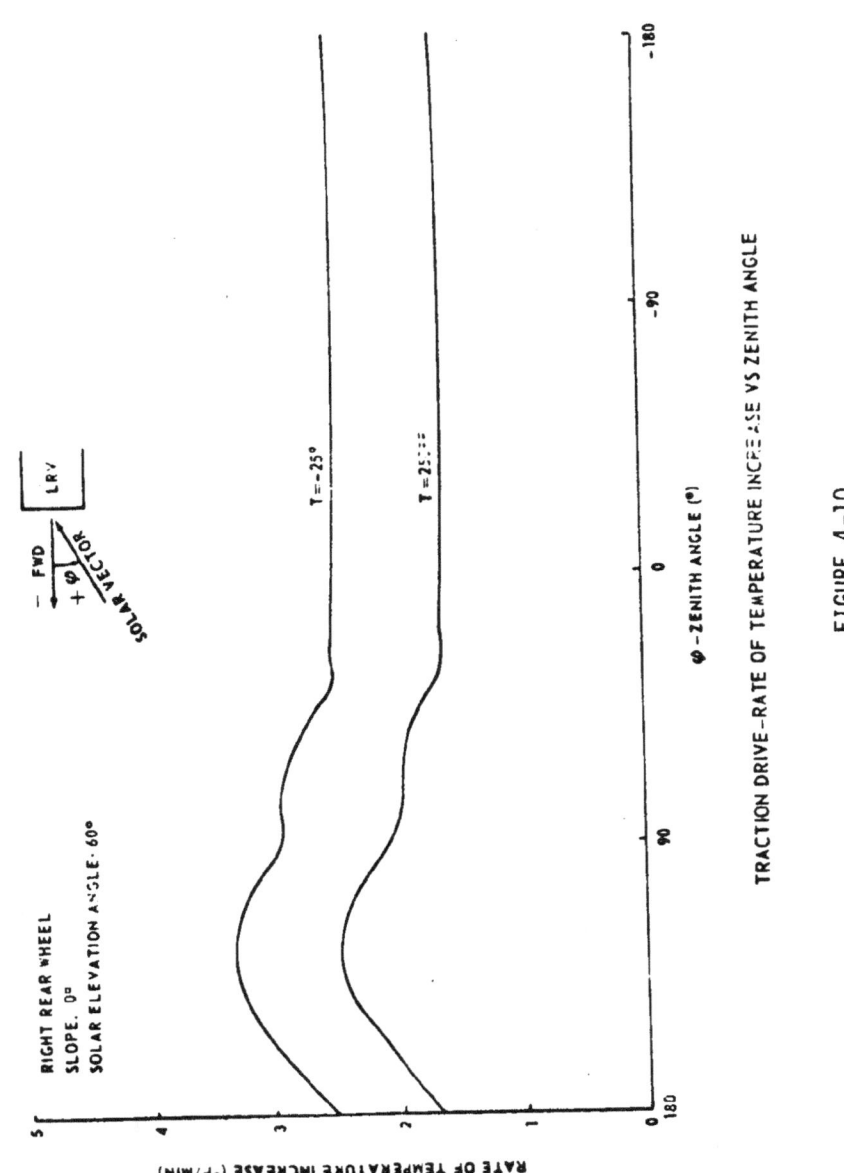

FIGURE 4-10
TRACTION DRIVE-RATE OF TEMPERATURE INCREASE VS ZENITH ANGLE

LS006-002-2H
LUNAR ROVING VEHICLE
OPERATIONS HANDBOOK
APPENDIX A

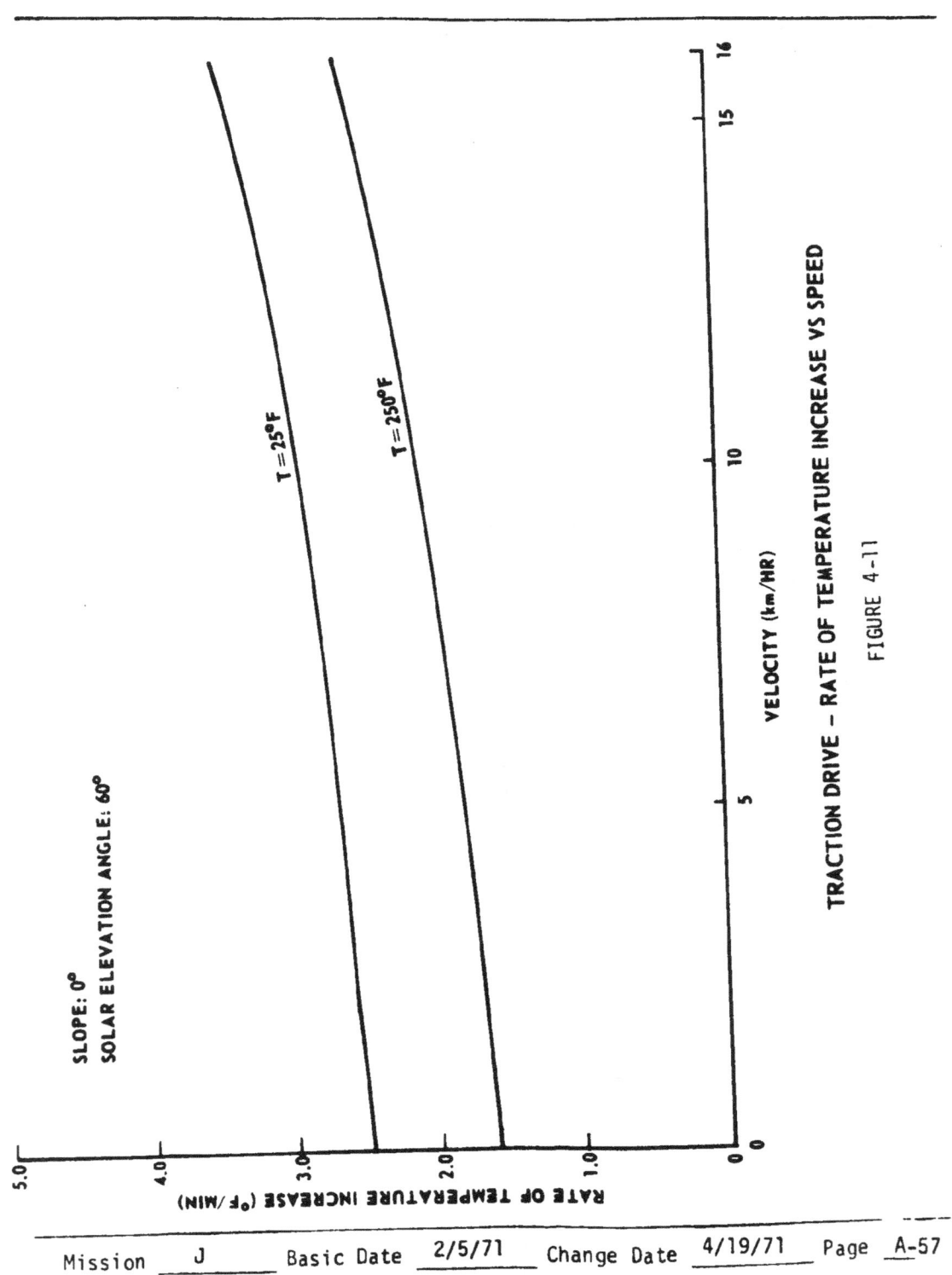

TRACTION DRIVE – RATE OF TEMPERATURE INCREASE VS SPEED

FIGURE 4-11

LS006-002-2H
LUNAR ROVING VEHICLE
OPERATIONS HANDBOOK
APPENDIX A

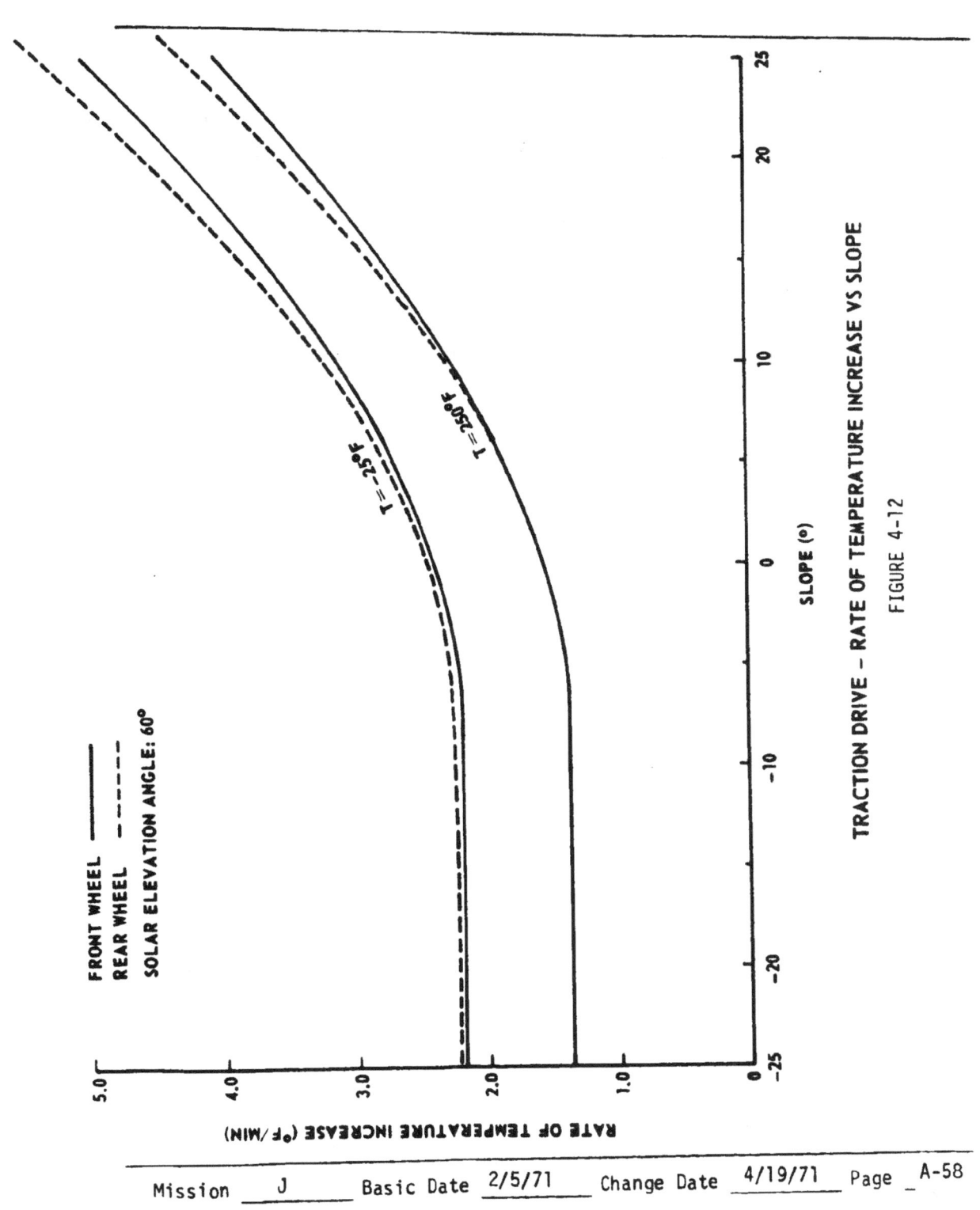

FIGURE 4-12
TRACTION DRIVE – RATE OF TEMPERATURE INCREASE VS SLOPE

4.2 ELECTRICAL SUBSYSTEM PERFORMANCE

The battery nominal voltage is $36 ^{+5}_{-3}$ VDC. Voltage for various current drains between 0 and 62 amperes is shown in Figure 4-13. Voltage vs. state of charge at a final discharge rate of 47 amperes is shown in Figure 4-14. The rate of battery temperature increase as a function of current flow and battery temperature is defined on Figures 4-15 and 4-16. Separate curves are provided for both left and right battery positions. The effect of battery temperature and solar reflector dust coverage upon battery temperature change for both left and right battery positions is defined on Figures 4-17 and 4-18. Rates of battery temperature increase with one battery providing all the required LRV power can be obtained from Figures 4-15 and 4-16.

Distribution losses in the power distribution system are shown in Figure 4-19. The 36 volt losses (I_E) are shown totaled for all four traction drive motors using the motor current scale. Motor controller cable losses (I_m) are shown in Figures 4-19 using the total battery current scale.

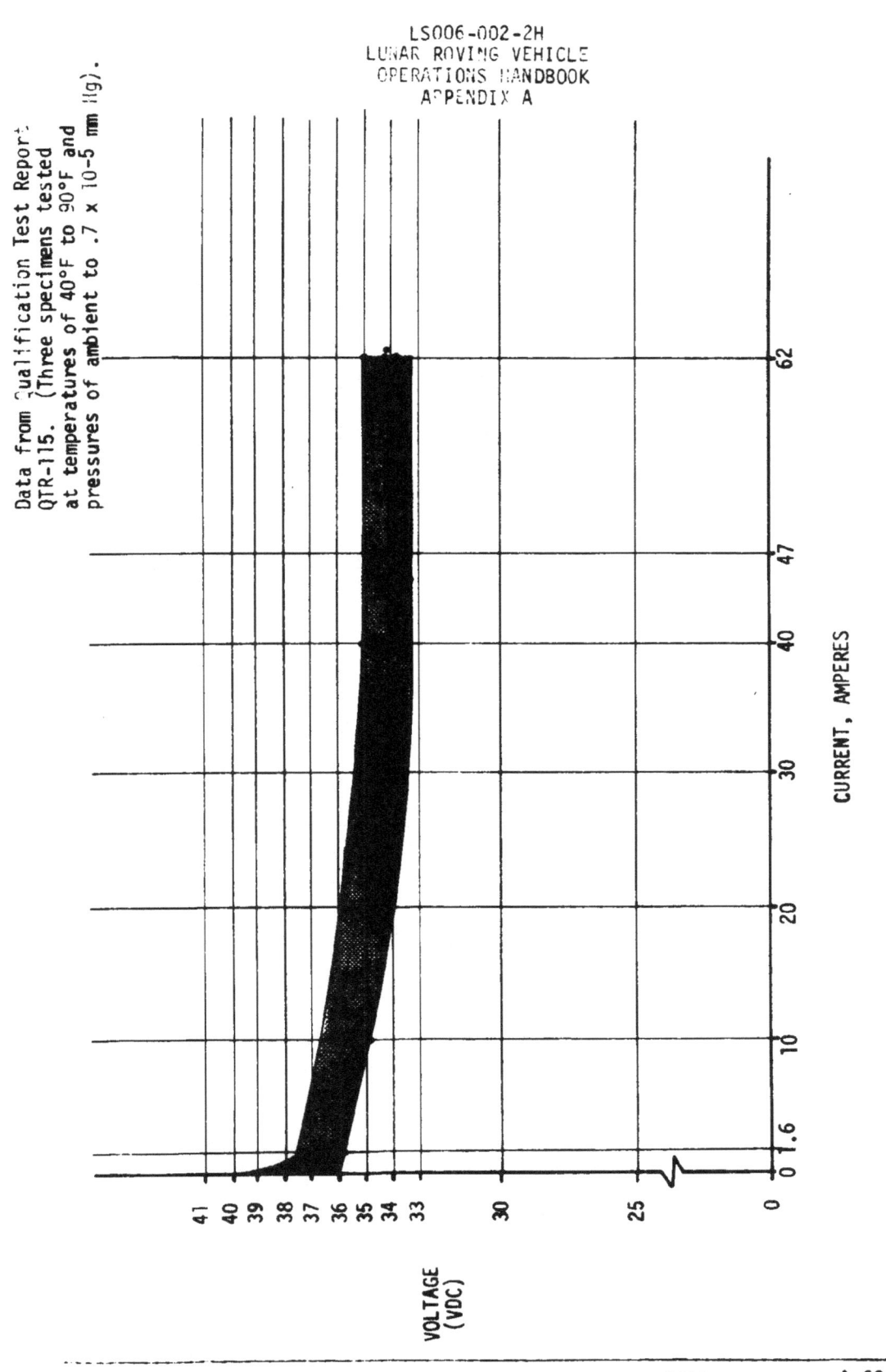

FIGURE 4-13 VOLTAGE VS CURRENT DRAW PER BATTERY

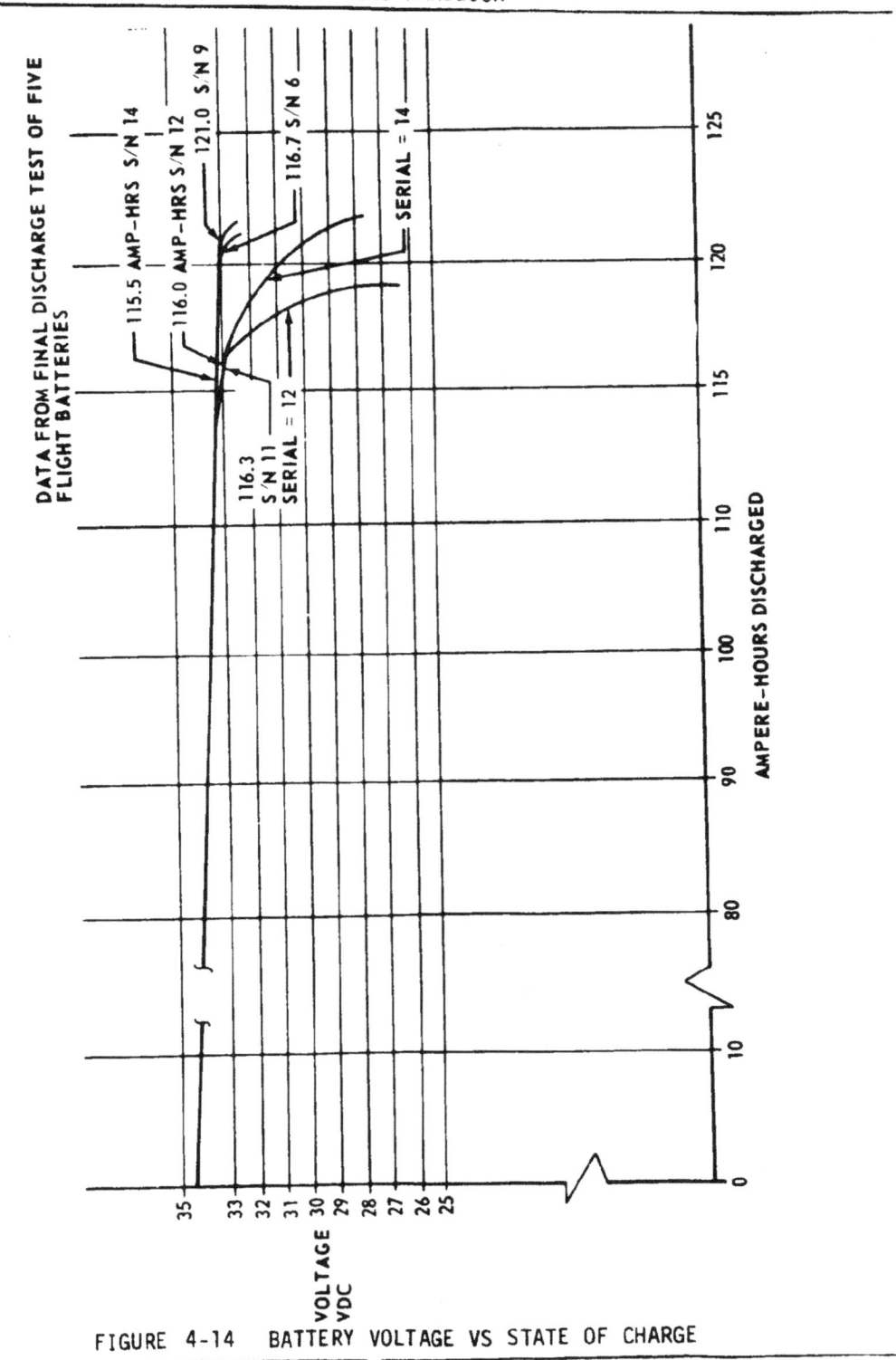

FIGURE 4-14 BATTERY VOLTAGE VS STATE OF CHARGE

LS006-002-2H
LUNAR ROVING VEHICLE
OPERATIONS HANDBOOK
APPENDIX A

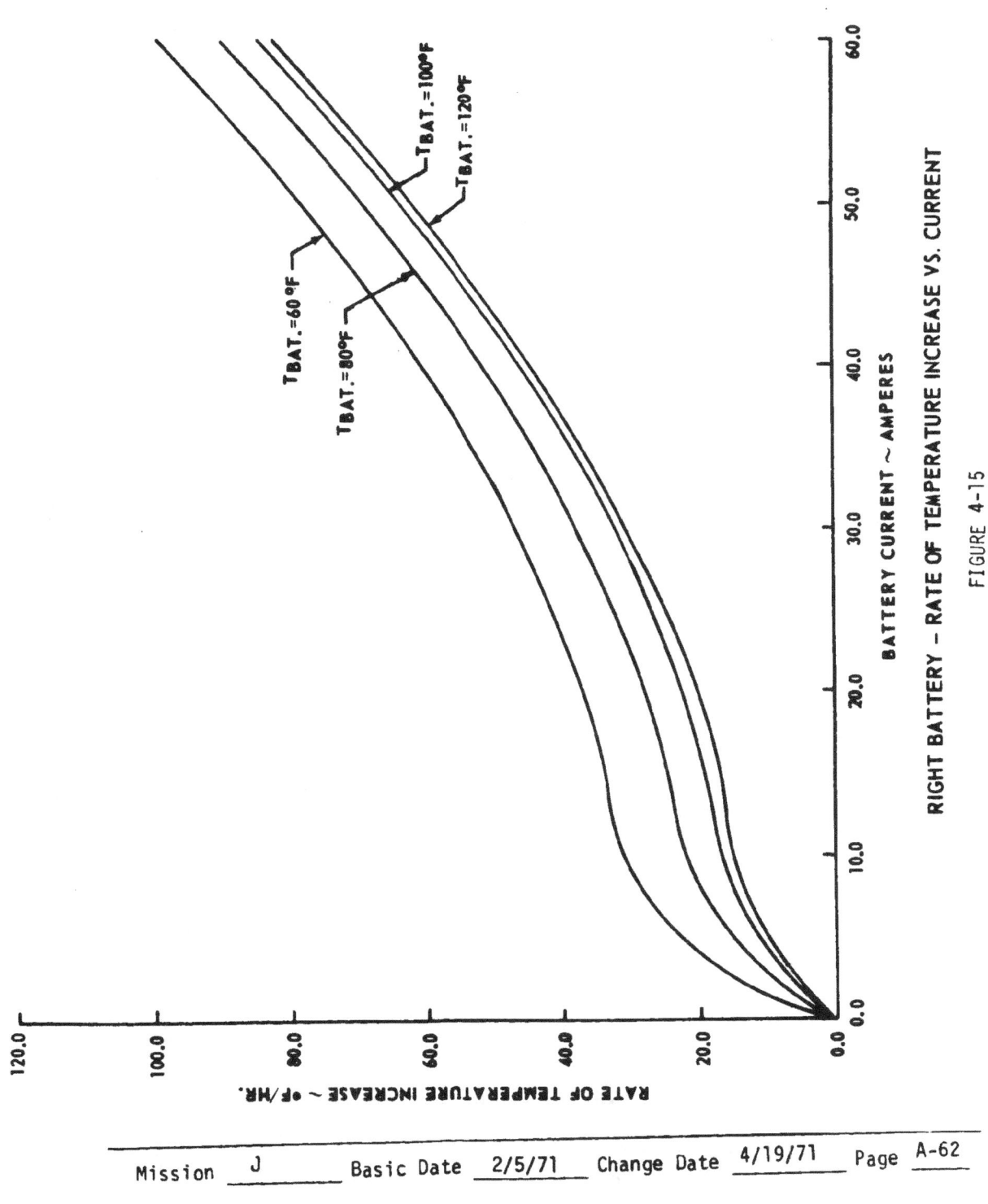

RIGHT BATTERY – RATE OF TEMPERATURE INCREASE VS. CURRENT

FIGURE 4-15

LS006-002-2H
LUNAR ROVING VEHICLE
OPERATIONS HANDBOOK
APPENDIX A

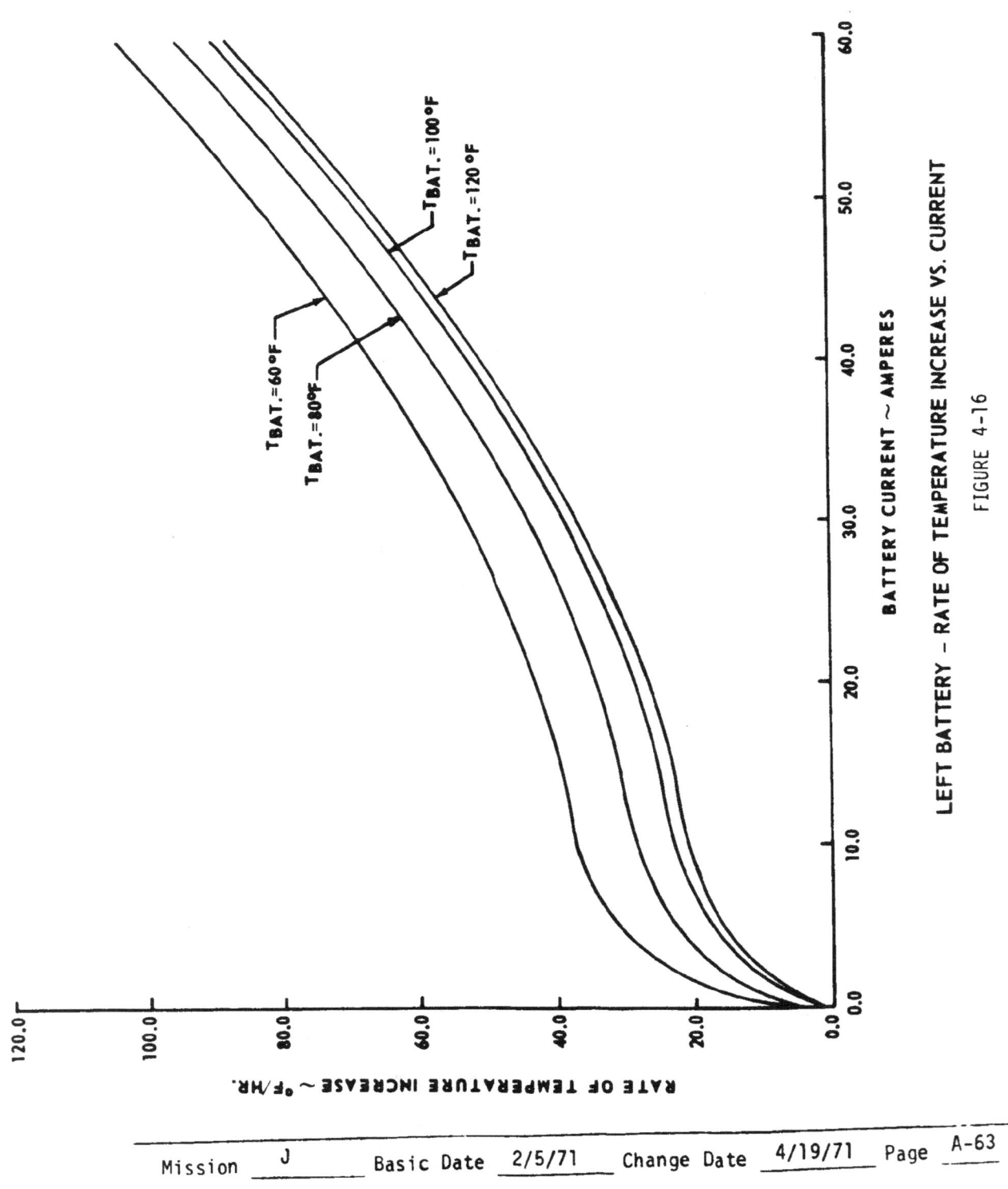

LEFT BATTERY - RATE OF TEMPERATURE INCREASE VS. CURRENT

FIGURE 4-16

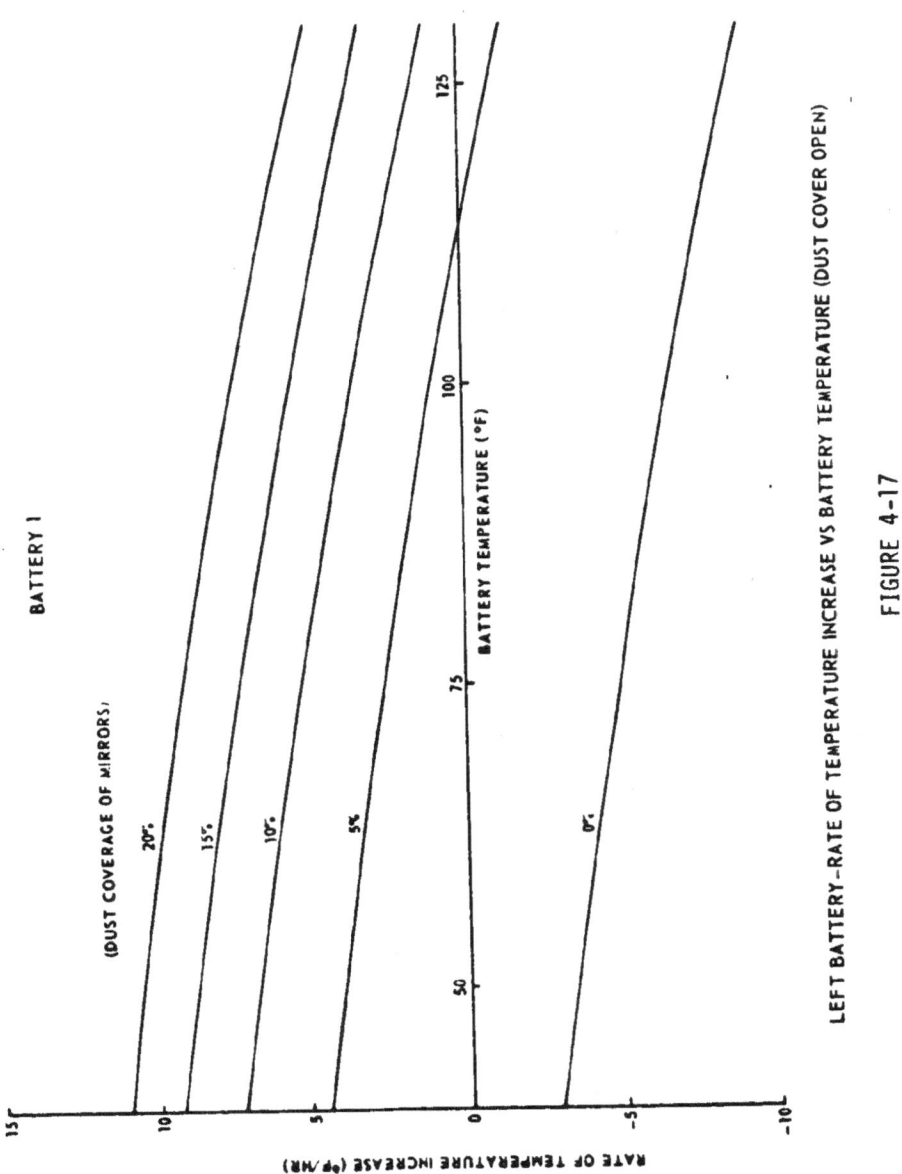

FIGURE 4-17

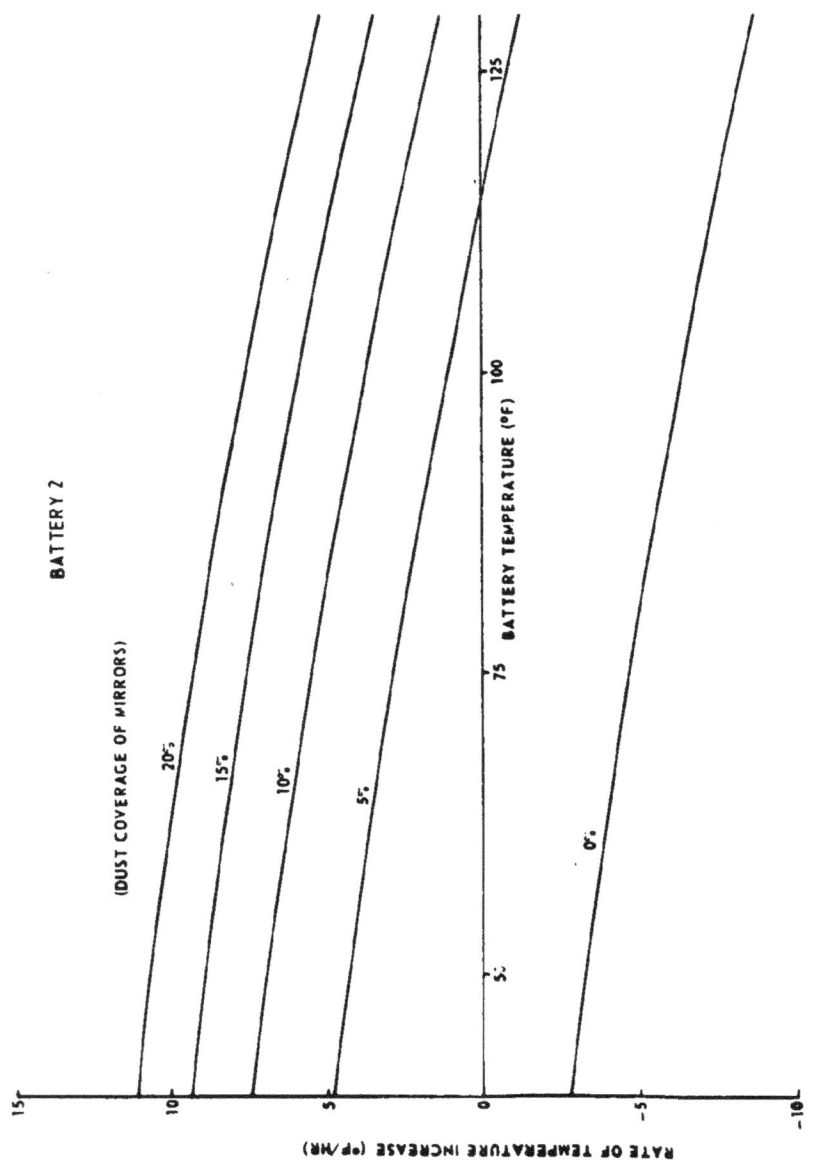

FIGURE 4-18
RIGHT BATTERY-RATE OF TEMPERATURE INCREASE VS BATTERY TEMPERATURE (DUST COVER OPEN)

LS006-002-2H
LUNAR ROVING VEHICLE
OPERATIONS HANDBOOK
APPENDIX A

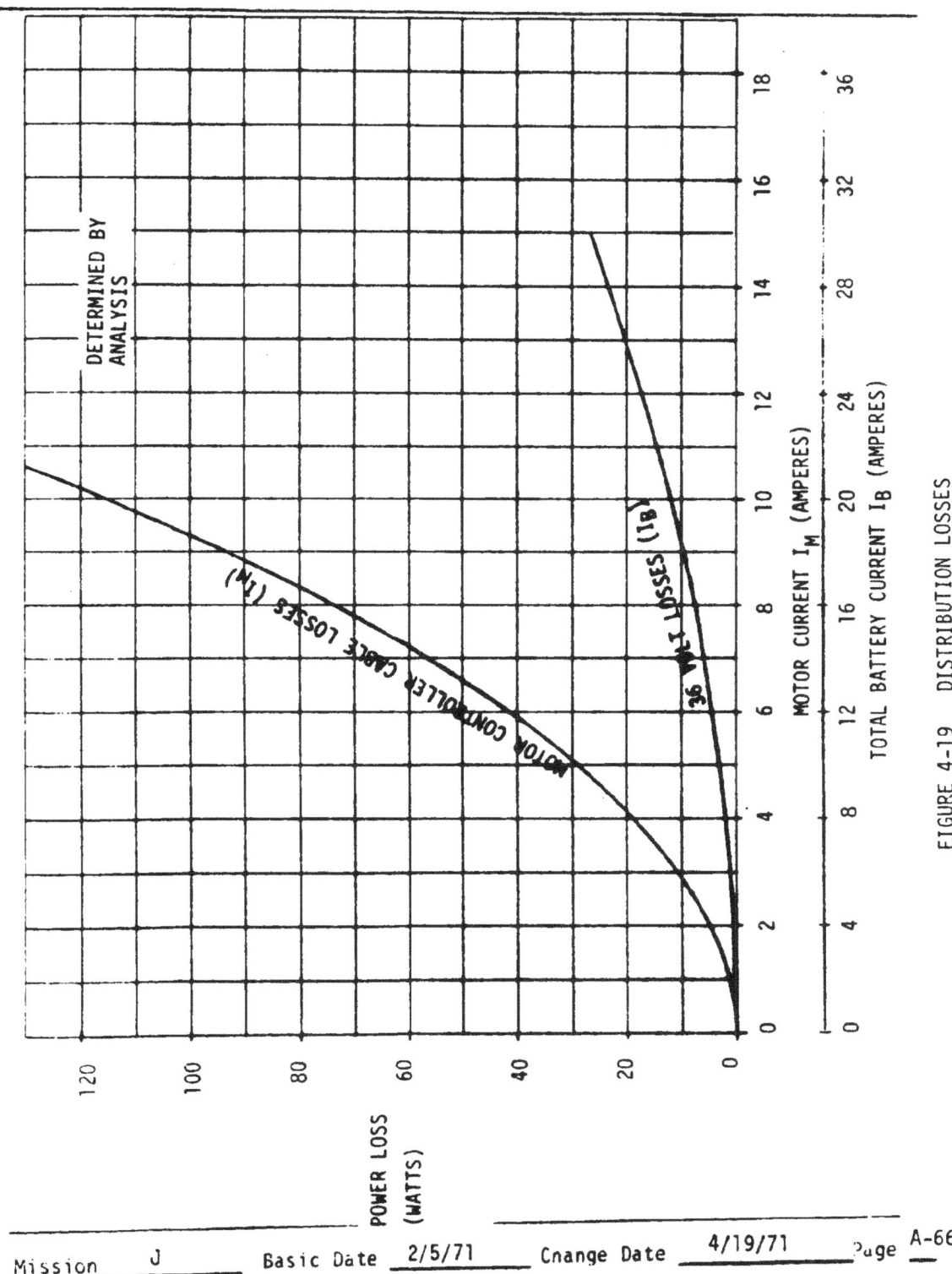

FIGURE 4-19 DISTRIBUTION LOSSES

LS006-002-2H
LUNAR ROVING VEHICLE
OPERATIONS HANDBOOK
APPENDIX A

4.3 NAVIGATION SUBSYSTEM PERFORMANCE

Directional Gyro Unit power consumption as a function of temperature is defined on Figure 4-20.

Directional Gyro Unit drift rate as a function of temperature is shown in Figure 4-21.

Bearing error as a function of gyro drift rate and navigation subsystem update period is shown on Figure 4-22.

Sun Shadow Device Azimuth corrections as a function of vehicle roll angle and sun elevation angle are defined on Figure 4-23.

Sun Shadow Device Azimuth corrections as a function of Sun Shadow Device Reading and pitch angle for sun elevation angles of 10, 25 and 40 degrees are shown in Figures 4-24, 4-25, and 4-26.

LS006-002-2H
LUNAR ROVING VEHICLE
OPERATIONS HANDBOOK
APPENDIX A

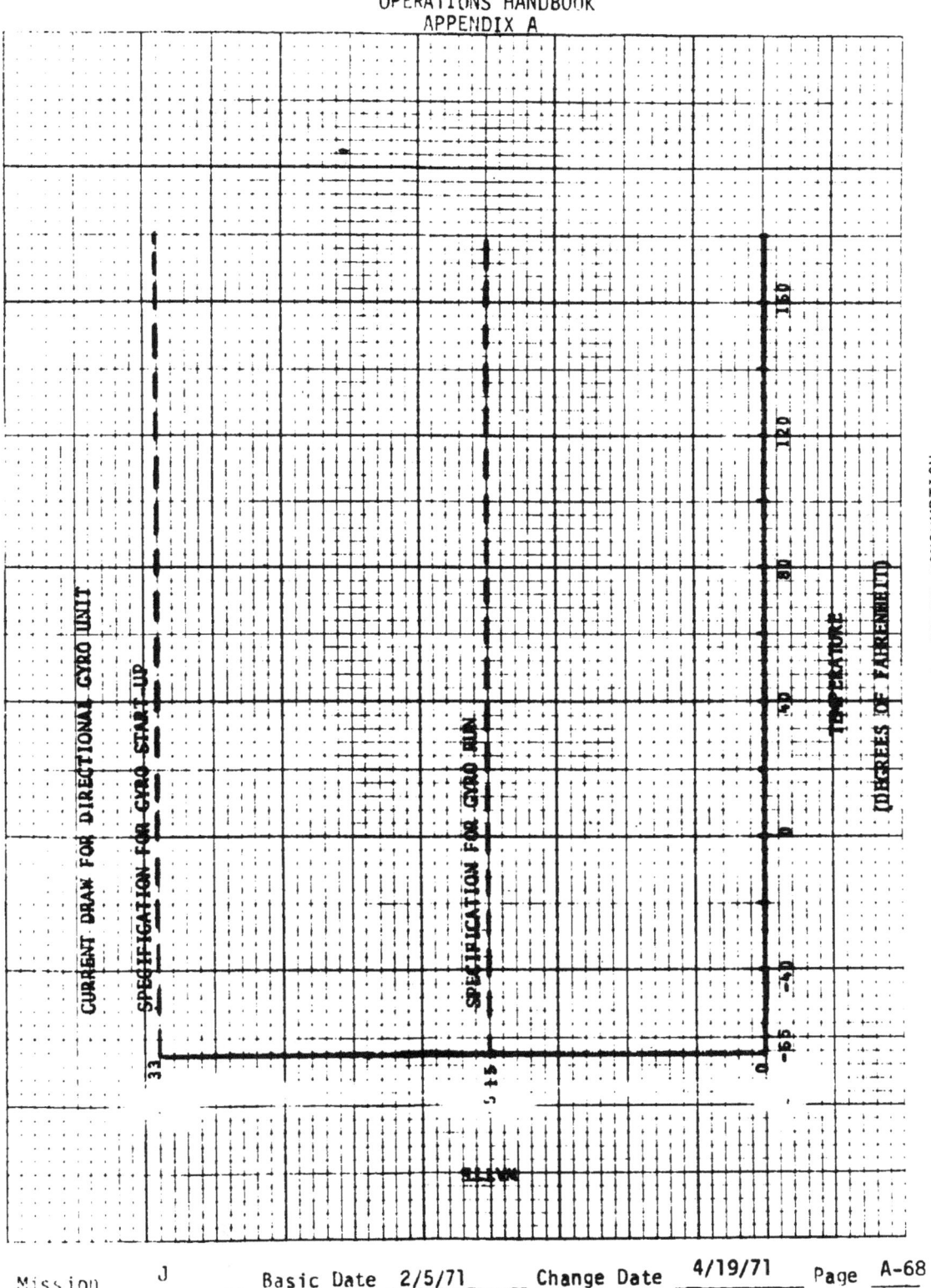

FIGURE 4-20 DGU POWER CONSUMPTION

LS006-002-2H
LUNAR ROVING VEHICLE
OPERATIONS HANDBOOK
APPENDIX A

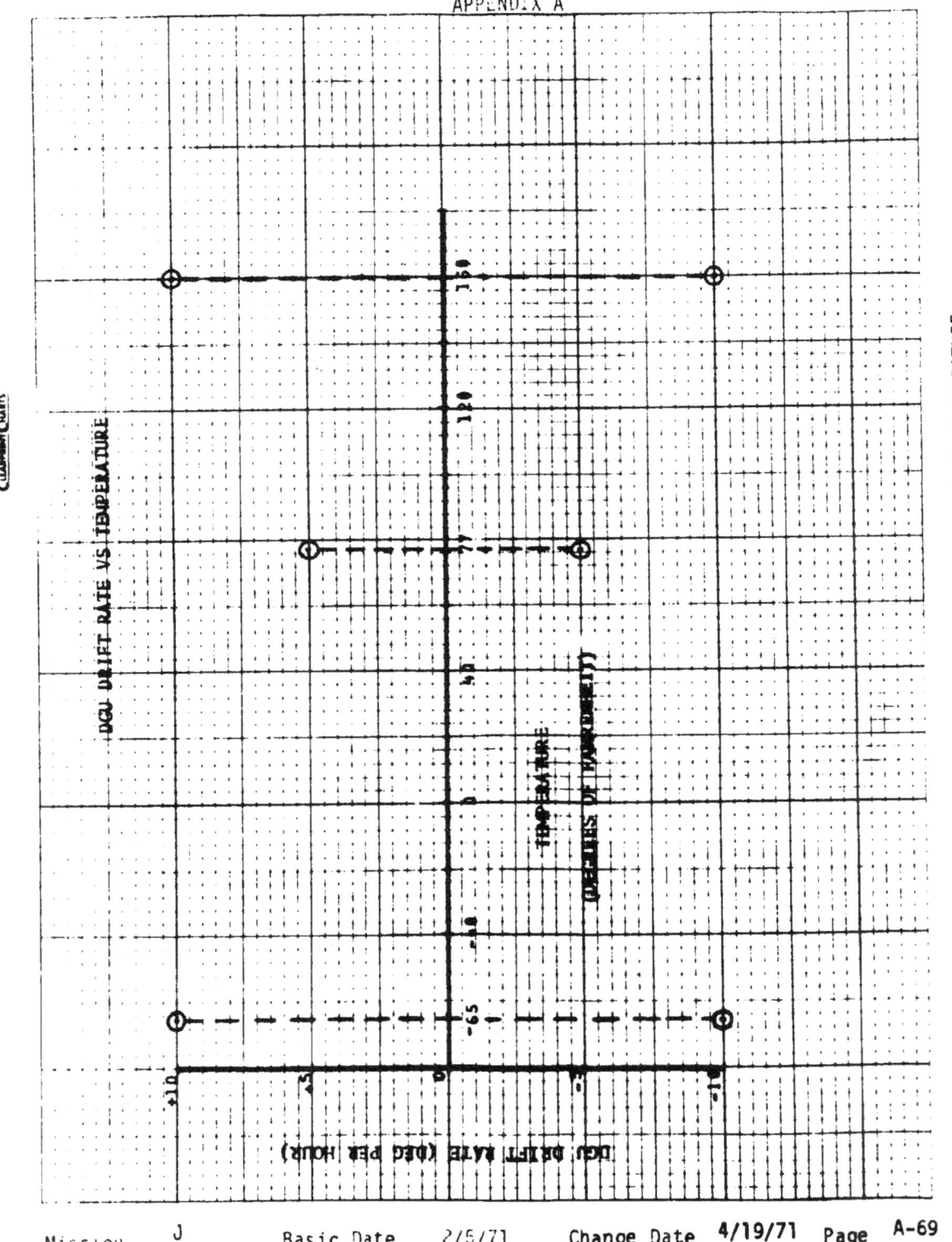

FIGURE 4-21 DGU DRIFT RATE VS TEMPERATURE

LS006-002-2H
LUNAR ROVING VEHICLE
OPERATIONS HANDBOOK
APPENDIX A

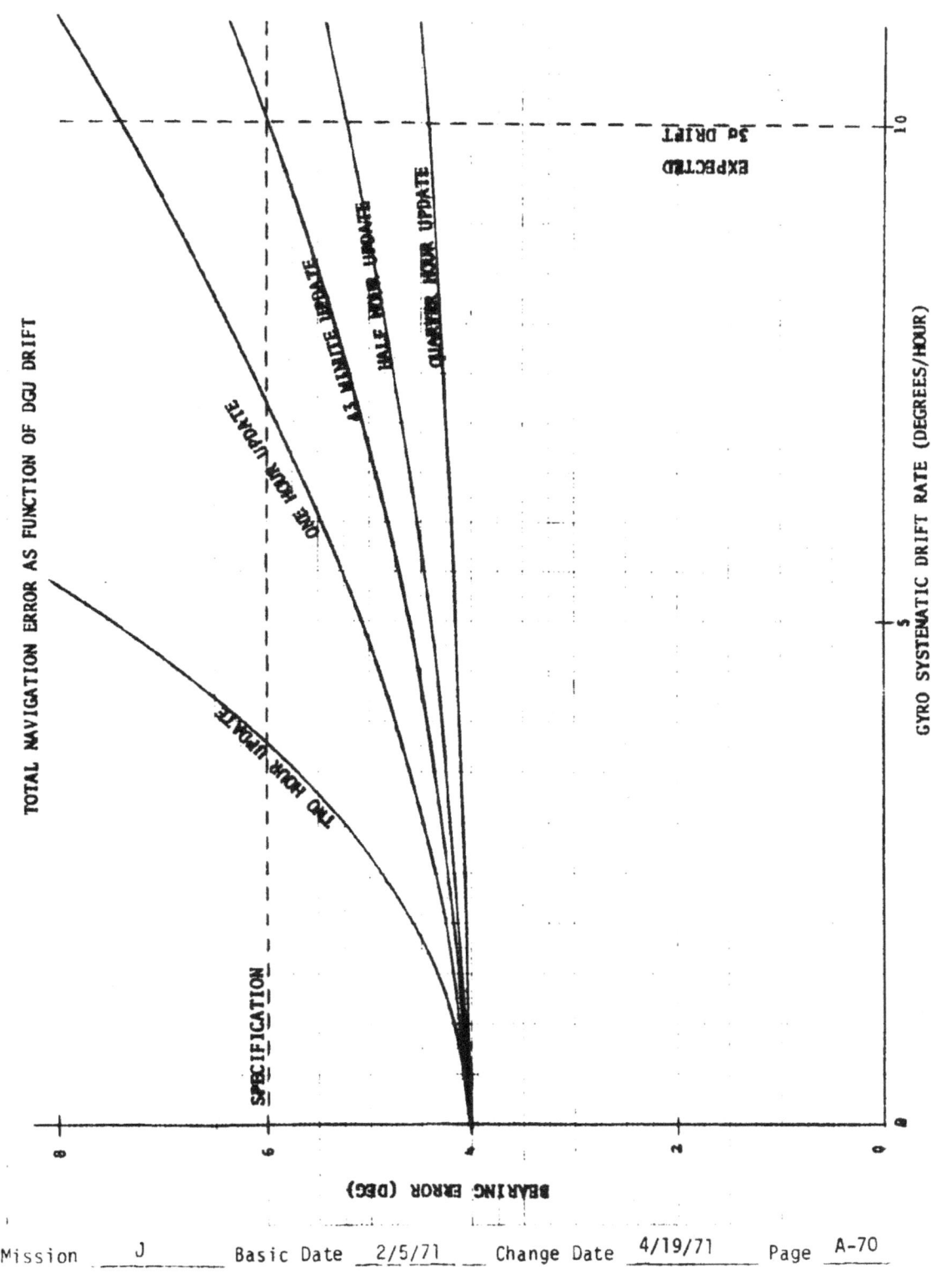

FIGURE 4-22 NAVIGATION BEARING ERROR AS A FUNCTION OF DGU DRIFT

Mission __J__ Basic Date __2/5/71__ Change Date __4/19/71__ Page __A-70__

LS006-002-2H
LUNAR ROVING VEHICLE
OPERATIONS HANDBOOK
APPENDIX A

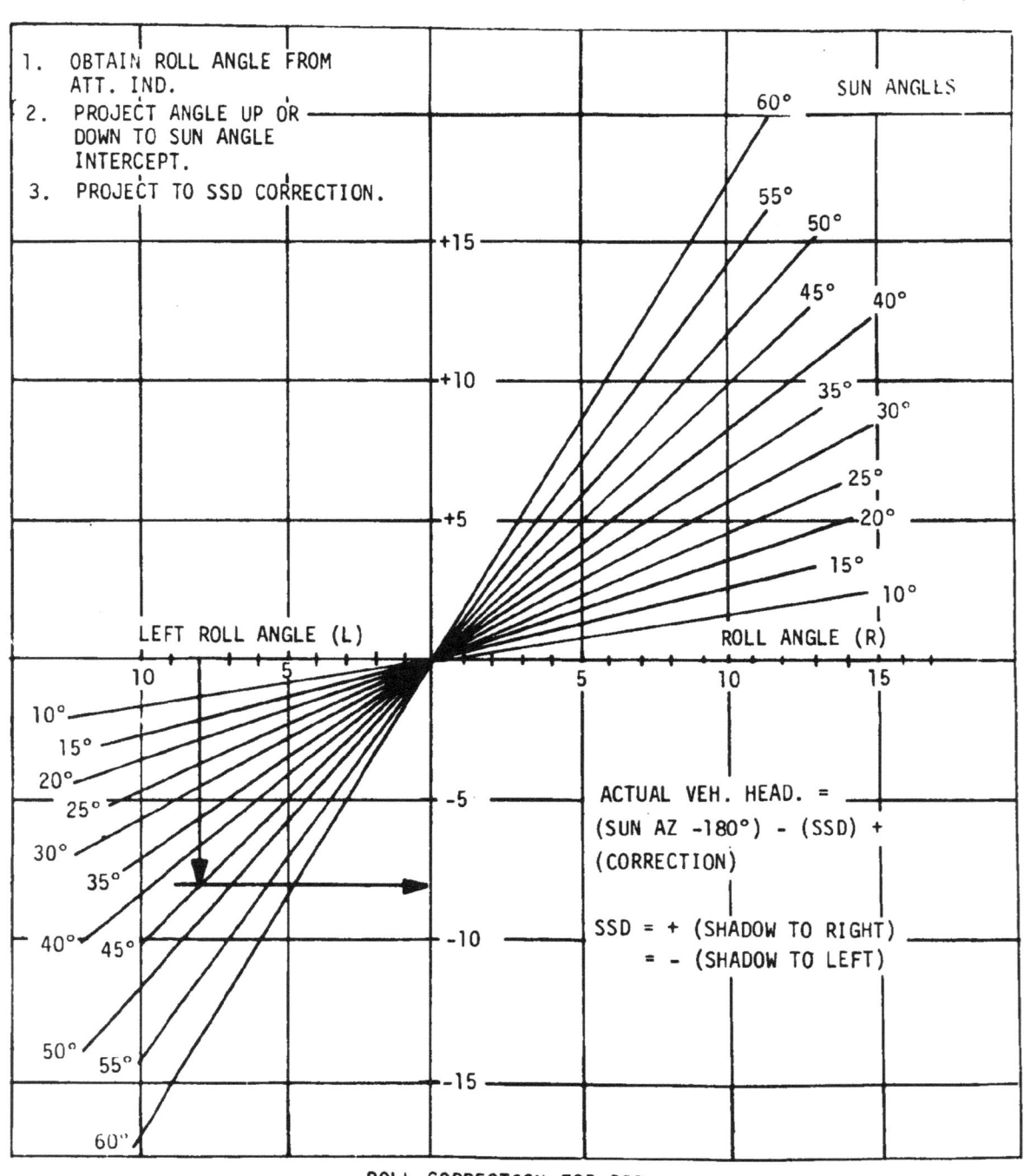

FIGURE 4-23 SSD AZIMUTH CORRECTION FOR VEHICLE ROLL

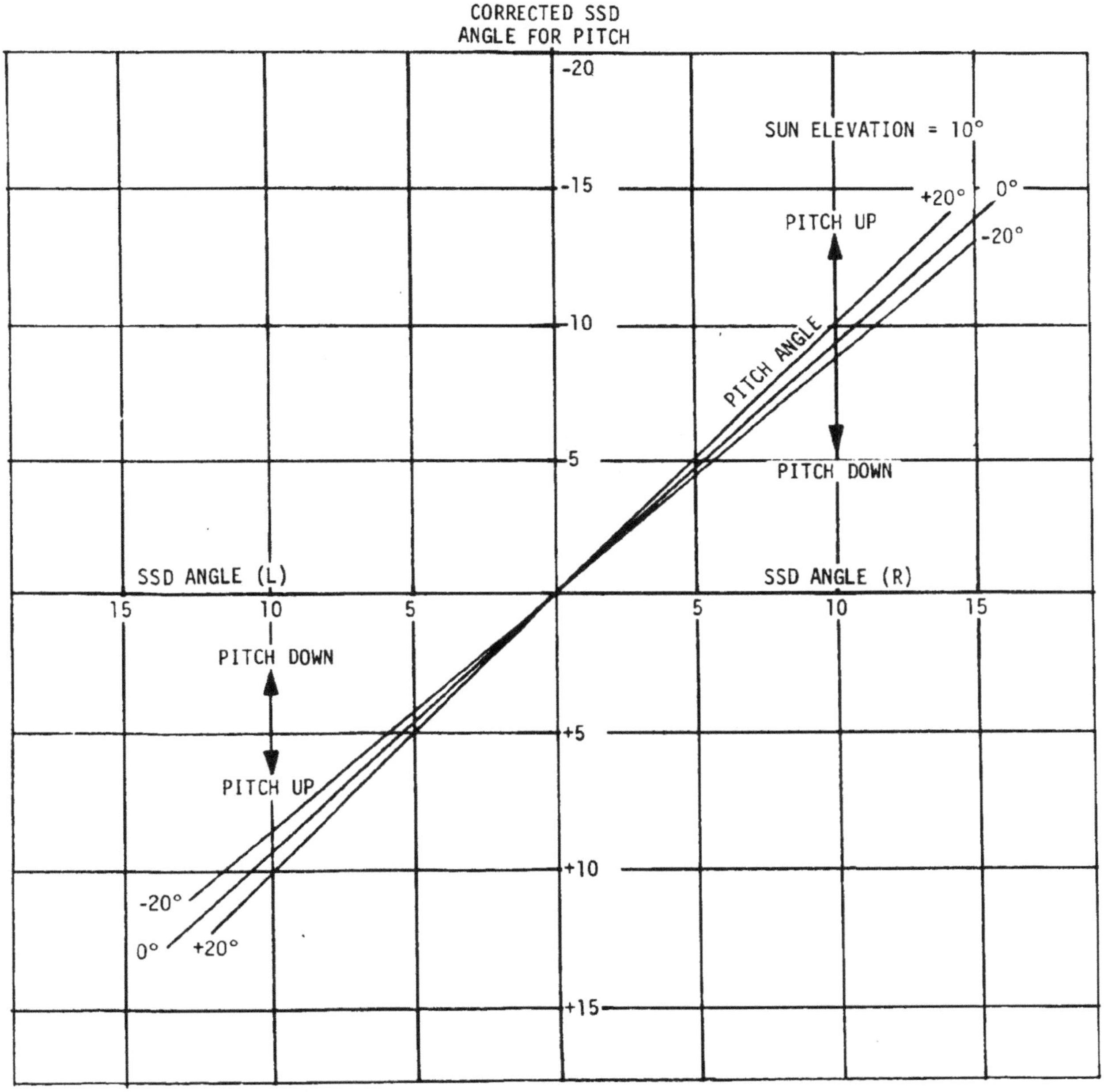

LS006-002-2H
LUNAR ROVING VEHICLE
OPERATIONS HANDBOOK
APPENDIX A

FIGURE 4-24. CORRECTED SSD ANGLE FOR VEHICLE PITCH (10° SUN ELEVATION)

LS006-002-211
LUNAR ROVING VEHICLE
OPERATIONS HANDBOOK
APPENDIX A

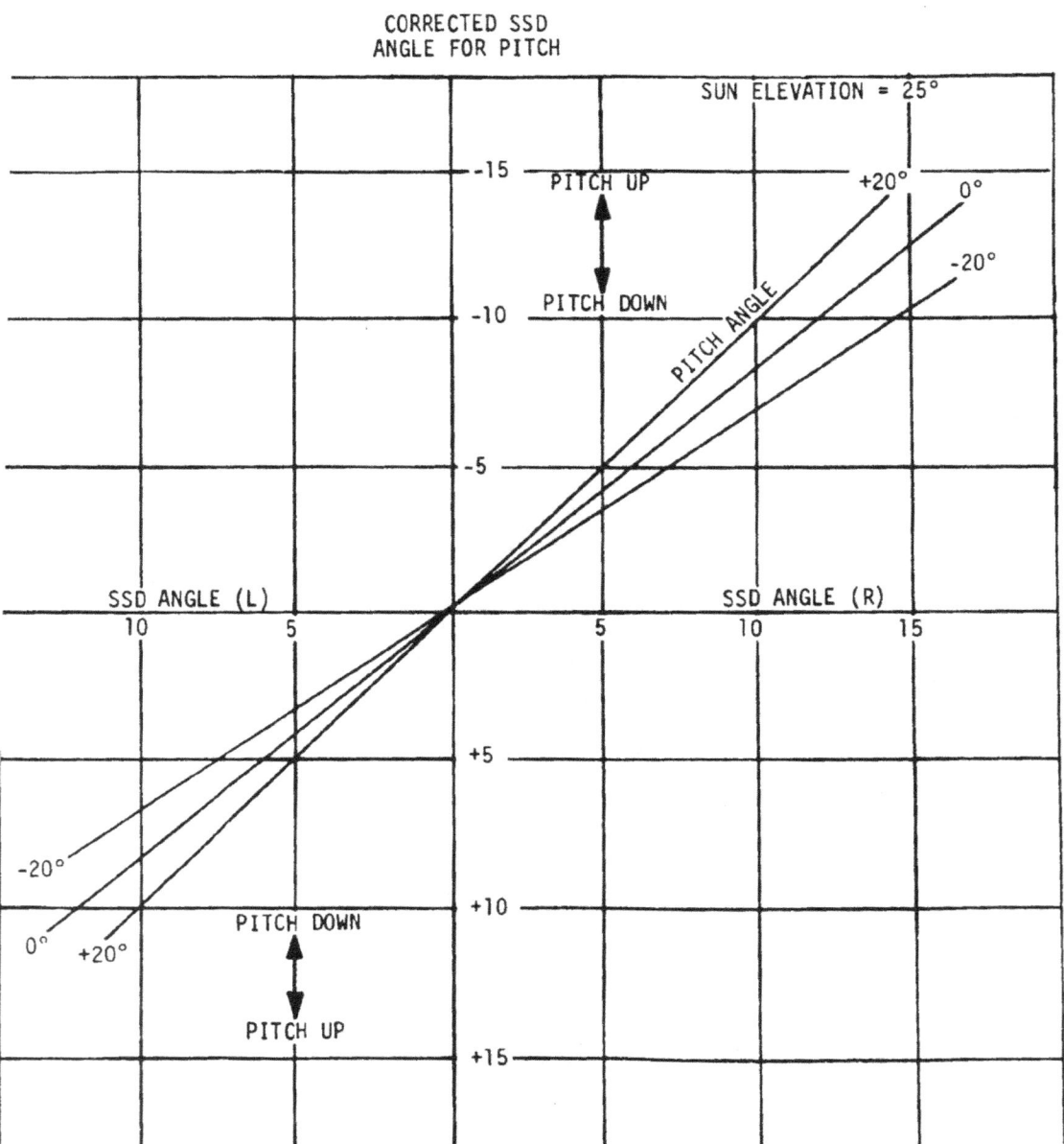

FIGURE 4-25   CORRECTED SSD ANGLE FOR VEHICLE PITCH (25° SUN ELEVATION)

Mission __J__   Basic Date __2/5/71__   Change Date __4/19/71__   Page __A-73__

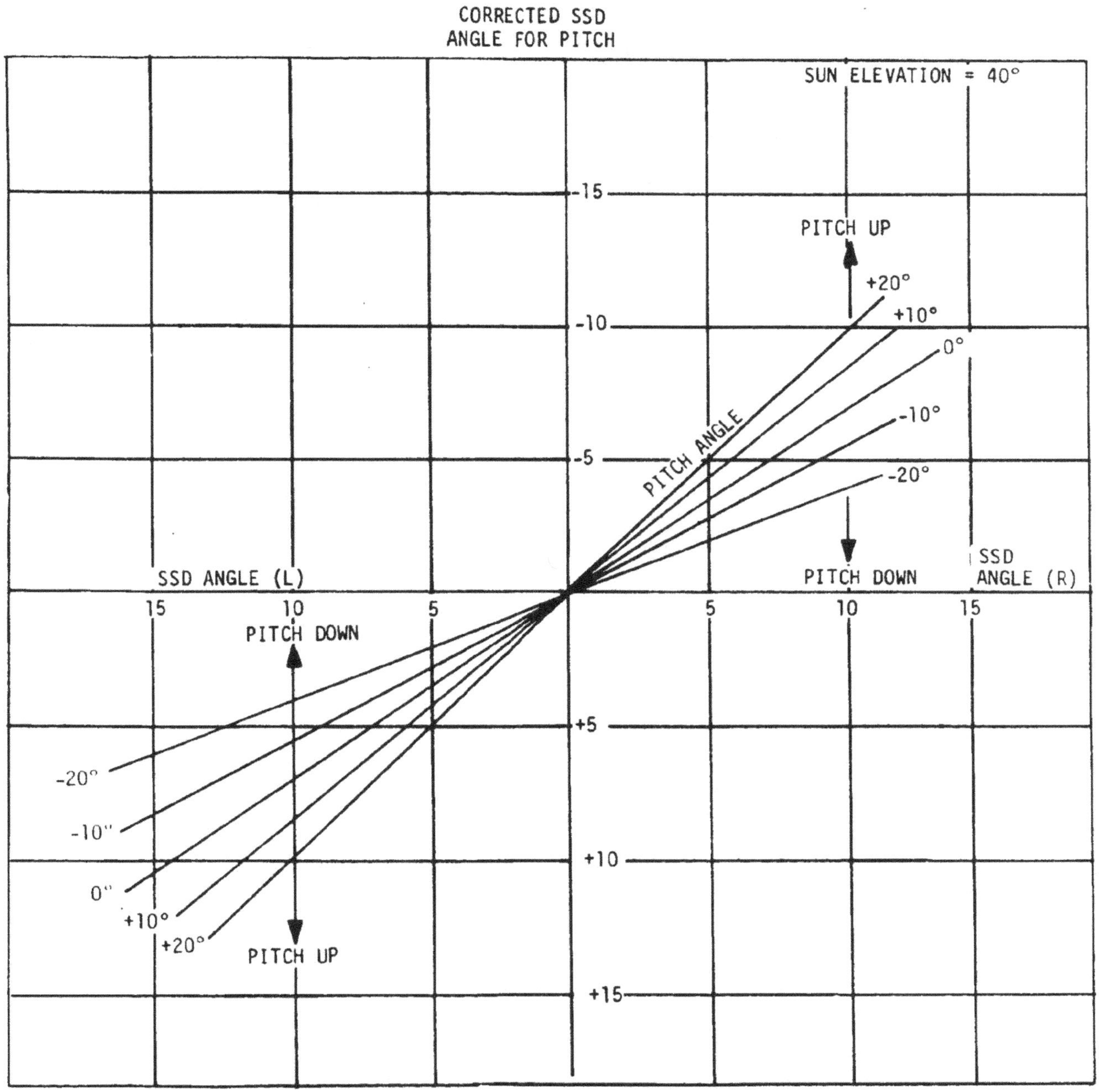

LS006-002-2H
LUNAR ROVING VEHICLE
OPERATIONS HANDBOOK
APPENDIX A

FIGURE 4-26   CORRECTED SSD ANGLE FOR VEHICLE PITCH (40° SUN ELEVATION)

LS006-002-2H
LUNAR ROVING VEHICLE
OPERATIONS HANDBOOK
APPENDIX A

4.4   DISPLAYS AND CONTROLS SUBSYSTEM PERFORMANCE

Torque required to provide throttle control at various hand controller angular displacements is defined on Figure 4-27.

Torque required to effect steering control at various hand controller angular displacements is defined on Figure 4-28.

Hand controller forces to accomplish vehicle braking as a function of hand controller linear and angular displacement is defined on Figure 4-29.

LS006-002-2H
LUNAR ROVING VEHICLE
OPERATIONS HANDBOOK
APPENDIX A

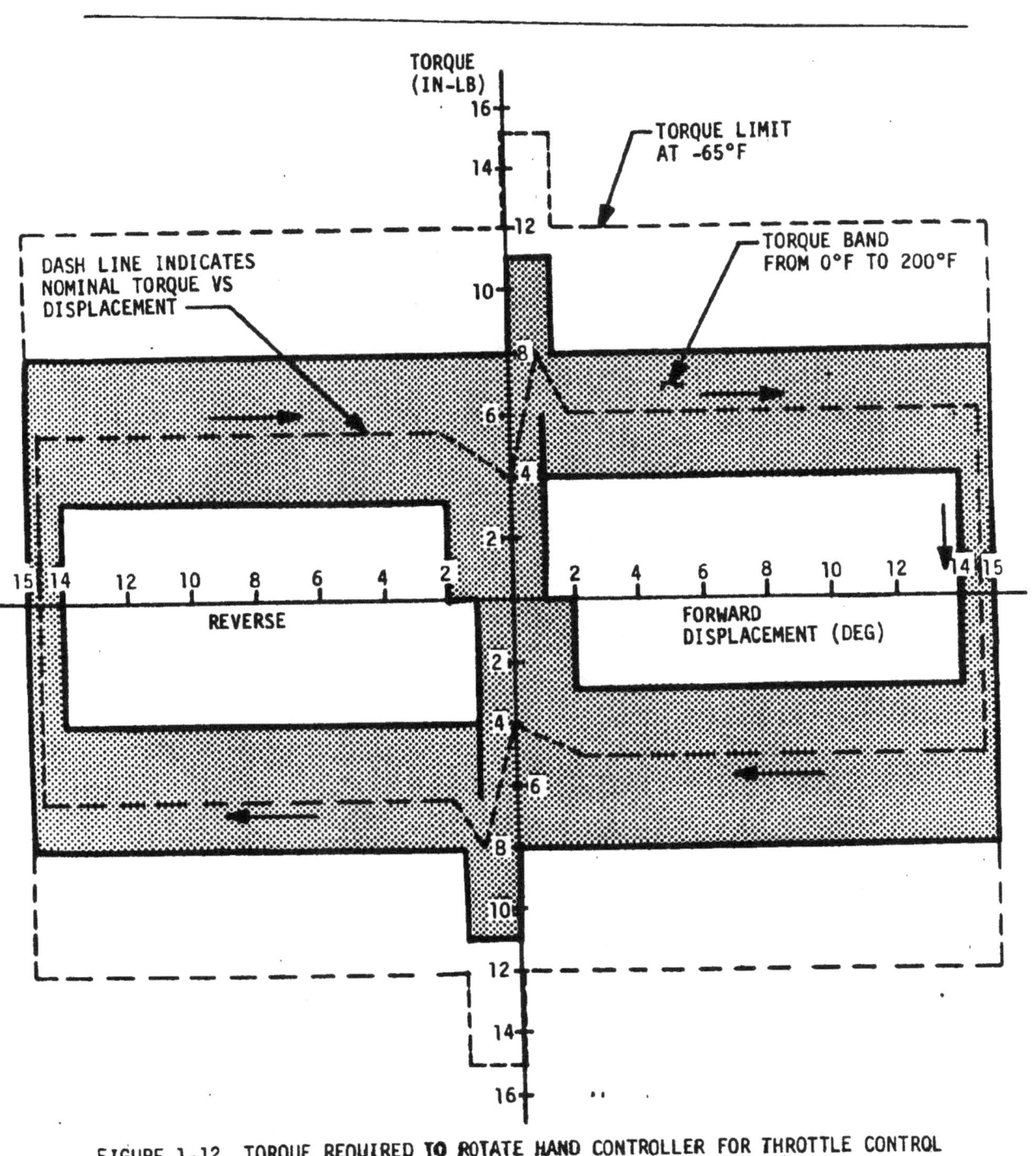

FIGURE 1-12 TORQUE REQUIRED TO ROTATE HAND CONTROLLER FOR THROTTLE CONTROL

LS006-002-2H
LUNAR ROVING VEHICLE
OPERATIONS HANDBOOK
APPENDIX A

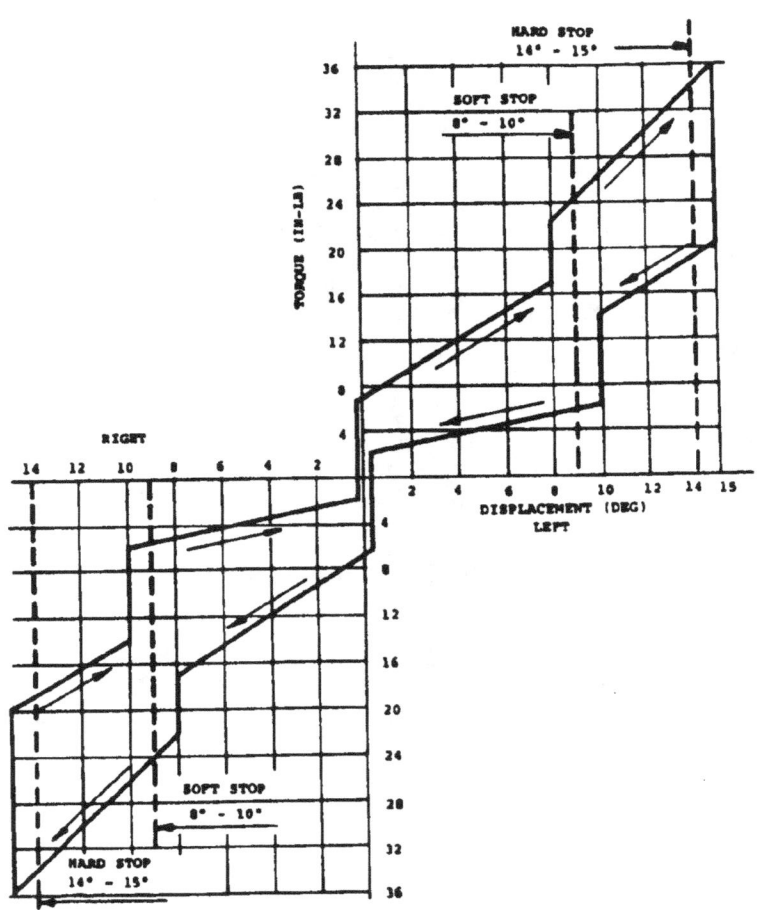

FIGURE 4-28 TORQUE REQUIRED TO ROTATE HAND CONTROLLER FOR STEERING CONTROL

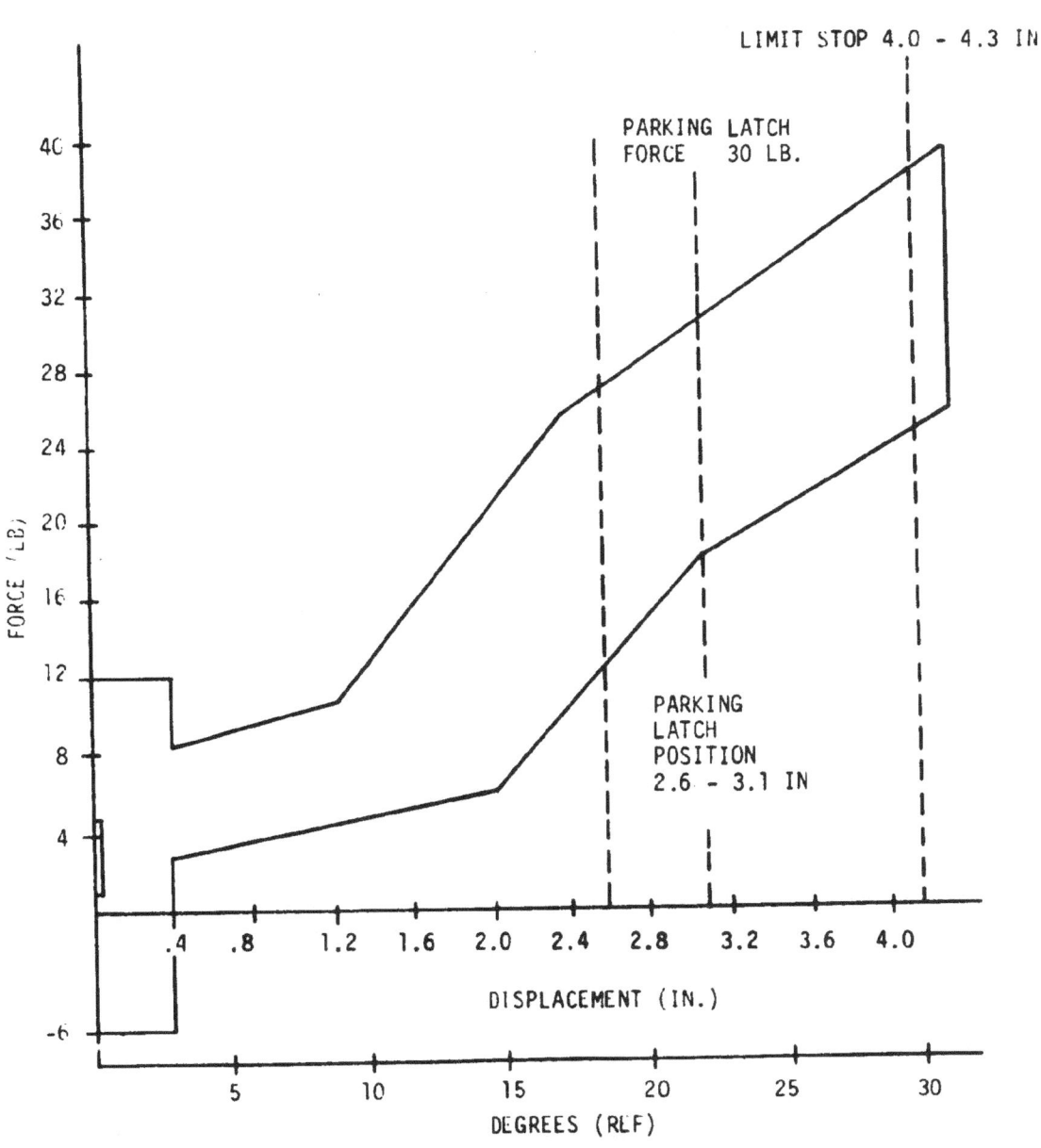

FIGURE 4-29 BRAKE CONTROL FORCE VS DISPLACEMENT

5.0 VEHICLE PERFORMANCE DATA

5.1 VEHICLE DYNAMIC RESPONSE

The LRV will exhibit dynamic characteristics dependent on the surface traversed and the speed at which the surface is traversed.

The vertical acceleration at the LRV seats for various speeds over various surface types is shown in Figure 5-1.

The vibration environments induced into the various LRV payload zones as a result of traversing the lunar surface are shown in Table 5-I.

LS006-002-2H
LUNAR ROVING VEHICLE
OPERATIONS HANDBOOK
APPENDIX A

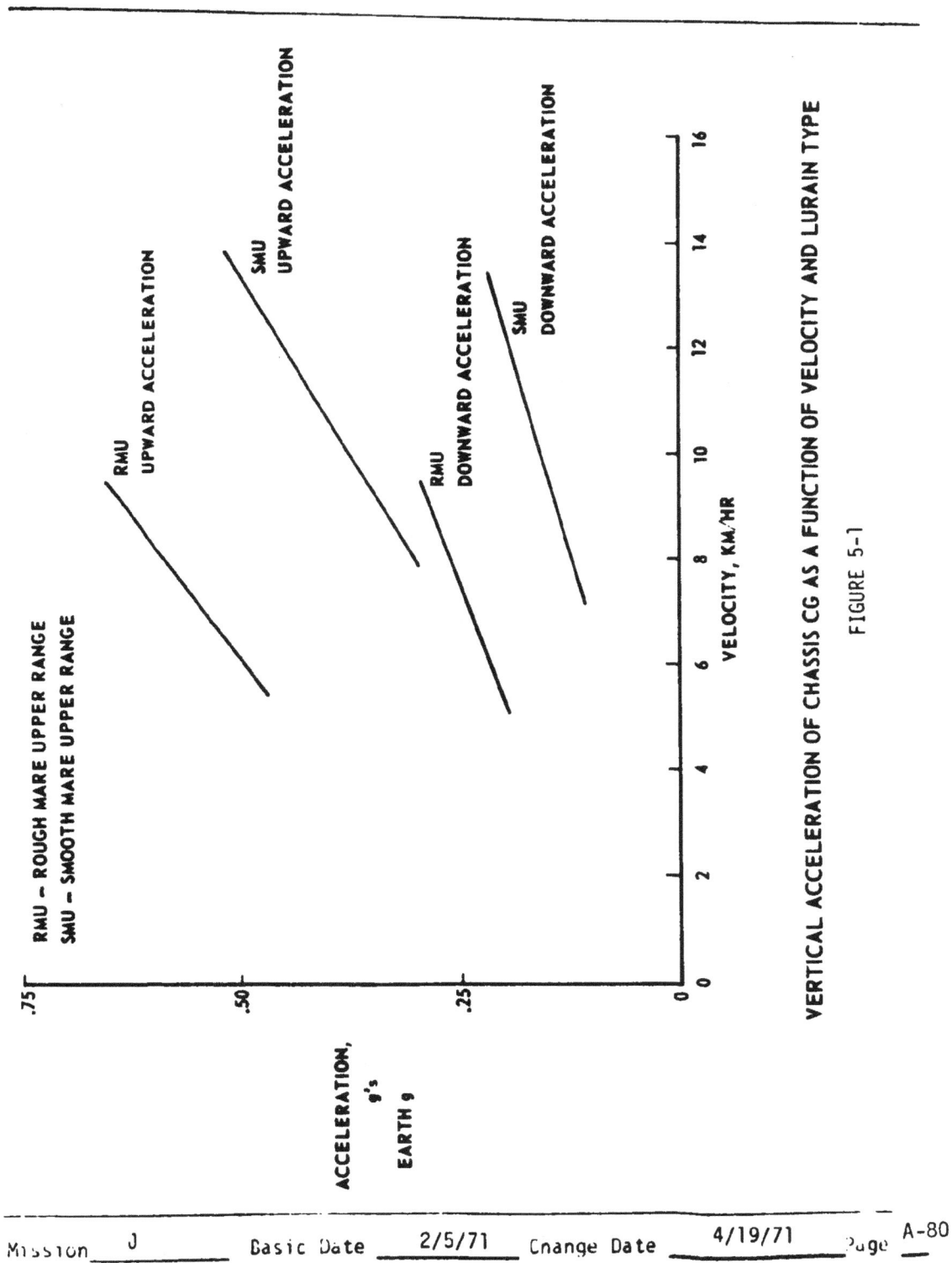

VERTICAL ACCELERATION OF CHASSIS CG AS A FUNCTION OF VELOCITY AND LURAIN TYPE

FIGURE 5-1

LS006-002-2H
LUNAR ROVING VEHICLE
OPERATIONS HANDBOOK
APPENDIX A

RANDOM VIBRATION

PAYLOAD ZONE	VERTICAL	FORE/AFT & LATERAL
FORWARD CHASSIS AT LCRU, TV, HI-GAIN ANTENNA MOUNTS	0.2 to 0.5 Hz @ +9 db/oct 0.5 to 1.0 Hz @ 0.3 g^2/Hz 1.0 to 10.0 Hz @ -6 db/oct Overall RMS Accel = 0.67g	0.2 to 0.5 Hz @ +9 db/oct 0.5 to 1.0 Hz @ 0.042 g^2/Hz 1.0 to 10.0 Hz @ -6db/oct Overall RMS Accel = 0.25 g
AFT CHASSIS AT PALLET ADAPTER ATTACH POINTS	0.2 to 0.5 Hz @ +9 db/oct 0.5 to 1.0 Hz @ 0.21 g^2/Hz 1.0 to 10.0 Hz @ -6 db/oct Overall RMS Accel = 0.56 g	0.2 to 0.5 Hz @ +9 db/oct 0.5 to 1.0 Hz @ 0.042 g^2/Hz 1.0 to 10.0 Hz @ -6 db/oct Overall RMS Accel = 0.25 g
CENTER CHASSIS FLOOR	0.2 to 0.5 Hz @ 9 db/oct 0.5 to 1.0 Hz @ 0.072 g^2/Hz 1.0 to 10.0 Hz @ -6 db/oct Overall RMS Accel = 0.33 g	0.2 to 0.5 Hz @ +9 db/oct 0.5 to 1.0 Hz @ 0.042 g^2/Hz 1.0 to 10.0 Hz @ -6 db/oct Overall RMS Accel = 0.25 g
TOP OF INBOARD HANDHOLDS AT 16 MM DAC AND LO-GAIN ANTENNA CONNECT POINTS	0.2 to 0.5 Hz @ 9 db/oct 0.5 to 1.6 Hz @ .072 g^2/Hz 1.6 to 2.0 Hz @ 9 db/oct 2.0 to 5.0 Hz @ .15 g^2/Hz 5.0 to 10.0 Hz @ -9 db/oct Overall RMS Accel = .93g	SAME AS VERTICAL

TABLE 5-I   VIBRATION ENVIRONMENT AT LRV PAYLOAD INTERFACES

Mission J   Basic Date 2/5/71   Change Date 4/19/71   Page A-81

LS006-002-2H
LUNAR ROVING VEHICLE
OPERATIONS HANDBOOK
APPENDIX A

5.2　　　　RANGE, SPEED, LURAIN CAPABILITY

5.2.1　　　Range

Power for the LRV is supplied by two 36 VDC primary batteries having a total capacity of 8280 watt-hours. LRV range is dependent on the way these watt-hours are consumed, e.g. the ratio of driving time to standby time, the length of time the LCRU is operated from LRV power, the length of time spent traversing slopes and speed selected.

To calculate range, all power consumption must be determined, including both variable and constant power levels. The variable power consumption for motor/gear box inefficiencies, steering motor inefficiency and drive controller losses can be calculated from the data presented in Section 4.1. Variable power use due to electrical distribution losses are contained in Figure 4-19. Variable damping power losses are shown in Figure 5-2.

The other power users operate independent of the mobility subsystem and have constant power levels regardless of LRV mobility activities. These constant power levels are defined in Table 5-II.

Range sensitivity to speed and weight are shown on Figure 5-4.

COMPONENT	POWER	TIME
CONTROL & DISPLAY	10 WATTS	ENTIRE SORTIE
NAVIGATION (WARM UP)	90 WATTS	3 MINUTES
NAVIGATION (AFTER WARM UP)	40 WATTS	ENTIRE SORTIE AFTER WARMUP
DRIVE CONTROLLER (STANDBY)	23 WATTS	DURING PARKED PERIOD WITH DRIVE MOTORS ON

TABLE 5-II　LRV STEADY STATE POWER CONSUMPTION

5.2.2　　　Speed

LRV maximum speed capability is dependent on the slope being traversed, the type of soil encountered, the surface roughness of the lurain and the LRV gross weight. Maximum speed capability is shown in Table 5-III. Effects of weight on maximum speed is shown in Figure 5-5.

Mission　J　Basic Date 2/5/71　Change Date 4/19/71　Page A-82

LS006-002-2H
LUNAR ROVING VEHICLE
OPERATIONS HANDBOOK
APPENDIX A

SURFACE ROUGHNESS

	SLOPE	MID-RANGE PSD
	0°	9.5
SMOOTH MARE	5°	8.2
	10°	7.3
	0°	10.5
ROUGH UPLANDS	5°	8.7
	10°	7.6

TABLE 5-III   LRV SPEED CAPABILITY

5.2.3   Lurain Capability

The LRV is designed with the following capabilities of traversing the lunar surface:

Crevasse Crossing Capability	70 cm width
Step Obstacle Climbing Capability	35 cm high
Clearance Under Chassis	35 cm

Slope climbing capability of the LRV is shown in Figure 5-6.

5.3   THERMAL PERFORMANCE

Thermal performance characteristics of the LRV are defined in Section 4 under the appropriate subsystem performance paragraph.

5.4   CONTROLLABILITY

5.4.1   Steering

The LRV has a turning radius of 122 inches utilizing the four wheel steering capability. The steering rate is such that lock-to-lock steering angle change can be accomplished in 6 seconds. Crewman forces required to control steering by use of the hand controller are shown in Figure 4-28.

5.4.2   Braking

Stopping distances for various initial speeds and different slopes are shown in Figure 5-7.

Hand Controller forces required to effect braking are shown in Figure 4-29.

5.4.3 Speed

At high speeds the LRV can become uncontrollable. The controllability speed limit is based on wheel slip angle considerations and tends to be independent of lurain type. The controllability speed limit should be utilized for mission planning purposes. Double Ackerman steering was used.

Actual driving experience on the moon's surface should dictate the final controllability speed restrictions.

CONTROLLABILITY SPEED LIMIT

ALL LURAIN TYPES	10 KM/HR

TABLE 5-IV   LRV CONTROLLABILITY SPEED LIMITS

LS006-002-2H
LUNAR ROVING VEHICLE
OPERATIONS HANDBOOK
APPENDIX A

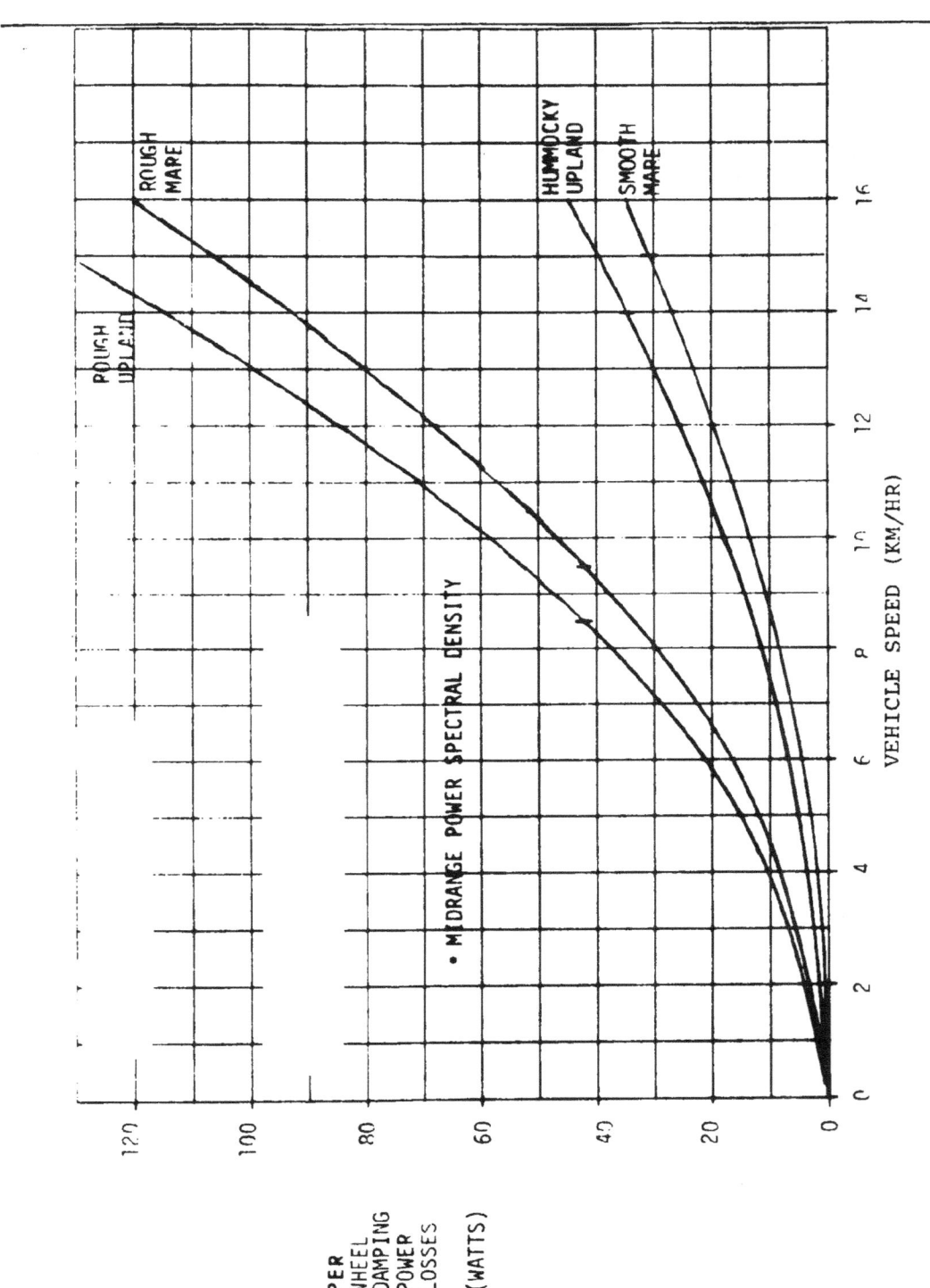

FIGURE 5-2 LRV DAMPING POWER LOSSES

LS006-002-2H
LUNAR ROVING VEHICLE
OPERATIONS HANDBOOK
APPENDIX A

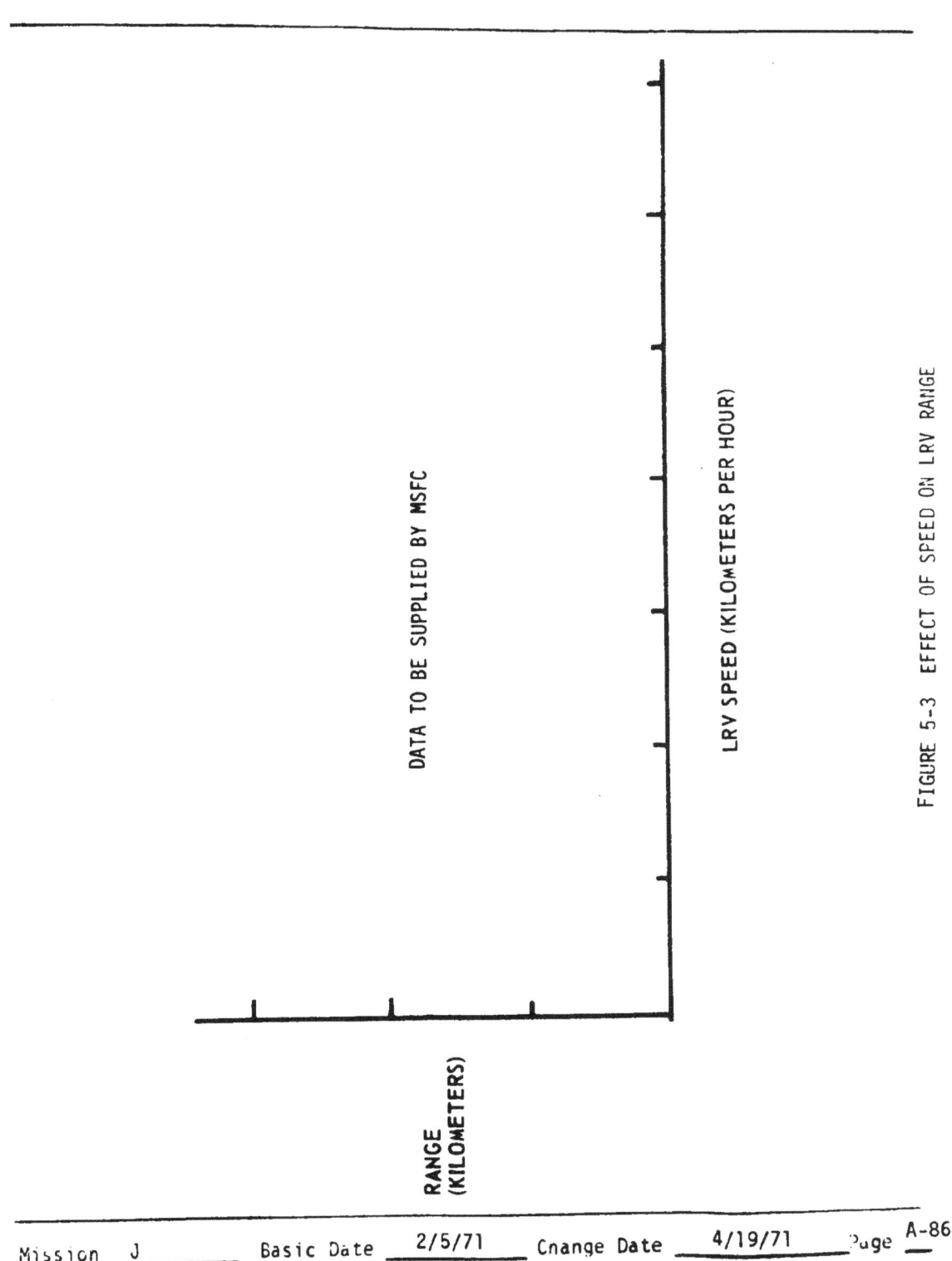

FIGURE 5-3 EFFECT OF SPEED ON LRV RANGE

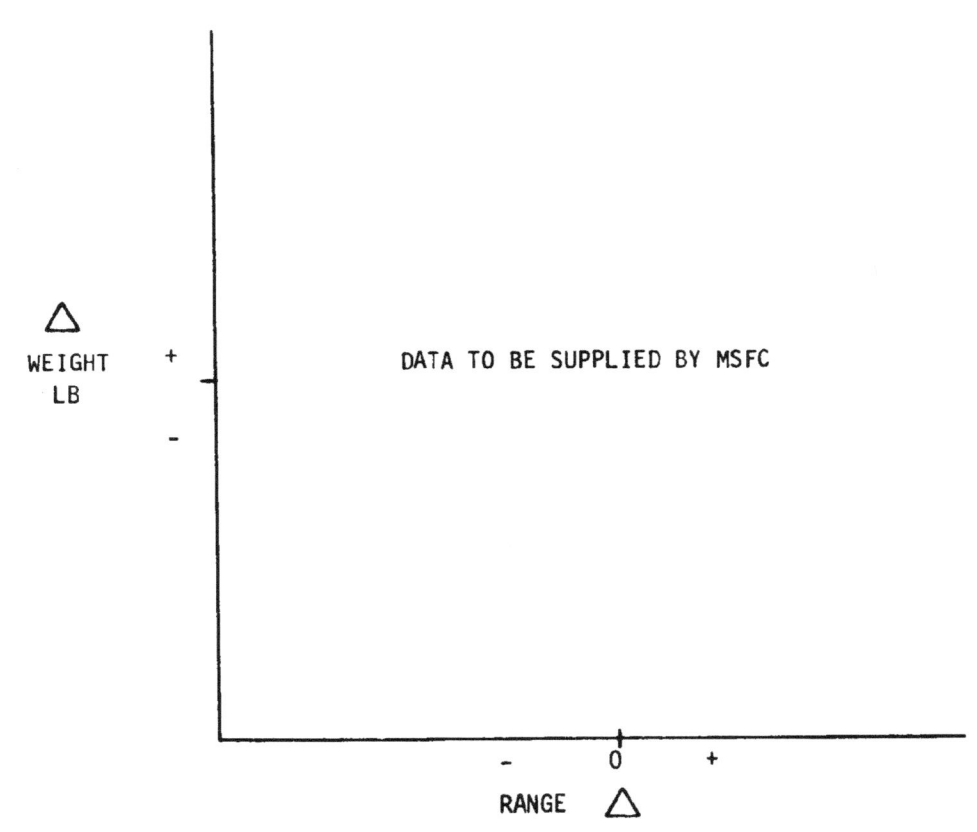

FIGURE 5-4 EFFECT OF WEIGHT CHANGE ON RANGE FOR FOUR LURAIN TYPES

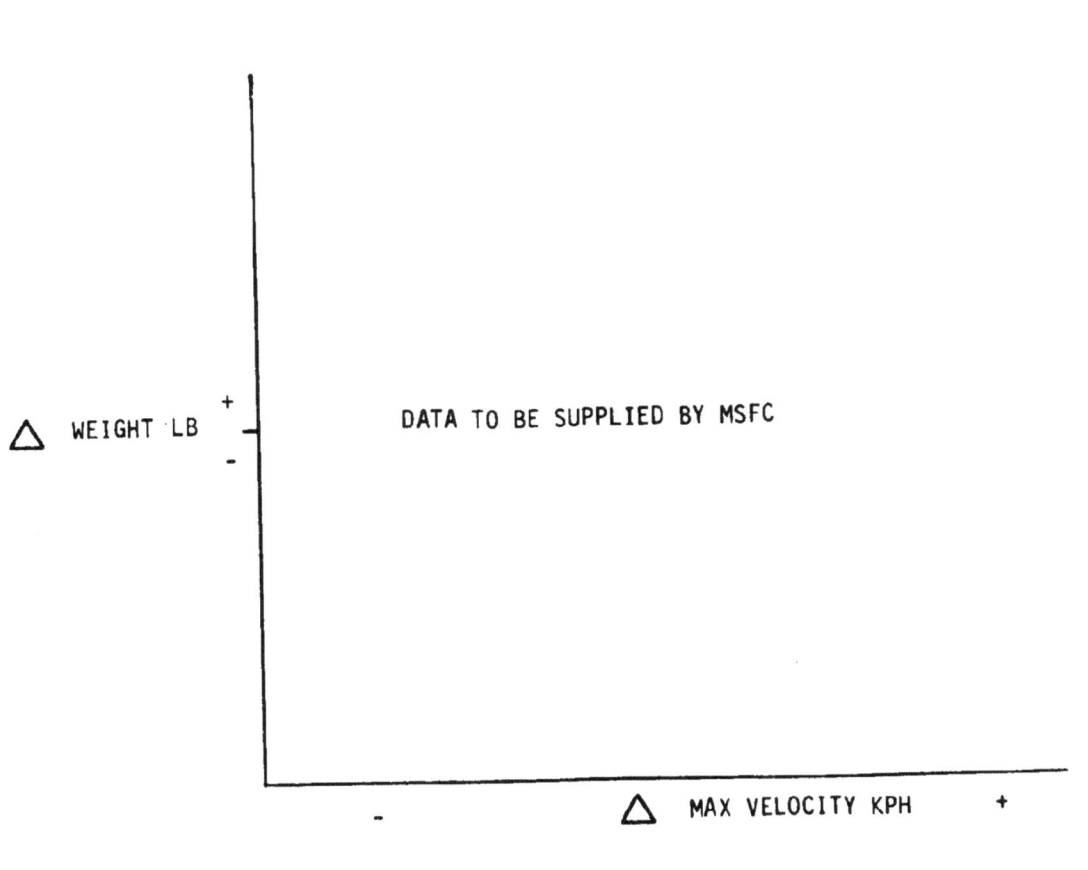

FIGURE 5-5 EFFECT OF WEIGHT CHANGE ON MAXIMUM VELOCITY FOR FOUR LURAIN TYPES

LS006-002-2H
LUNAR ROVING VEHICLE
OPERATIONS HANDBOOK
APPENDIX A

DATA TO BE PROVIDED BY MSFC

FIGURE 5-6

SLOPE CLIMBING CAPABILITY FOR VARYING C.G. LOCATION

LS006-002-2H
LUNAR ROVING VEHICLE
OPERATIONS HANDBOOK
APPENDIX A

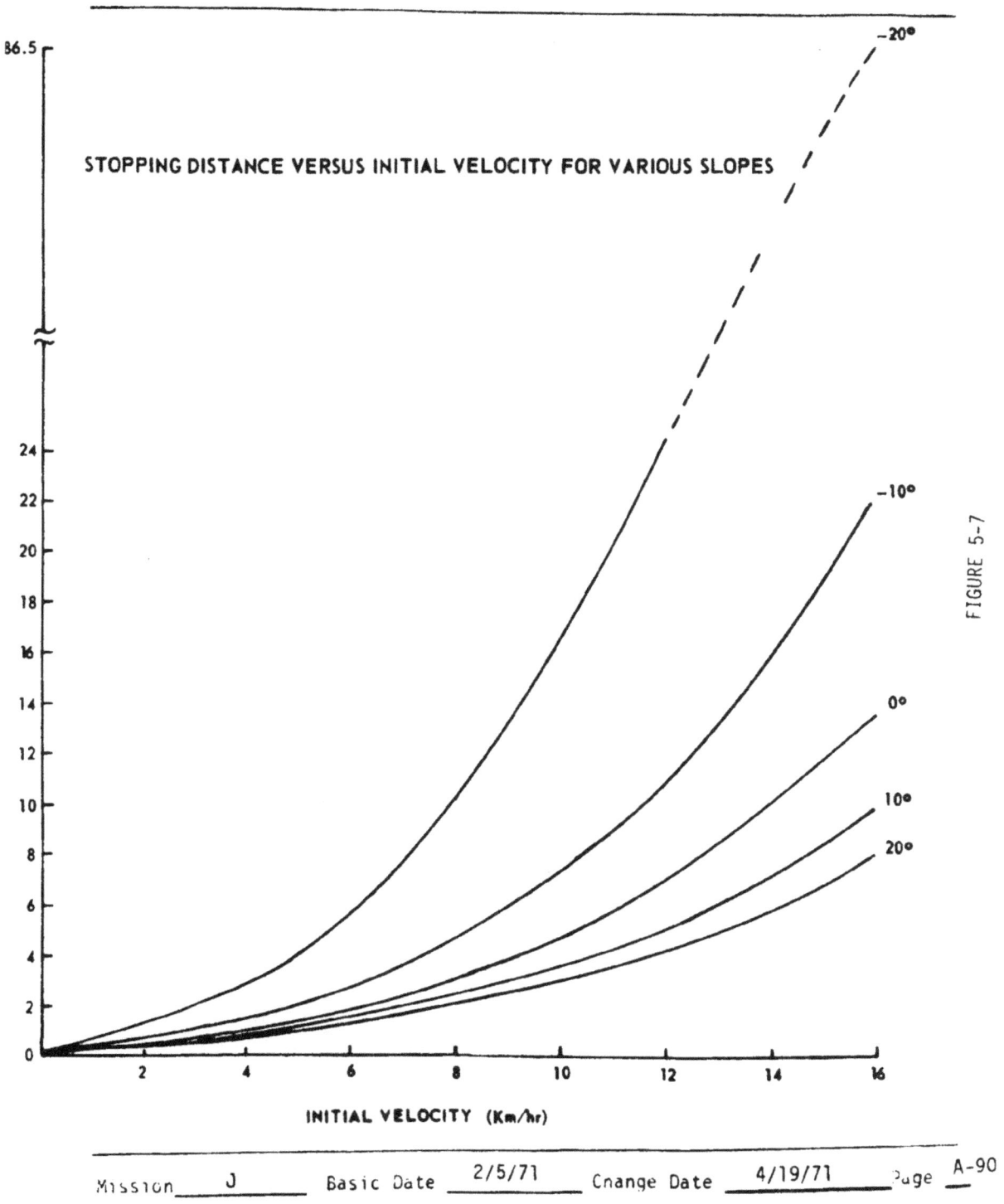

FIGURE 5-7

Mission __J__ Basic Date __2/5/71__ Change Date __4/19/71__ Page __A-90__

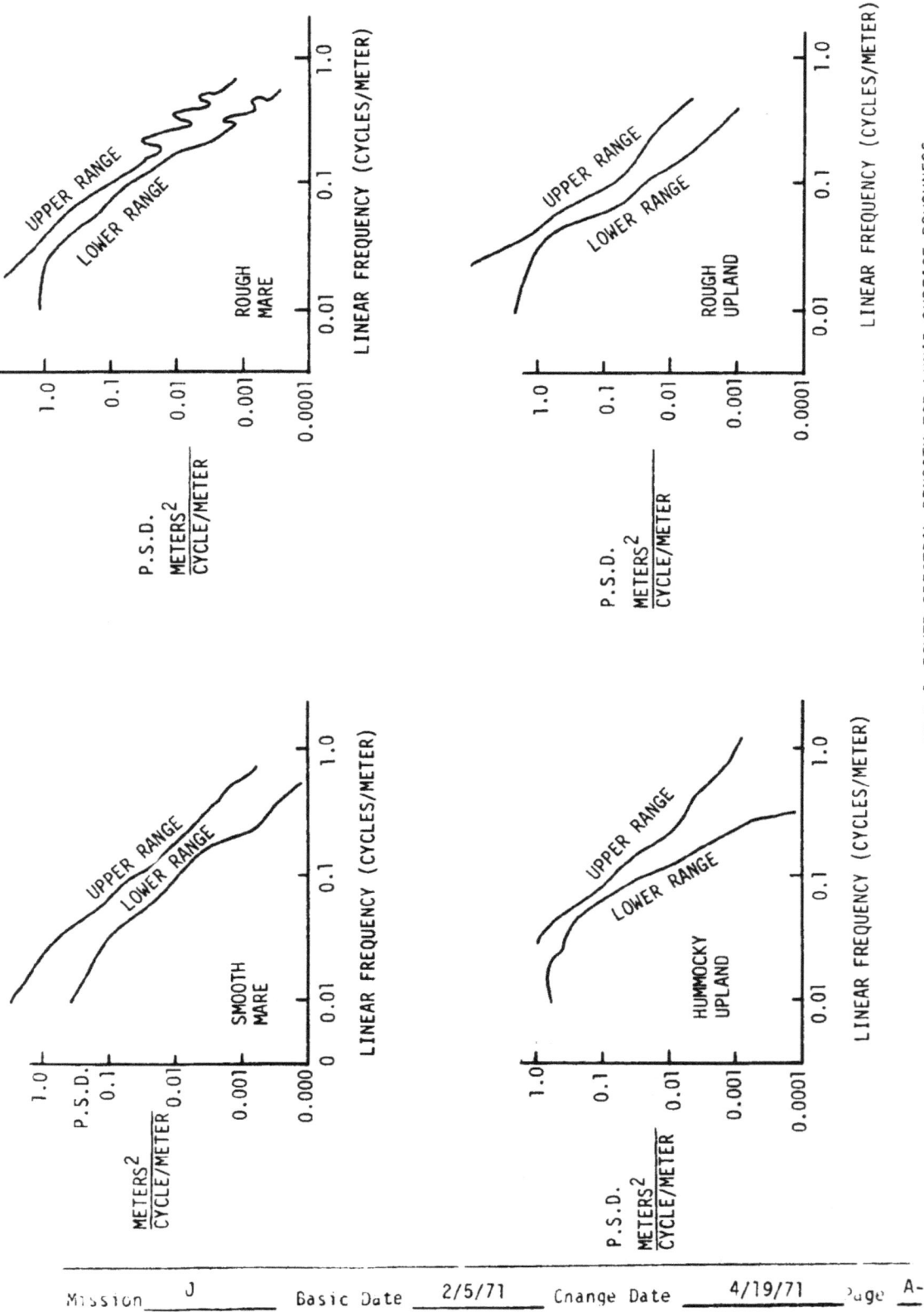

FIGURE 5-8 POWER SPECTRAL DENSITY FOR LUNAR SURFACE ROUGHNESS

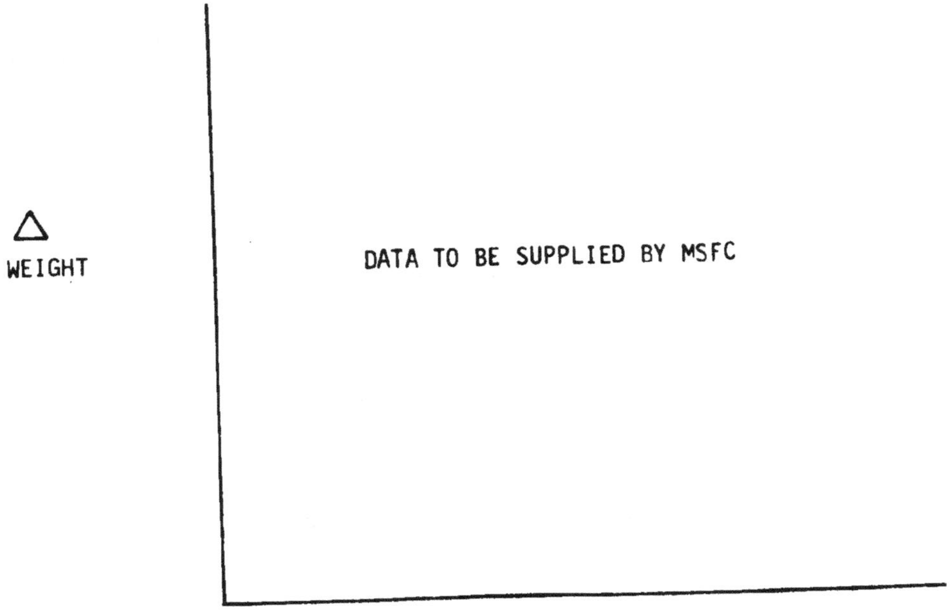

Δ WEIGHT

DATA TO BE SUPPLIED BY MSFC

FIGURE 5-9 EFFECT OF WEIGHT CHANGE ON CONTROLLABILITY FOR FOUR LURAIN TYPES

LS006-002-2H
LUNAR ROVING VEHICLE
OPERATIONS HANDBOOK
APPENDIX A

6.0   SPECIFIC VEHICLE DATA

6.1   LRV-1

6.1.1   Constraints

No constraints other than those shown in Section 3 are applicable.

6.1.2   Subsystem Performance

The traction drive characteristics for each of the four traction drives on LRV-1 are shown in Figure 6-1 through 6-4.

Forward and rear steering motor maximum current measured during acceptance test was 1.50 amps for the front and 1.45 amps for the rear against a torque of 200 in-lbs.

6.1.3   Vehicle Performance Data (Acceptance Test Data)

Power consumption characteristics of the navigation system are shown on Table 6-I.

Chassis ground clearance under full load (1520 lbs.) is 35.1 cm.

6.1.4   Controllability (Acceptance Test Data)

The turning radius measured during acceptance test is 3.1 meters.

Lock-to-lock steering required $5.0 ^{+0.1}_{-0}$ seconds, against a torque of 200 in-lbs.

6.1.5   Weight and Center of Gravity

The weight and C.G. of LRV-1 in all flight and deployed configurations is shown in Tables 6-II and 6-III.

6.1.6   Navigation Odometer Factor

The LRV-1 navigation system is set up such that the odometer will register 0.245 meters per count. There are 9 counts per wheel revolution.

6.1.7   Meter Calibration

The final meter calibration curves for each of the four meters on LRV-1 are shown in Figures 6-5 through 6-8.

Mission   J   Basic Date 2/5/71   Change Date 4/19/71

LS006-002-2H
LUNAR ROVING VEHICLE
OPERATIONS HANDBOOK
APPENDIX A

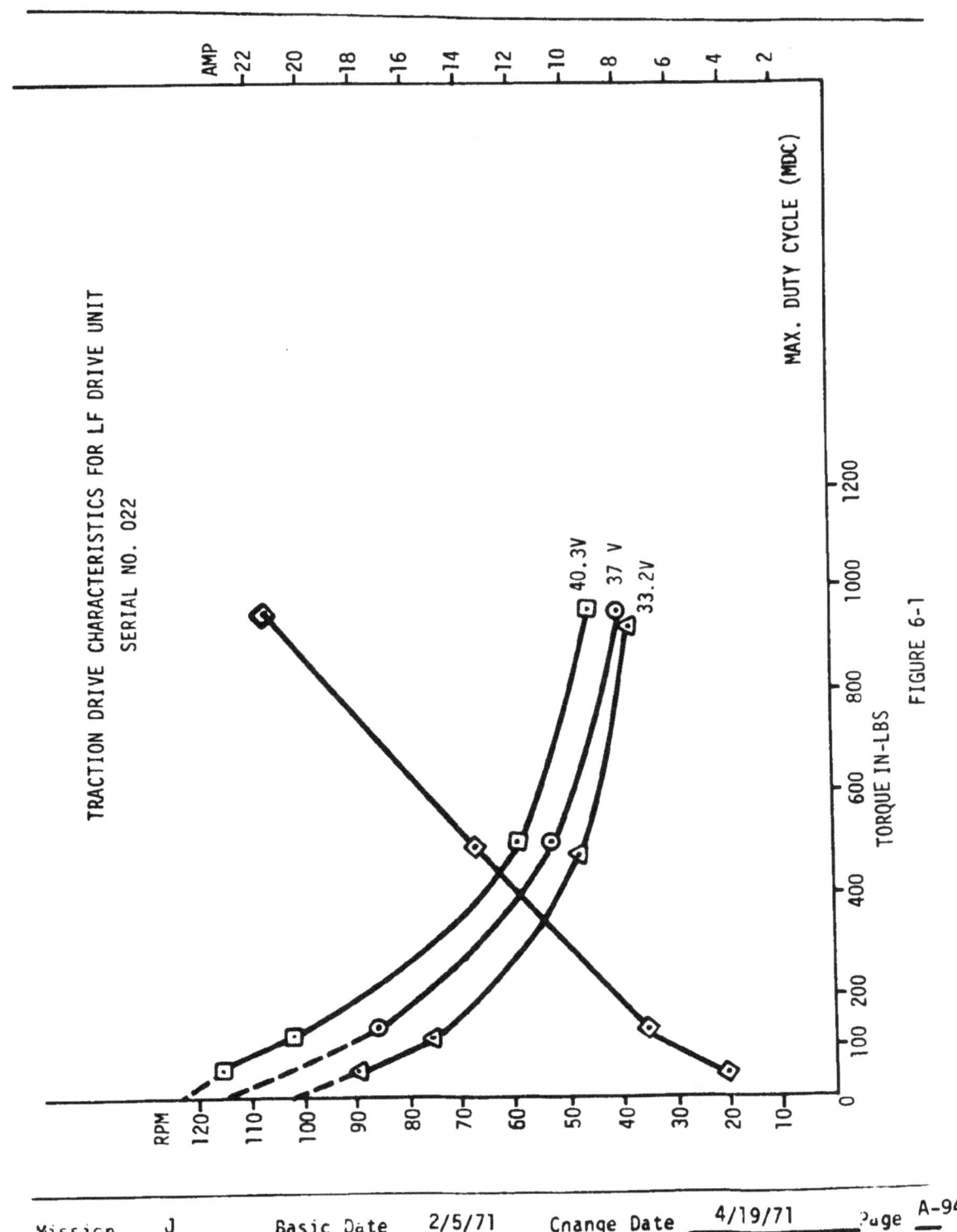

FIGURE 6-1

Mission __J__   Basic Date __2/5/71__   Change Date __4/19/71__   Page __A-94__

LS006-002-2H
LUNAR ROVING VEHICLE
OPERATIONS HANDBOOK
APPENDIX A

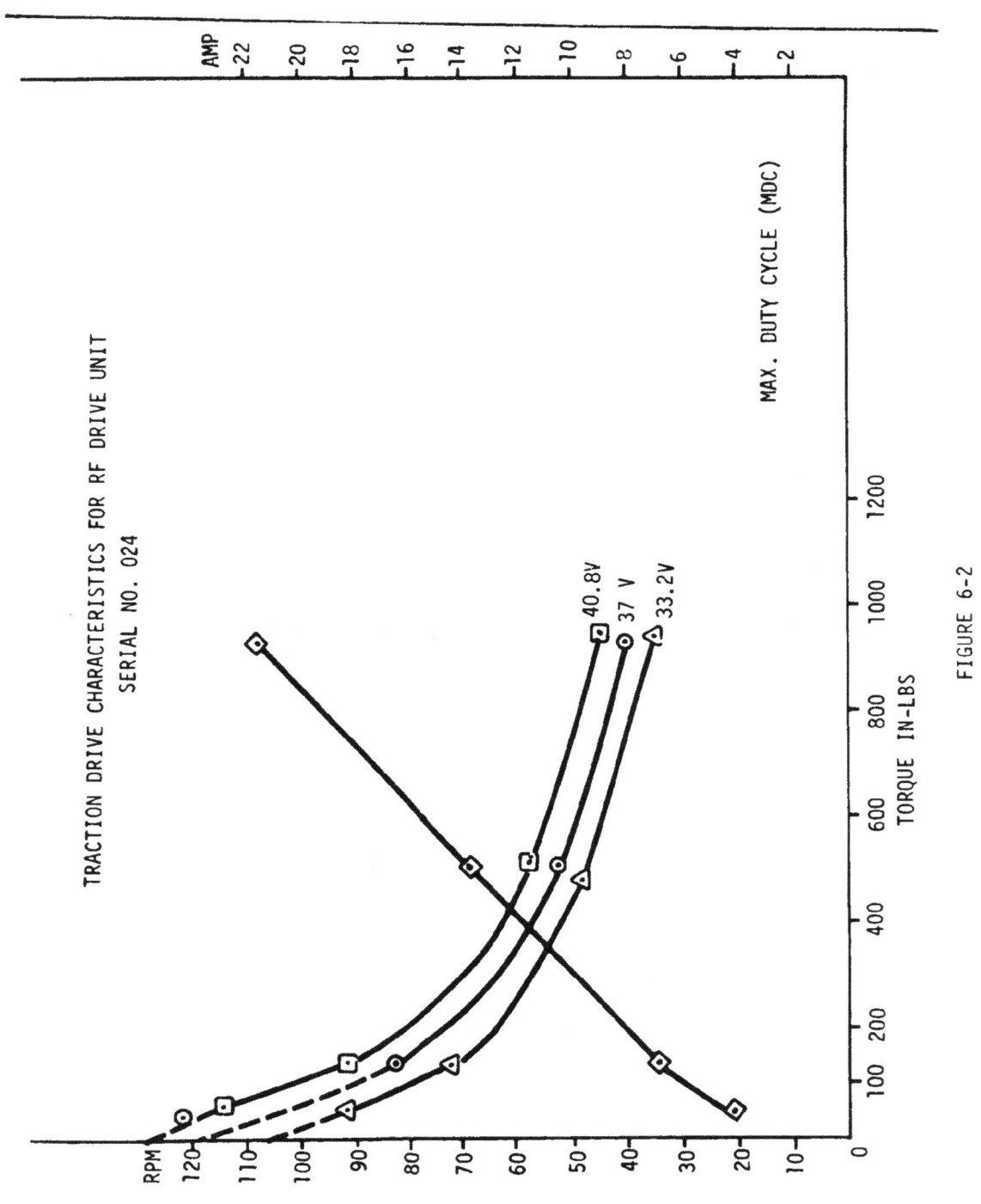

FIGURE 6-2

LS006-002-2H
LUNAR ROVING VEHICLE
OPERATIONS HANDBOOK
APPENDIX A

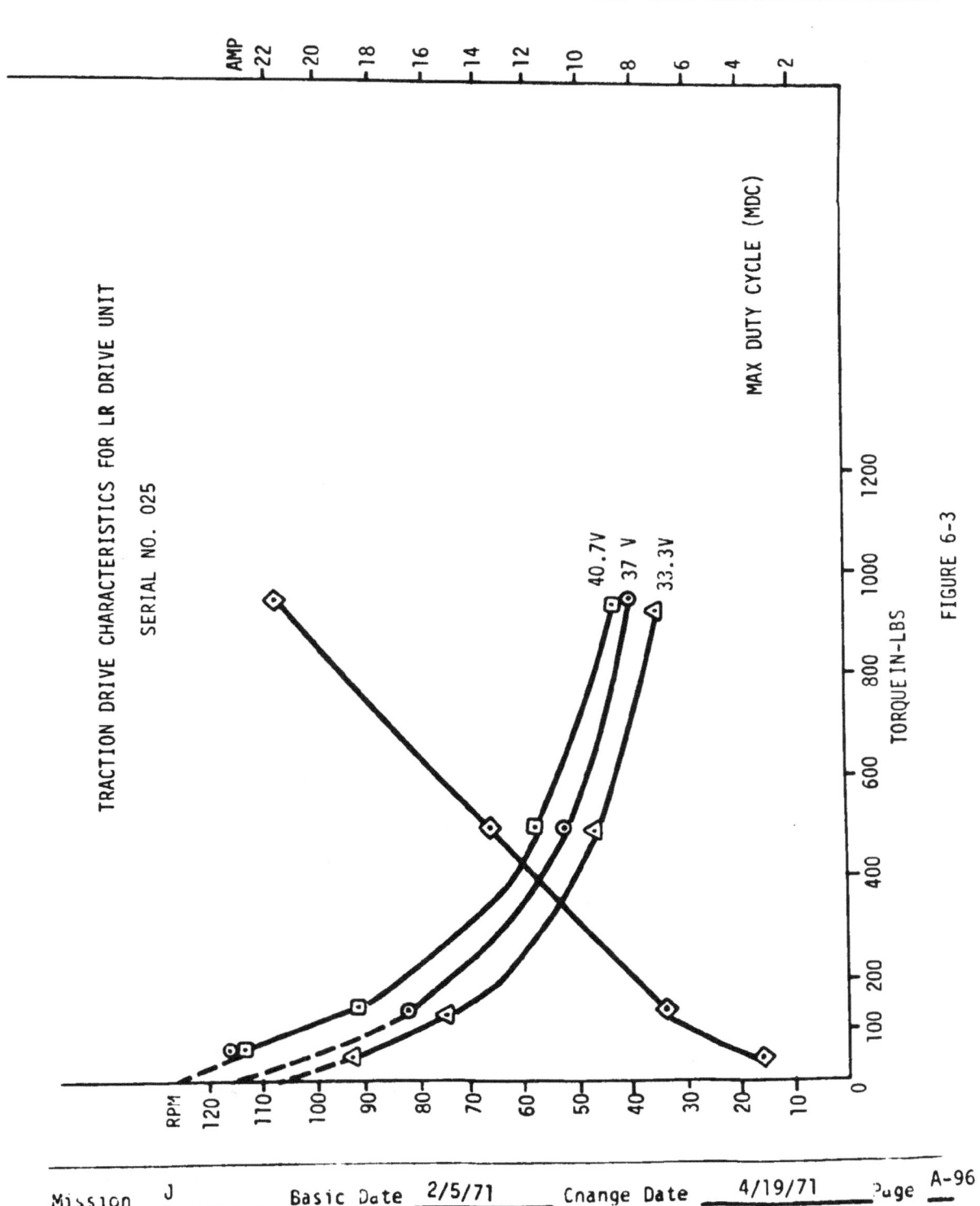

FIGURE 6-3

LS006-002-2H
LUNAR ROVING VEHICLE
OPERATIONS HANDBOOK
APPENDIX A

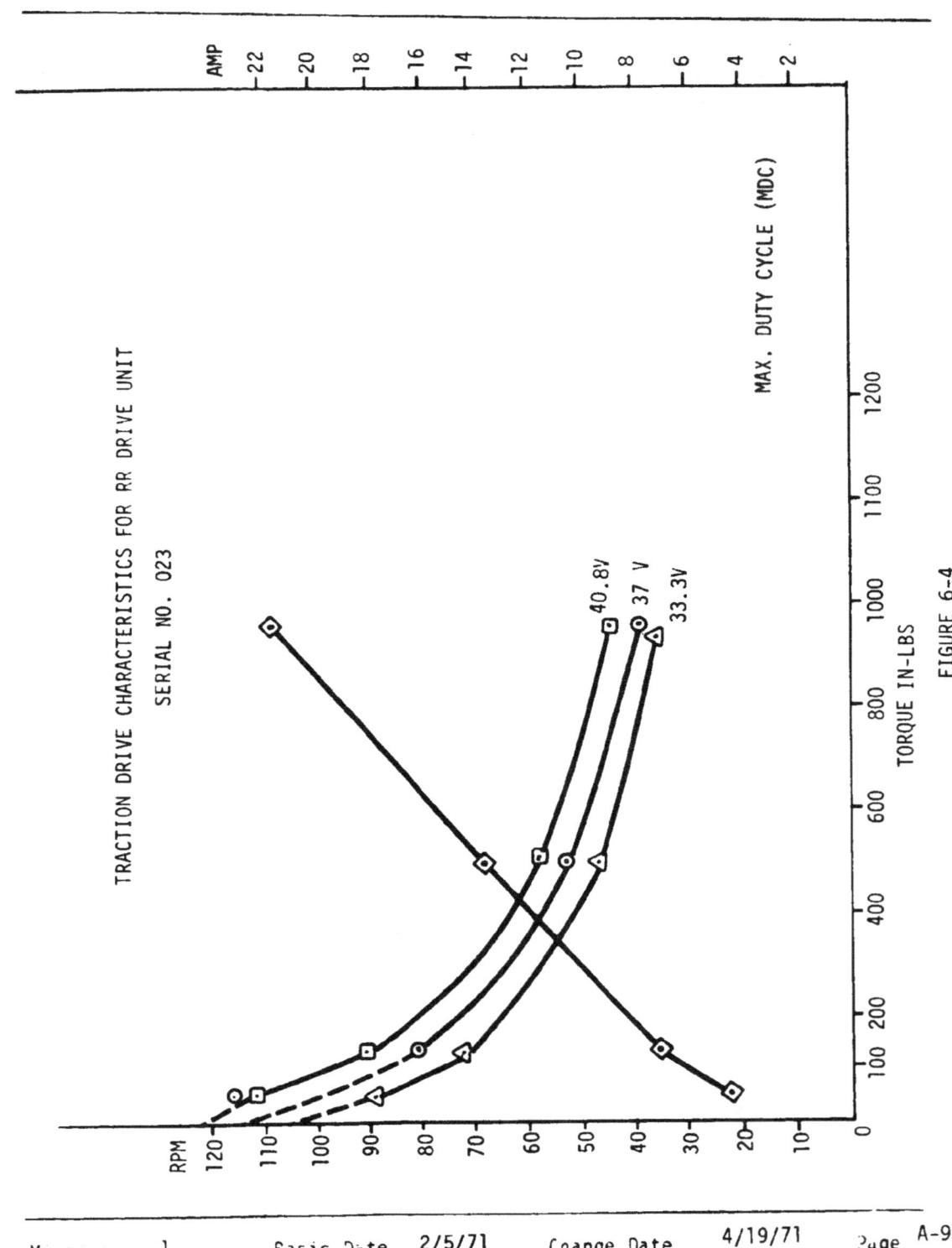

FIGURE 6-4

NAVIGATION SYSTEM RESET AND WARMUP

NSS CURRENT (AMPS)	BATTERY VOLTAGE TO SPU (VOLTS)	NSS POWER CONSUMED (WATTS)
2.36	34.8	82.1

LRV-1 NAVIGATION SYSTEM POWER CONSUMPTION

TABLE 6-I

LS006-002-2H
LUNAR ROVING VEHICLE
OPERATIONS HANDBOOK
APPENDIX A

CONFIGURATION	WEIGHT (POUNDS)	CENTER OF GRAVITY (LRV STATIONS)			MOMENT OF INERTIA (LB-IN^2)		
		X	Y	Z	Ix	Iy	Iz
BASIC LRV WITH SSE (FOLDED)	494.25	66.74	-0.54	111.34	149,745	171,079	231,426
LCRU SUPPORT FITTINGS	+3.00						
INBOARD HANDHOLD	+1.00						
UNDERSEAT ITEMS	+2.60						
AFT PALLET SUPPORT STRUCTURE	+3.10						
LCRU POWER CABLE, LCRU SUPPORT LEGS, ANTENNA & TV ADAPTERS	+3.80						
INSTALLED LRV IN LM	508.75	66.95	-0.42	111.21	151,167	173,662	223,771
SSE ATTACHED TO LM	-37.88						
WHEEL LOCK STRUTS	-0.97						
LRV SUPPORT TRIPODS	-3.56						
LRV DEPLOYMENT FITTING	-1.03						
SSE THERMAL PROTECTION	-0.60						
LRV DEPLOYED (EMPTY OPERATIONAL)	464.71	53.69	-0.24	104.33	209,476	645,261	799,786
SCIENCE EQUIPMENT	41.70						
PHOTOGRAPHIC EQUIPMENT	35.20						
SUPPORTING EQUIPMENT	116.60						
CREW SYSTEMS	802.20						
LUNAR SAMPLES	60.00						
LRV OPERATIONAL - LOADED	1520.41	72.96	-0.21	116.99	687,279	1,901,693	2,037,460

TABLE 6-II LRV-1 WEIGHT, C.G., AND MOMENTS OF INERTIA

LS006-002-2H
LUNAR ROVING VEHICLE
OPERATIONS HANDBOOK
APPENDIX A

LRV-1 LOADED WEIGHT DISTRIBUTION: FRONT WHEELS 48.4%
                                    REAR WHEELS   51.6%

LRV-1 LOADED WHEEL LOADING: RIGHT FRONT 365.5 LBS (24.0%)
                             LEFT FRONT   369.9 LBS (24.3%)
                             RIGHT REAR   390.2 LBS (25.7%)
                             LEFT REAR   394.8 LBS (26.0%)

TABLE 6-III LRV-1 LUNAR OPERATIONAL WEIGHT
             DISTRIBUTION - STATIC, LEVEL
             LURAIN CONDITION

LS006-002-2H
LUNAR ROVING VEHICLE
OPERATIONS HANDBOOK
APPENDIX A

PARAMETER	
1. GROSS VEHICLE MASS	47.2 SLUGS (1520 LBS)
2. SUSPENDED VEHICLE MASS	44.2 SLUGS (1424 LBS)
3. WHEEL MASS	.745 SLUG (24 LBS)
4. WHEEL ROTATIONAL MOMENT OF INERTIA	2.2 SL-FT^2
5. VEHICLE MOMENTS OF INERTIA	See Table 6-I
6. CG LOCATION	See Table 6-I
7. VERTICAL SUSPENSION RATE	14 LB/IN (0-9 INCHES) 500 LB/IN ($<$ 0 OR $>$ 9 IN)
8. VERTICAL DAMPING RATE	17.3 LB-SEC^2/FT^2
9. HORIZINTAL SUSPENSION RATE	51,000 LB/FT
10. HORIZONTAL SUSPENSION DAMPING RATE	2420 LB/(FT/SEC)
11. WHEEL RADIAL SPRING RATE	400 LB/FT (0-1.5 IN) 680 LB/FT (1.5-3 IN) 7300 LB/FT (3 IN)
12. WHEEL DAMPING RATE	2.5 LB/(FT/SEC)
13. WHEEL DIAMETER	32 INCHES
14. VEHICLE WHEEL BASE	90 INCHES

TABLE 6-IV   SUMMARY OF LRV-1 MOBILITY PARAMETERS

Mission J   Basic Date 2/5/71   Change Date 4/19/71   Page A-101

LS006-002-2H
LUNAR ROVING VEHICLE
OPERATIONS HANDBOOK
APPENDIX A

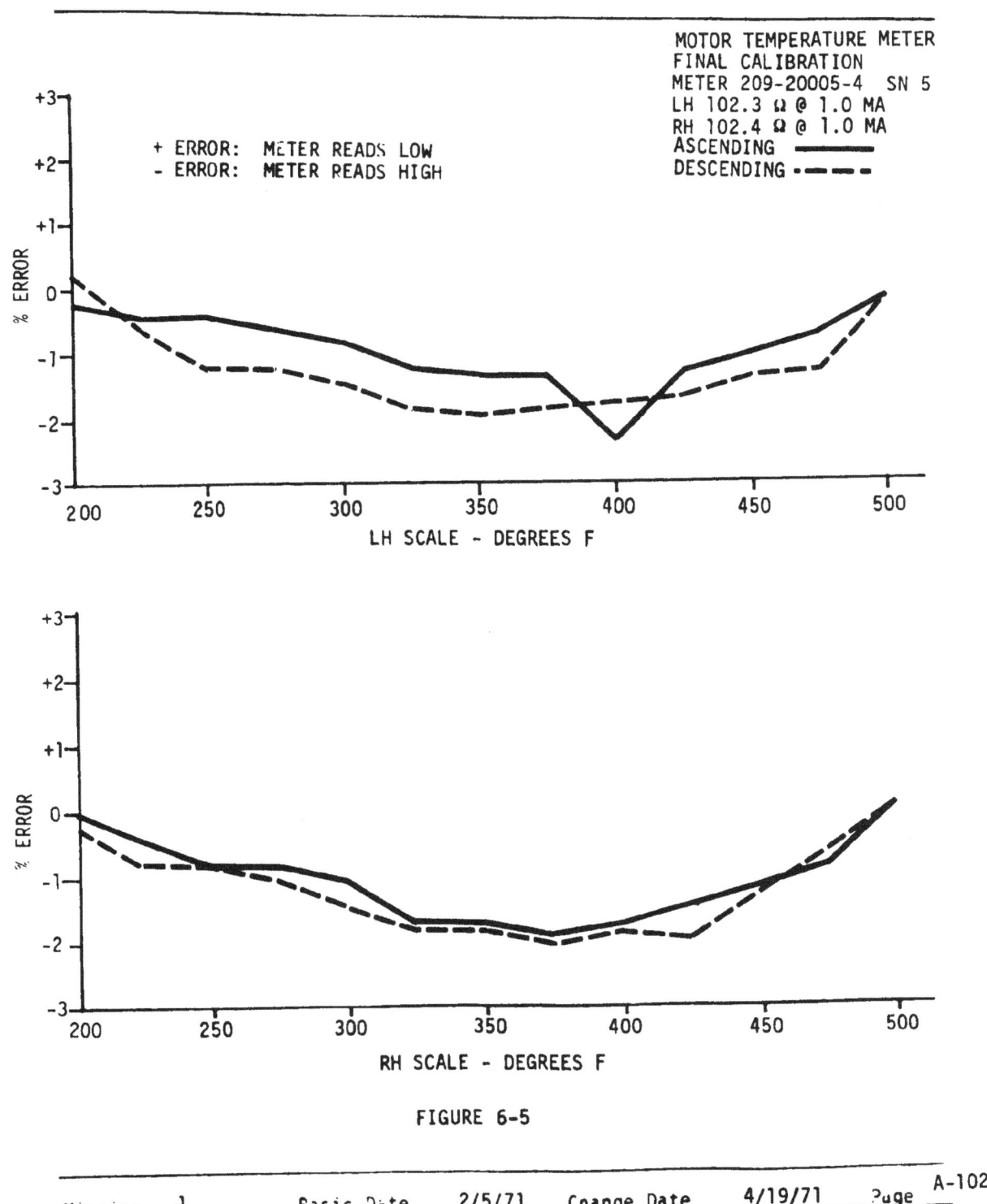

FIGURE 6-5

Mission **J**   Basic Date **2/5/71**   Change Date **4/19/71**   Page **A-102**

LS006-002-2H
LUNAR ROVING VEHICLE
OPERATIONS HANDBOOK
APPENDIX A

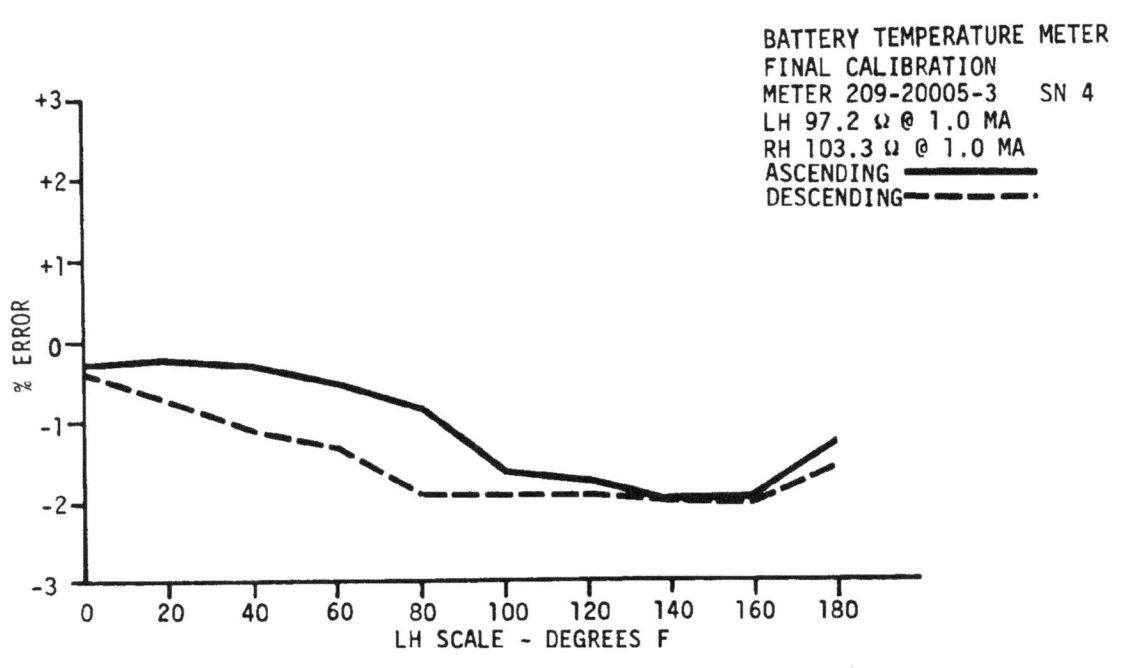

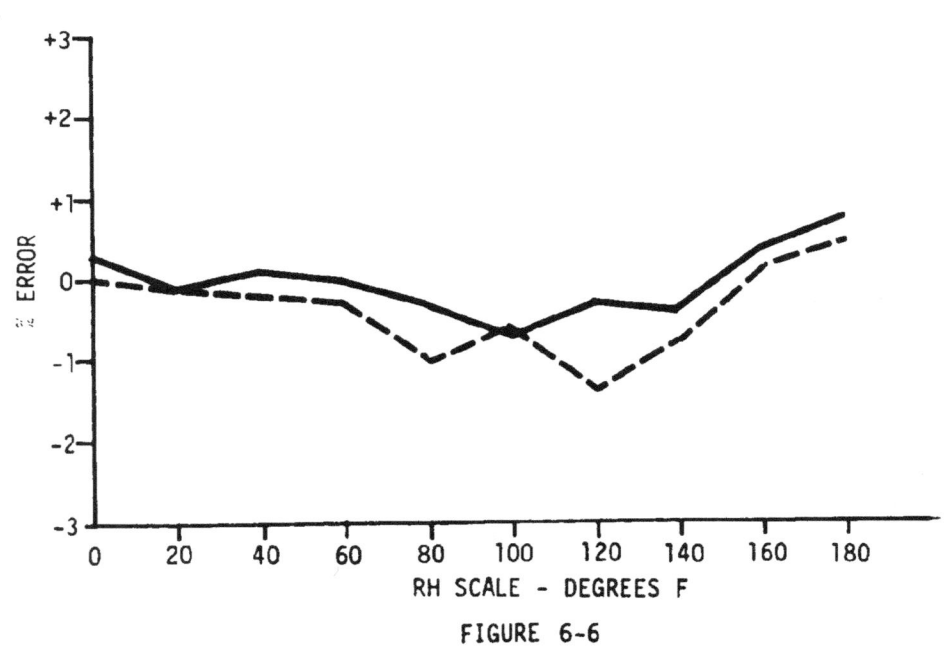

FIGURE 6-6

LS006-002-2H
LUNAR ROVING VEHICLE
OPERATIONS HANDBOOK
APPENDIX A

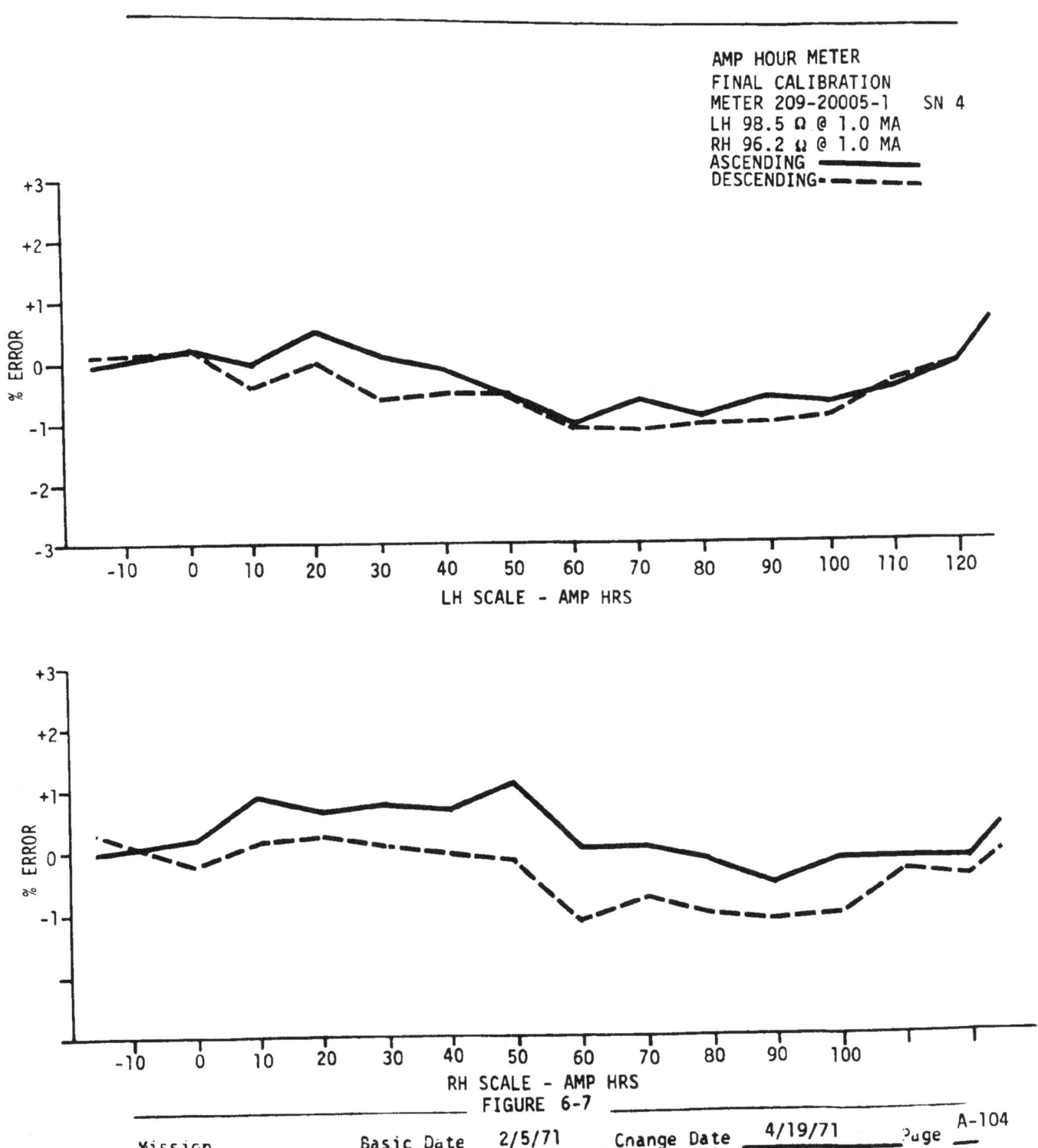

FIGURE 6-7

Mission _____ Basic Date 2/5/71 Change Date 4/19/71 Page A-104

LS006-002-2H
LUNAR ROVING VEHICLE
OPERATIONS HANDBOOK
APPENDIX A

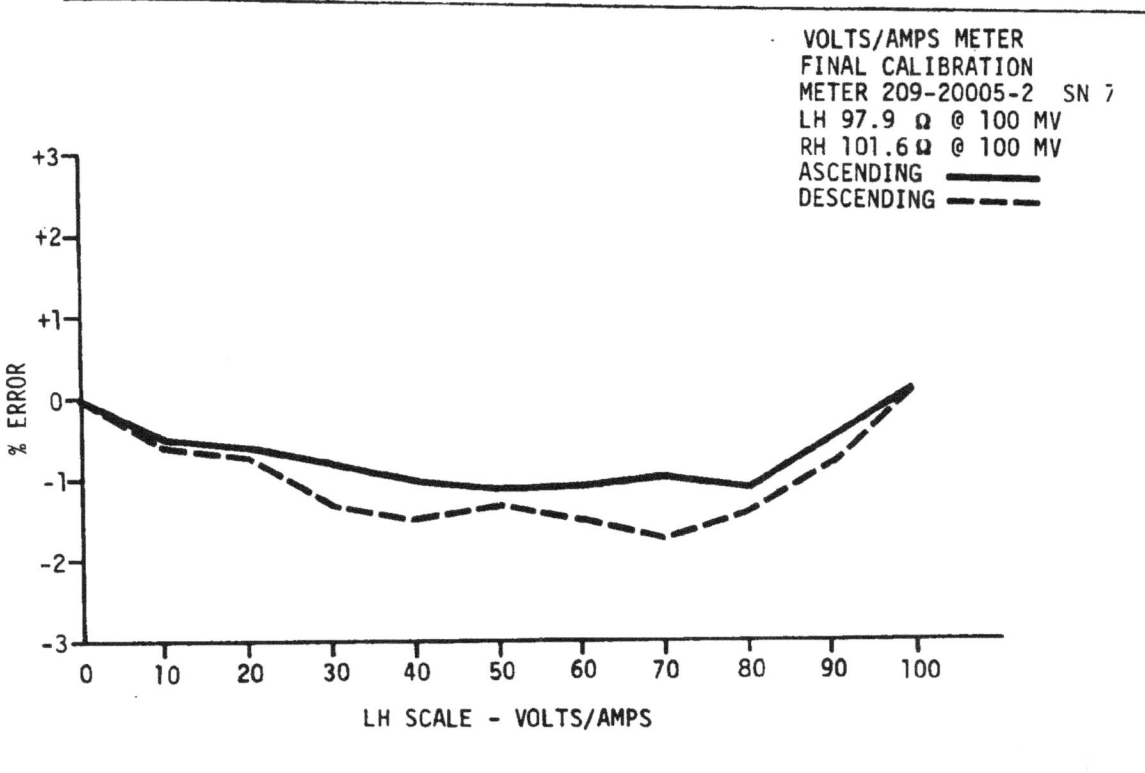

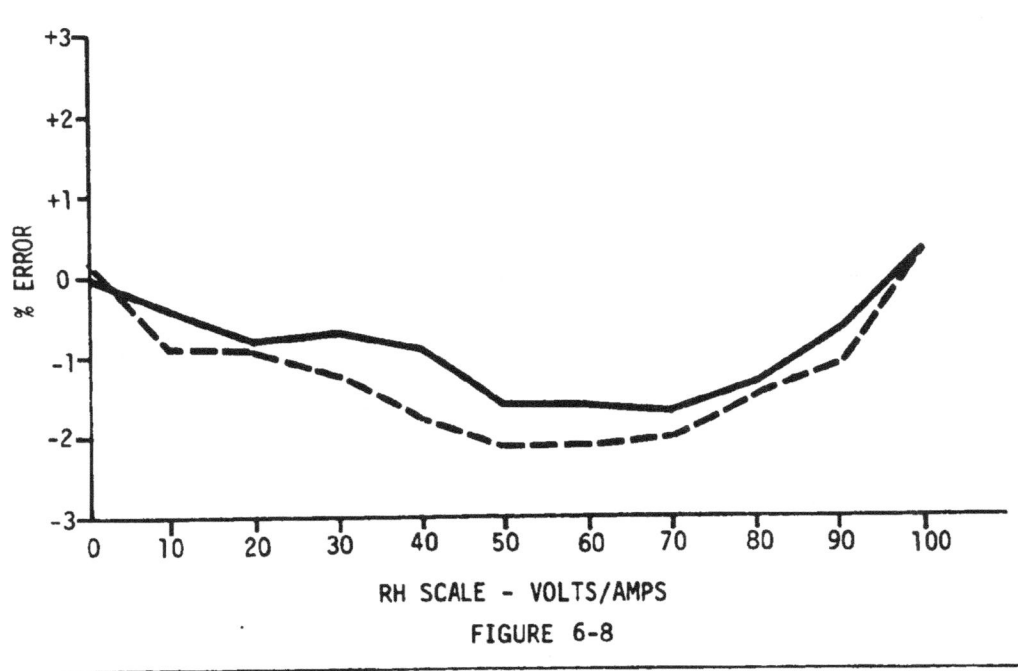

FIGURE 6-8

Mission J   Basic Date 2/5/71   Change Date 4/19/71   Page A-105

MMS SUBCOURSE NUMBER 151   EDITION CODE 3

NIKE MISSILE
and Test Equipment

NIKE HERCULES

DECLASSIFIED

by U.S. Army Missile and Munitions Center and School
Periscope Film LLC

NASA
PROJECT GEMINI

FAMILIARIZATION MANUAL
Manned Satellite Capsule

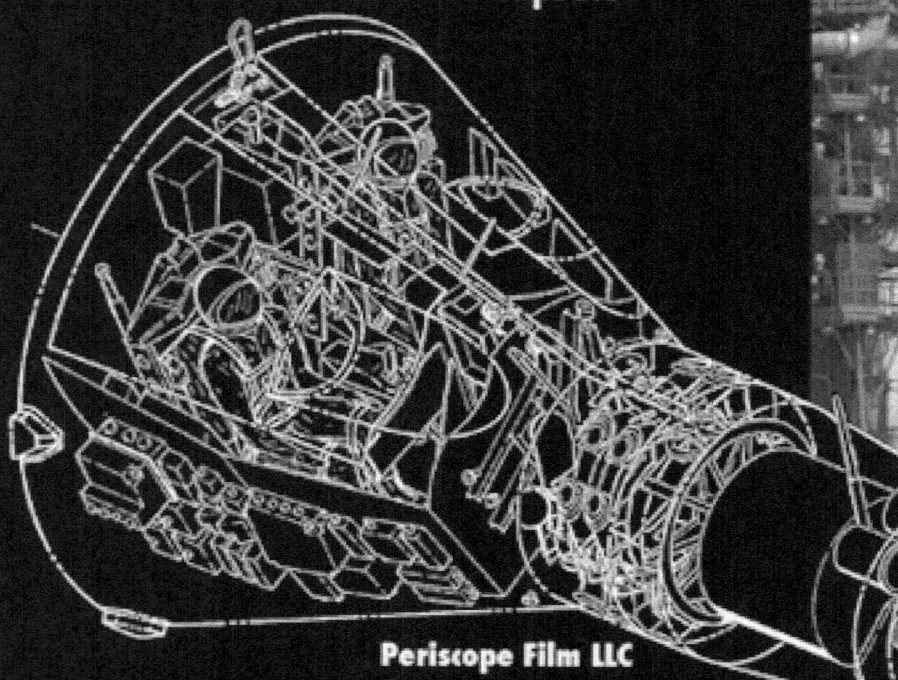

Periscope Film LLC

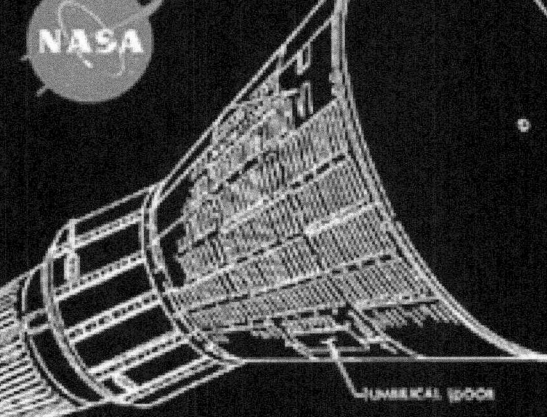

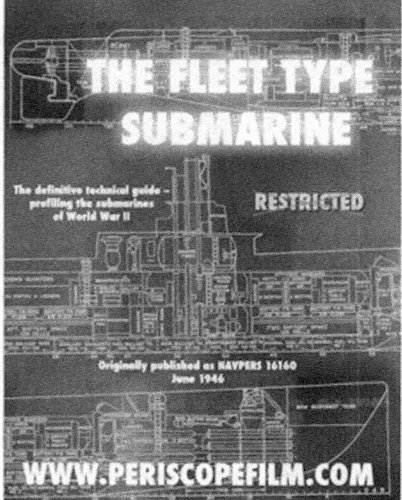
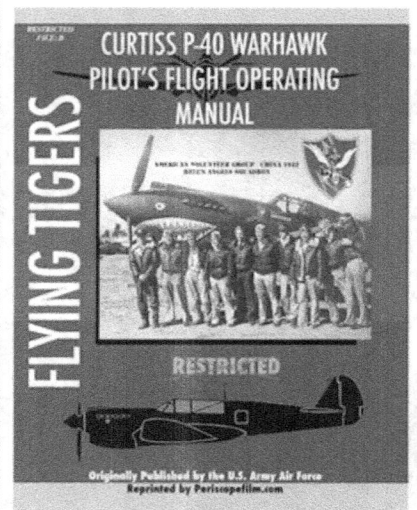

ALSO NOW AVAILABLE FROM PERISCOPEFILM.COM

©2012 Periscope Film LLC
ALL RIGHTS RESERVED
ISBN #978-1-937684-89-1

www.ingramcontent.com/pod-product-compliance
Lightning Source LLC
Chambersburg PA
CBHW080724230426
43665CB00020B/2603